PLANT PHYSIOLOGY

Molecular, Biochemical, and Physiological
Fundamentals of Metabolism and Development

PLANT PHYSIOLOGY

Molecular, Biochemical, and Physiological
Fundamentals of Metabolism and Development

Dieter Hess

Springer-Verlag New York · Heidelberg · Berlin 1975

DIETER HESS

Professor, Botanische
Entwicklungsphysiologie
Universitaet Hohenheim
Postfach 106
D-7000, Stuttgart 70
West Germany

Library of Congress Cataloging in Publication Data

Hess, Dieter.
Plant Physiology

Translation of Pflanzenphysiologie.
Bibliography: p.
1. Plant physiology. I. Title
QK711.2.H4713   581.1   73-22278

ISBN 0-387-06643-8 Springer-Verlag New York ● Heidelberg ● Berlin
ISBN 3-540-06643-8 Springer-Verlag Berlin ● Heidelberg ● New York

To my mother

PREFACE

In recent years, molecular biology has infiltrated into all branches of botany. This is particularly true of plant physiology. This book attempts to provide an introduction to the metabolic and developmental physiology of higher plants from a molecular biological point of view. Starting from the heterocatalytic function of DNA the first ten chapters deal with metabolism; development is presented in the last nine, starting from the autocatalytic functions of DNA and including certain topics oriented more toward metabolic physiology. Both fields of plant physiology are so closely linked that an integrated presentation of this kind seemed not only possible but desirable.

In contrast to other accounts, an attempt has been made to give *equal weight* to metabolism and development. In particular, the so-called "secondary" plant materials, which are of considerable interest to the pharmacist, the nutrition technologist, the plant breeder, and the agriculturalist, as well as to the biologist, are treated sufficiently. It is obvious that the wealth of material made an illustrative style of presentation necessary.

The book is intended for beginners, and so it has had, in part, to be simplified. Even so it has not been possible to write it without mentioning hypotheses that anticipate much more research. The beginner ought also to learn how working hypotheses are first postulated on the basis of certain facts and then must either be proved or refuted.

In this connection it may be asked how much the "beginner" is expected to know. He should be provided with a good textbook of general botany as well as an introduction to biochemistry. The latter is all the more necessary because of the material that is treated in every textbook of biochemistry, such as biological oxidation, only the basic concepts are presented here. In compensation for this the metabolic processes that are specific for higher plants have been given more than usual emphasis.

Apart from brief references, an outline of methodology had to be omitted in the interests of the size and price of this book. Information of this nature can, however, be found in any biochemically oriented textbook and there are also abridged introductory texts, e.g. E. S. Lenhoff's *Tools of Biology.*

The author hopes that his book will be of use not only to students of biology and related disciplines at the beginning of their studies but also to teachers at the high school and grammar school level. They may wish to keep themselves informed of recent developments in the field of plant physiology and also to take this or that fact into account in teaching.

The author wishes to thank his publisher, Herrn Roland Ulmer, and his staff for their kind cooperation and Herrn Ekkehart Volk for his care in preparing the figures. Thanks are also due to his wife and daughter who showed full understanding of the many hours of "overtime" the book entailed.

It is a real pleasure to thank Dr. Derek Jarvis for the English translation and for his efforts in reading the galleys and finishing the index. Should the book be well accepted in the English speaking countries, it will be due to his interest and cooperation.

The German editions of this book received a warm reception from the reading public. It is the author's hope that the same may prove true of the English edition.

Stuttgart-Hohenheim   Dieter Hess

CONTENTS

Preface vii

CONTROL OF CHARACTER FORMATION BY
NUCLEIC ACIDS   1

A. The Chemical Constitution of the Nucleic Acids   1
    1. The Building Blocks of the Nucleic Acids   1
    2. Nucleosides, Nucleotides, Polynucleotides   4
    3. The Watson-Crick Model of DNA   5
B. Direct Evidence for the Role of the Nucleic Acids as Carriers
    of Genetic Information   7
    1. Transformation   7
    2. Transfection   9
C. The Heterocatalytic Function of DNA: Transcription and
    Translation   10
    1. The Concept of Molecular Genetics   10
    2. The Genetic Code   11
    3. Transcription   12
    4. Translation   14
    5. Antimetabolites of Transcription and Translation   20
    6. Evidence for mRNA in Higher Plants   24
    7. Transcription and Translation in a Cell-Free System   29
    8. One Gene–One Polypeptide   30

PHOTOSYNTHESIS   35

A. Division of Photosynthesis into Primary and Secondary
    Processes   35
B. Primary Processes of Photosynthesis   36
    1. Electron Transport Chains   36
    2. Redox Systems in the Primary Processes of Photosynthesis   38
    3. Pigment Systems I and II of Photosynthesis   43
    4. Primary Processes of Photosynthesis   46
    5. Quantum Yield of Photosynthesis   48

C.   Secondary Processes of Photosynthesis.................... 49
    1. The CO_2 Acceptor...................................... 49
    2. The Connection with the Primary Processes............ 50
    3. The Calvin Cycle..................................... 51
    4. The C_4 dicarboxylic Acid Pathway...................... 53
D.   The Chloroplast: Site of Photosynthesis.................. 54

CARBOHYDRATES   57

A.   Monosaccharides 57
    1. Phosphorylation (Kinases) 57
    2. Intramolecular Migration of Phosphate (Mutases)........ 58
    3. Sugar Nucleotides (UDPG)............................. 59
    4. Inversion of an OH Group (Epimerases)................ 59
    5. Control of the Equilibrium Between Aldoses and Ketoses
       (Isomerases).. 59
    6. Oxidative Degradation of 1 C Atom (Hexose-pentose
       Transition)... 60
    7. The Pentose Phosphate Cycle......................... 62
B.   Oligosaccharides and Polysaccharides 63
    1. Glycosides 63
    2. Oligosaccharides..................................... 64
    3. Polysaccharides...................................... 67

BIOLOGICAL OXIDATION   74

A.   Glycosis 75
B.   Oxidative Decarboxylstion of Pyruvate, Formation of Active
    Acetate... 78
C.   Citric Acid Cycle 81
D.   The Respiratory Chain 83
E.   Mitochondria as Power Plants............................ 86

FATS   89

A.   Chemical Constitution of the Fatty Acids 89
B.   Biosynthesis of the Fatty Acids 90
    1. Formation of Malonyl CoA............................. 90
    2. Fatty Acid Synthesis Proper 91
C.   Biosynthesis of the Neutral Fats 93

D. Degradation of the Fats 94
   1. β-Oxidation 95
   2. α-Oxidation 96
E. The Glyoxylate Cycle 96

TERPENOIDS **99**

A. Chemical Constitution................................ 99
B. Secondary Plant Substances.......................... 100
C. Volatile Oils.. 101
D. Biosynthesis (General).............................. 102
E. Biosynthesis (Particular) 104
   1. Monoterpenes...................................... 104
   2. Sesquiterpenes.................................... 105
   3. Triterpenes 106
   4. Diterpenes.. 110
   5. Tetraterpenes: Carotenoids 111
   6. Polyterpenes 115

PHENOLS **117**

A. Chemical Constitution................................ 117
B. Biosynthesis (General).............................. 118
   1. The Shikimic Acid Pathway......................... 118
   2. The Acetate-Malonate Pathway 120
   3. Precursors and Intermediates 120
C. Biosynthesis (Particular) 121
   1. Cinnamic Acids 121
   2. Coumarins.. 122
   3. Lignin... 124
   4. Phenol Carboxylic Acids and Simple Phenols........ 128
   5. Flavan Derivatives............................... 129
   6. Flower Pigmentation.............................. 136

AMINO ACIDS **138**

A. The Reduction of Nitrogen........................... 138
B. Reductive Amination 140
C. The Formation of Glutamine 140
D. Transamination...................................... 141
E. The Origin of the C Skeleton of the Amino Acids..... 142

ALKALOIDS   144

A. Derivatives of the Aliphatic Amino Acids, Ornithine and
    Lysine. 146
    1. Quinolizidine Alkaloids. 146
    2. Nicotiana Alkaloids and Nicotinic Acid 147
    3. Tropane Alkaloids 149

B. Derivatives of the Aromatic Amino Acids, Phenlalanine and
    Tyrosine. 150
    1. Amaryllidaceae Alkaloids and Colchicine. 150
    2. Betacyanins and Betaxanthins. 152
    3. Isoquinoline Alkaloids (Benzylisoquinoline Alkaloids). . . . 153

C. Derivative of the Amino Acid Tryptophan: Indole Alkaloids
    and Derivatives. 154

D. Purine Alkaloids. 156

E. Biochemical Systematics. 157

PORPHYRINS   160

CELL DIVISION   163

A. Development — Growth and Differentiation 163

B. Cell Division. 163
    1. The Mitotic Cycle 163
    2. The Autocatalytic Function of DNA: Replication 165
    3. Plant Tumors: Crown Galls. 169

DIFFERENTIAL GENE ACTIVITY AS PRINCIPLE OF
DIFFERENTIATION   172

A. Totipotency. 172

B. Differential Gene Activity: The Phenomenon 174
    1. RNA Synthesis on Giant Chromosomes. 175
    2. Phase-specific mRNA 176
    3. Phase- and Tissue-specific Protein Patterns 178

REGULATION   181

A. States of Activity of the Gene. 181

B. Regulation: Point of Departure............................ 182

C. Regulation by Internal Factors........................... 183
1. Intracellular Regulation 183
2. Intercellular Regulation: Phytohormones.............. 194

D. Regulation by External Factors.......................... 219
1. Temperature 219
2. Light... 220

POLARITY AND UNEQUAL CELL DIVISION AS FUNDAMENTALS OF DIFFERENTIATION   225

A. Polarity.. 225

B. Unequal Cell Division.................................. 227
1. Development of Stomata 228
2. Root Hair Formation................................ 228
3. Pollen Mitosis...................................... 229

CELL ELONGATION   231

A. The Phenomenon 231

B. The Process of Elongation Within a Cell 232
1. The Suction Pressure Equation of the Cell............ 232
2. The Stages of Cell Elongation 234

C. Regulation 235
1. Adjustment of the Equilibrium Between Division and
   Elongation 235
2. Regulation by IAA 235

THE FORMATION OF SEEDS AND FRUITS   241

A. Complex Developmental Processes and Their Regulation ... 241

B. Formation of Seeds and Fruit........................... 242
1. The Process of Formation 242
2. Regulation 244

GERMINATION   250

A. Dormancy 250
1. Incomplete Embryos................................ 250
2. Maturation by Drying............................... 250

    3. Impermeability to Water and/or Gases   251
    4. Inhibitors   251
B. Conditions for Germination   252
    1. Water   253
    2. Oxygen   253
    3. Temperature   253
    4. Light   256
C. Mobilization of Reserve Materials   256
D. Assembly of the Photosynthetic Apparatus   257
E. Regulation of Germination by Photohormones   258
F. Regulation of Germination and Evolution   259

                THE VASCULAR SYSTEM   263

A. The Elements   264
B. Differentiation   264
C. Function   268
    1. Transport in Both Directions   268
    2. Transport of the Xylem   269
    3. Transport in the Phloem   277

                FLOWER FORMATION   286

A. Definitions   286
B. Temperature and Flower Induction: Vernalization   287
    1. Petkus Rye   288
    2. Henbane (Hyoscyamus Niger)   290
    3. Streptocarpus Wendlandii   292
    4. A Hypothesis Concerning Vernalization   296
C. Length of Day and Flower Induction: Photoperiodism   297
    1. Long and Short Day Plants, Neutral Day Plants   298
    2. Analysis of Photoperiodism in Flower Induction   298
    3. Photoperiodism in Flower Induction as a Sign of
       Adaptation   307
    4. Light and Circadian Rhythms   308

BIBLIOGRAPHY   313
SOURCES OF ILLUSTRATION   319
INDEX   322

PLANT PHYSIOLOGY

Molecular, Biochemical, and Physiological
Fundamentals of Metabolism and Development

Control of Character Formation
by Nucleic Acids

The development of a plant consists of the formation of a succession of characters: a seed germinates, puts forth a root and seed leaves, forms shoots that unfurl their leaves, blossoms and, finally, bears fruit and seeds. (Fig. 1) The formation of each character, in turn, consists of the intricate interlocking of many sequences of chemical reactions. All of these reaction sequences, and thus the formation of the characters, are genetically controlled. For that reason, let us inquire first into which substances form the genetic material and how they are involved in the formation of characters.

A. The Chemical Constitution of the Nucleic Acids

Much painstaking experimental work during the first 30–40 years of this century led to its being established that the genetic material of nucleated higher organisms is located primarily in the chromosomes of the cell nucleus. Genetic material is also found in some organelles of the cytoplasm, such as the plastids of plants and the mitochondria, though in rather small amounts.

Accordingly, we shall concentrate our attention on the genetic material that is localized in the chromosomes of the nucleus. Chromosomes consist predominantly of proteins and nucleic acids. The proteins can be divided into histones, which are basic proteins, and non-histones. In addition, there are enzyme proteins that have specific functions. The second, most important group of chromosome constituents, the nucleic acids, were discovered about 100 years ago by the Swiss scientist Miescher in Tubingen. Nucleic acids are the carriers of genetic information. For the moment, we shall not consider the evidence for this assertion, and instead concern ourselves with the chemical structure of the nucleic acids.

1. The Building Blocks of the Nucleic Acids (Fig. 2, Table 1)

All nucleic acids are built up, in principle, from three groups of substances: nitrogen-containing cyclic bases, pentoses, which are sugars con-

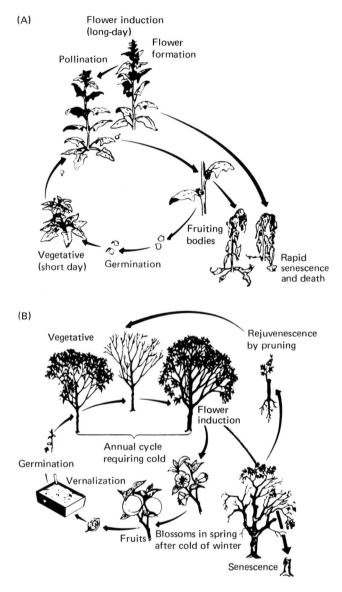

Fig. 1. Diagrams showing the development of a herbaceous (A) and a woody (B) plant. The spinach seed (A) germinates under certain light conditions, develops into a vegetative plant but blossoms only when the days become longer. After fruiting, the plant dies. The peach seed (B) germinates after vernalization and develops into a tree. After 1–3 years, the vegetative phase is complete and the plant has matured to the flowering and fruit-bearing stage. Both of these processes occur after a period of wintry cold. The cycle is repeated annually, each time a period of cold being a precondition of flowering. Gradually, the tree grows old and, after a rather long period of senescence, dies. Rejuvenescence is possible by pruning (Janick et al. 1969).

taining 5 carbon atoms, and inorganic phosphate. The two large groups of nucleic acids, deoxyribonucleic acids (DNA) and ribonucleic acids (RNA), can be distinguished from each other by the kinds of bases and the kind of sugar which participate in their structures. Each of these large groups is itself heterogeneous. There is not simply one DNA, there are innumerable different kinds of DNA. Similarly, there is not one RNA but three large subgroups of RNA, each of which can be subdivided into a large number of different kinds.

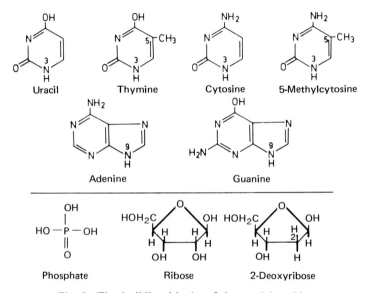

Fig. 2. The building blocks of the nucleic acids.

The bases occurring in DNA are the purine bases adenine and guanine and the pyrimidine bases cytosine and thymine. The pentose sugar of DNA is 2-deoxyribose.

The purine bases adenine and guanine and the pyrimidine base cytosine also occur in RNA. The second pyrimidine base, however, is not thymine, but uracil. The pentose sugar is also different: RNA contains ribose.

Table 1. Designations of the Nucleosides and Nucleotides in DNA and RNA
ph = phosphate, d = deoxy-(d- is often omitted when it is obvious that a deoxy compound is being referred to)

Base	Abbreviation	RNA Nucleoside	RNA Nucleotide	DNA Nucleoside	DNA Nucleotide
Thymine	T	—	—	d-Thymidine	d-Thymidine-5'-ph
Cytosine	C	Cytidine	Cytodine-5'-ph	d-Cytidine	d-Cytidine-5'-ph
Uracil	U	Uridine	Uridine-5'-ph	—	—
Adenine	A	Adenosine	Adenosine-5'-ph	d-Adenosine	d-Adenosine-5'-ph
Guanine	G	Guanosine	Guanosine-5'-ph	d-Guanosine	d-Guanosine-5'-ph

In addition to the bases already mentioned there is a number of less commonly occurring bases. Of these, 5-methyl cytosine must be mentioned, since it is one of the less common bases that occurs relatively often in the DNA of higher plants.

Let us stress again those building blocks with respect to which DNA and RNA differ: DNA contains the pyrimidine base thymine in place of which RNA contains the pyrimidine base uracil. The pentose sugar of DNA is 2-deoxyribose, that of RNA is ribose.

2. Nucleosides, Nucleotides and Polynucleotides (Fig. 3)

The three types of building blocks already mentioned, the bases, the pentose sugars, and phosphate, are assembled to form the nucleic acids according to certain structural principles. A base linked to a pentose sugar through one of its nitrogen atoms is called a nucleoside. If inorganic phosphate is attached to a hydroxyl group of the pentose sugar, a nucleotide is obtained. Finally, many nucleotides may be linked together to form a polynucleotide; the linkage between the individual nucleotides is through their phosphate groups.

We must now consider which functional groups are involved in the assembly of these compounds. In the case of the formation of a nucleoside, the nitrogen atom number 3 of a pyrimidine base or the nitrogen atom number 9 of a purine base becomes linked to the carbon atom number 1 of the pentose sugar, with the elimination of water. According to the conventions of sugar chemistry the carbon atom number 1

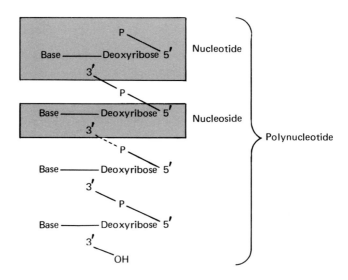

Fig. 3. The assembly of nucleosides, nucleotides, and polynucleotides.

is a glycosidic carbon atom. Correspondingly, a nucleoside is a N-glycoside.

The required nucleotide is formed by esterification of the hydroxyl group attached to carbon atom number 5 of the pentose with phosphate. Thus, the nucleotide is the phosphoric acid ester of the corresponding nucleoside. Now we are left with the linkages between the individual nucleotides. These are formed by the hydroxyl group on carbon atom number 3 of the pentose of one nucleotide reacting with the phosphate group of the next nucleotide, with elimination of water.

Polynucleotides are long strands with a pronounced polarity. This polarity is due to the repeated sequence, pentose-3'-phosphate-5'-pentose-3' or vice versa, depending on which end one starts from. At the beginning of such a chain, there is a nucleotide with a phosphate residue attached to carbon atom number 5 and at the end, a nucleotide with a free hydroxyl group at carbon atom number 3.

3. The Watson-Crick Model of DNA (Fig. 4)

The polynucleotide strands of DNA seldom occur as single strands. Usually, two DNA single strands are found wound together in a double-stranded spiral. That this is so was realized by Watson and Crick in 1953. They proposed a model of the structure of DNA which bears their names, a proposal which was inspired, to some extent, by the x-ray photographs of DNA obtained by Wilkins.

According to the Watson-Crick model, DNA consists of a double coil, a double helix. Two DNA strands of opposite polarity take part in the formation of this helix. Thus, the 3'-hydroxyl end of the one DNA strand and the 5'-phosphate end of the other DNA strand lie at the same end of the helix. The two strands are held together by hydrogen bonds between the purine and pyrimidine bases. The bases project from the backbone of the helix, which is formed by the sugar-phosphate sequence, toward the inside. Hence, there is an opportunity for hydrogen bonds to be formed between the bases of both strands. This phenomenon is known as base pairing and it follows a strict rule, which is called the base pairing rule. Thus, a cytosine or 5-methyl cytosine of one strand always pairs with a guanine of the other, and a thymine of one strand always pairs with an adenine of the other. Three hydrogen bonds can be formed between cytosine or 5-methyl cytosine and guanine, and two between thymine and adenine. As a consequence of the base pairing rule, the base sequences of the two strands of DNA are complementary to each other.

The important points may be summarized as follows: according to the Watson-Crick model, DNA usually consists of two complementary, single strands of DNA of opposite polarity that are coiled around each other and held together by hydrogen bonds between specific base pairs. This double coil is called the DNA double helix.

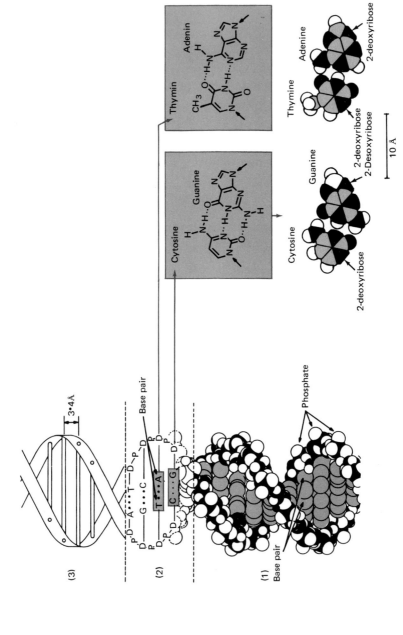

Fig. 4. The Watson-Crick model of DNA. P = phosphate, D = deoxyribose. For abbreviations of the bases, see Table 1 (modified from Bennett 1970).

B. Direct Evidence for the Role of the Nucleic Acids as Carriers of Genetic Information

Let us now turn to the experimental evidence that nucleic acids can be the carriers of genetic information. Direct evidence can be obtained from experiments with isolated nucleic acids, as in transformation and transfection.

1. Transformation

The age of molecular genetics began with the transformation experiments of Avery in 1944. In his experiments he used pneumococci, bacteria, which, among other things, are known to cause pneumonia. There are strains of pneumococci in which the bacterium is surrounded by a capsule of polysaccharide. The capsule protects the bacterium against attack by enzymes of the invaded organism. For this reason, bacteria with capsules can multiply in the host organism and thus become pathogenic, or virulent. When such capsulated pneumococci are grown in a Petrie dish, they form colonies with a smooth surface. Hence, they are known as S-bacteria or S-strains.

Other pneumococcal strains do not possess a protective capsule and, as a result, are not virulent. Their colonies have a rough surface and so they are called R-bacteria or R-strains.

Avery transferred DNA isolated from S-strains to cultures of R-strains. A small precentage (in these experiments less than 1%) of R-bacteria that had been treated in this way developed capsules (Fig. 5). Furthermore, the capacity to form a capsule, once acquired, was maintained by later generations. Thus, the gene responsible for capsule formation had been successfully transferred and this gene must have been in the DNA with which the R-strains were treated.

This experiment clearly showed that DNA can function as genetic material. This kind of gene transplantation with the help of isolated DNA is called transformation and the genetically altered bacteria or organisms (see below) are known as transformants. In the last 25 years transforma-

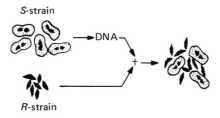

Fig. 5. Transformation in pneumococci. DNA from the S-strain is transferred to a few bacteria of the R-strain. As a consequence, the latter form capsules just like the bacteria of the S-strain (from Kaudewitz 1958).

tion has been carried out with a variety of other bacteria. Many other characters, in addition to the capacity to form a capsule, have been successfully transferred, such as the capacity to synthesize certain amino acids or resistance to various antibiotics.

Attempts have also been made to achieve transformation in higher organisms. In principle, the procedure has been the same as that used with bacteria. Two pure lines were taken that differed from each other with respect to a specific character: one line possessed the character and thus the gene(s) for its expression, the other did not. DNA was isolated from the former, the DNA donor, and used to treat the latter, the DNA receptor.

Up to now transformation has been successfully carried out with animal, human, and plant cells in culture, and also with a few whole organisms (the fruit fly *Drosophila,* the silkworm *Bombyx,* the meal moth *Ephestia* and among plants *Petunia* and the cruciferous plant *Arabidopsis).* Let us consider briefly the experiments with petunia which have been performed repeatedly. The DNA receptor was a pure line of *Petunia hybrida* which had lost the capacity to synthesize red pigments (anthocyanins) as a result of a mutation and, consequently, had white blossoms. Seedlings of this white flowering mutant were treated with DNA from a red flowering, i.e. anthocyanin carrying, line. Some of the treated plants subsequently developed pale or deep red blossoms. It was noted that, once acquired, the capacity to synthesize anthocyanins was maintained throughout successive generations (Fig. 6). Apparently, a gene responsi-

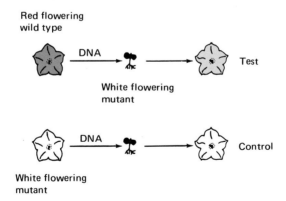

Fig. 6. Transformation in petunia. Seedlings of a white flowering mutant are treated with DNA from a red flowering wild type. Several of the treated mutants subsequently develop red blossoms. This capacity to form red pigment (anthocyanin) is maintained throughout subsequent sexual reproduction and vegetative growth. As a control, white flowering mutants were treated with their own DNA. They continue to produce white blossoms, except in a few cases where external factors, beyond experimental control, have induced slight anthocyanin synthesis. Synthesis of anthocyanin so induced is not maintained by subsequent generations.

ble for anthocyanin synthesis was transferred with the DNA from the line with red blossoms to that with white. This interpretation is supported by a number of additional experiments.

Experiments of this kind with isolated protoplasts, which take up DNA easily, or with tissue cultures are more promising than those with seedlings. It is also advantageous to work with episomes or phages rather than with the entire DNA of an organism since the former are already enriched with the genetic material which is of interest. In the case of phages the protective protein coat is an additional positive feature. Recently, genetic material for the utilization of lactose and galactose has been introduced, in the form of phages, into human fibroblasts and tissue cultures of *Arabidopsis,* the tomato and the mountain maple, which are not equipped with the genes in question. The phages themselves had acquired the genetic material from bacteria in the course of transduction processes.

Thus, direct experimental evidence for the role of DNA as genetic material in higher organisms, including higher plants, is still not plentiful. However, the findings with viruses and microorganisms and a wealth of indirect evidence leave no doubt that DNA can also function as the genetic material in higher plants.

2. Transfection

One speaks of transfection when viral infection is carried out using isolated viral nucleic acid. Transfections have been accomplished with both DNA from certain bacteriophages (bacteriophages are bacterial viruses) and the nucleic acids of other viruses that infect cells of higher organisms. As an example, transfection with the RNA of the tobacco mosaic virus will be described. The tobacco mosaic virus consists of RNA wrapped in a protein coat. Tobacco leaves infected with the virus are no longer their normal green color but develop a light green-green mosaic; hence the name of the virus. It is possible to prepare viral RNA by careful treatment of the isolated tobacco mosaic virus. When this RNA is rubbed into tobacco leaves, the RNA penetrates into the plant cells at sites where the tissue has been damaged, possibly through broken leaf hairs. In the cells the viral RNA not only replicates itself but also induces the synthesis of virus-specific coat protein. Finally, completed new virus particles arise from the RNA and coat protein that can infect neighboring cells. The characteristic mosaic can be seen to develop on the plant leaves. Thus, the RNA of the tobacco mosaic virus contains the genetic information for its own replication and the synthesis of virus-specific protein.

Such transfection experiments provide, in the first place, direct evidence that the nucleic acids used are the genetic material of their respective viruses. Moreover, these foreign nucleic acids take the place of the cell's own nucleic acid, be it DNA or RNA, in cellular metabolism. Instead of the genetic program encoded in the cell's own nucleic acid being expressed, that encoded in the viral nucleic acid now appears. Further, this is accomplished by the same mechanism as that employed by the cell's own program before transfection. That the cell can utilize the implanted nucleic acids at all is circumstantial evidence that, in normal cellular metabolism, nucleic acids serve as carriers of information.

C. The Heterocatalytic Function of DNA: Transcription and Translation

If DNA really is the genetic material, then it must, at some point, initiate its own reduplication. Only on the basis of its identical reduplication is it possible to understand how the genetic material can be transmitted unchanged from cell to cell and from organism to organism. In this connection one speaks of an *autocatalytic function* of DNA. We shall deal with this topic later when we discuss cell division.

For our present purpose it is more important to discuss another necessary property of DNA. If DNA is the genetic material, then it must play a regulatory role in the formation of characters. One speaks in this instance of the *heterocatalytic function* of DNA. Since, however, visible characters have their origin, ultimately, in specific chemical interconversions, we may expect DNA to control chemical reactions. Now it is well known that chemical interconversions in living organisms are catalyzed by enzymes. Here, then, is one of the key functions of DNA: DNA can initiate the formation of enzymes. We shall see later that the induction of enzymes is not the only way in which the genetic material is engaged in the formation of characters. However, since we shall be primarily concerned with the metabolism of plants in the next few chapters, we should now consider the question of the relationship between DNA and enzyme synthesis.

1. The Concept of Molecular Genetics

The concept of molecular genetics can be summarized as follows (Fig. 7): each gene is a specific segment of DNA. In higher organisms DNA is localized in the chromosomes of the cell nucleus. Such a segment of DNA can induce the synthesis of a RNA which is specific for that segment. To each segment of DNA, and thus to each gene, there is a corresponding, quite specific segment of RNA. This RNA migrates from the nucleus to the ribosomes of the cytoplasm where it serves as template for the synthesis of polypeptides. Thus, to each RNA there corresponds a quite specific polypeptide.

In other words, the genetic information contained in the DNA is first transcribed into RNA. This step from DNA to RNA is called *transcription*. Like a messenger, the RNA then carries the genetic information to the

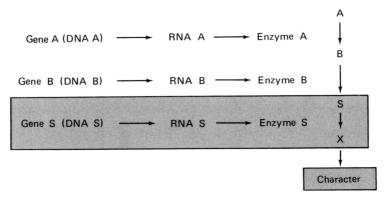

Fig. 7. The concept of molecular genetics.

ribosomes of the cytoplasm. For this reason it is called messenger or mRNA. The genetic information stored in the mRNA is then read off on the ribosomes and used for the synthesis of polypeptides. This second step, from mRNA to polypeptide, is called *translation*. The processes of transcription and translation therefore lead to the conclusion that for each gene there is, finally, a corresponding, specific polypeptide. This polypeptide may be—but need not be—an enzyme protein or a part of an enzyme protein that catalyzes a particular metabolic reaction. Usually, many such DNA-controlled enzyme proteins cooperate in the eventual appearance of a particular character. Genetic information thus flows from DNA via RNA to protein. That scientific discipline whose subject is this flow of information between macromolecules is called molecular genetics.

2. The Genetic Code

We spoke just now of the flow of genetic information between macromolecules. This immediately poses the question: in what form is genetic information contained in these macromolecules and what is the genetic code? The letters of the genetic code are the individual nucleotides. Any 3 nucleotides (a nucleotide triplet) constitute a code word or *codon*. We are thus concerned with a three-letter code. We have already mentioned two kinds of nucleic acids that are implicated in the expression of genetic information, DNA and mRNA. What, then, is a codon: a nucleotide triplet of DNA or of mRNA? The genetic code was deciphered, in part, by experiments with synthetic ribonucleic acids. From this, it is easy to see that a codon is designated as a nucleotide triplet in a mRNA which is responsible for the incorporation of a particular amino acid into a polypeptide. In the initial experiments of Nirenberg and Matthaei a synthetic mRNA which consisted entirely of uracil nucleotides, was introduced into a cell-free system which contained all other factors necessary for polypeptide synthesis. Under the direction of this "poly U," a polypeptide was synthesized that consisted almost entirely of phenylalanine. On the basis of a three-letter code, the code word for the incorporation of phenylalanine into a polypeptide had to be UUU. Furthermore, refined experiments in a cell-free system, particularly those of Khorana, led, ultimately, to the elucidation of the code words for all of the amino acids of proteins. In this way the RNA code was deciphered and, simultaneously, the DNA code too, as we shall see (Fig. 8).

3. Transcription

Let us now deal with the relation between DNA, RNA, and polypeptide in somewhat greater detail. This will require that we refine our rather crude concept of transcription and translation.

Of the two strands of the DNA double helix only one is "read" *in vivo*. How the choice between the strands is made is unknown. The transcrip-

1. Base	2. Base				3. Base
	A	G	T	C	
A	Phe	Ser	Tyr	Cys	A
	Phe	Ser	Tyr	Cys	G
	Leu	Ser	(PP-End)	—	T
	Leu	Ser	(PP-End)	Tyr	C
G	Leu	Pro	His	Arg	A
	Leu	Pro	His	Arg	G
	Leu	Pro	GluN	Arg	T
	Leu	Pro	GluN	Arg	C
T	Ileu	Thr	AspN	Ser	A
	Ileu	Thr	AspN	Ser	G
	Ileu	Thr	Lys	Arg	T
	Met (PP-Start)	Thr	Lys	Arg	C
C	Val	Ala	Asp	Gly	A
	Val	Ala	Asp	Gly	G
	Val	Ala	Glu	Gly	T
	Val (PP-Start)	Ala	Glu	Gly	C

Fig. 8. The genetic code. Experimentally, the mRNA code was first elucidated and from that the DNA code, which is presented here, could be inferred. Each code word consists of three bases. For example, AGC means "incorporation of serine into the corresponding position of the growing polypeptide." The code is degenerate in that more than one codon exists for the same amino acid. Special code words signify the beginning (PP-Start) and the end (PP-End) of the polypeptide chain (after Lynen 1969).

tion of genetic information takes place in such a way that the DNA strand serves as template for the synthesis of a complementary RNA, the mRNA. The ordering principle is base pairing. In the synthesis of RNA, a uracil in the mRNA corresponds to an adenine in the DNA and a cytosine in the mRNA to a guanine in the DNA, and vice versa. This means that we find base pairs just like those between the two strands of a DNA double helix, with the difference that in the mRNA uracil appears in place of thymine.

Nucleotide triplets are the important functional units in DNA too (Fig. 9). To a codon in mRNA there always corresponds a complementary *codogen* in DNA, whereby only uracil and thymine are interchanged. Thus, the deciphering of the RNA code implied the simultaneous elucidation of the DNA code. Let us now consider the mechanism of mRNA synthesis in somewhat greater detail. The starting materials are not nucleotides, let alone the free bases, but nucleoside-5'-triphosphates. They are the entities that are ordered along the DNA matrix according to the base pairing rule. A specific enzyme, the DNA-dependent *RNA polymerase,* cleaves pyrophosphate from each of the triphosphates and links the resulting nucleoside-5'-monophosphates-nucleotides simultaneously to RNA. The RNA polymerase from *E. coli* has been particularly well studied. In higher plants, the DNA-dependent RNA polymerase from pea and maize seedlings, for example, have been studied in some detail.

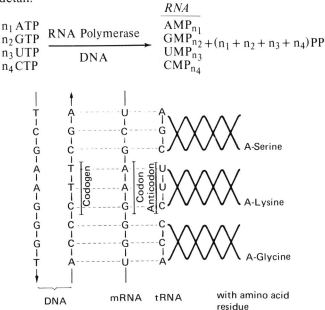

Fig. 9. The relationship between codogen on the DNA, codon on the mRNA and anticodon on the tRNA (modified from Hess 1968).

Transcription does not always lead directly to functional RNA. Rather, in some cases, larger transcription units of RNA are first formed. In plants clear evidence has been obtained for the existence of transcription units containing tRNA and rRNA. In these units the tRNA or certain kinds of rRNA is associated with RNA which corresponds to neither. Only after cleavage of the accompanying RNA is the functional tRNA or rRNA set free. This process of cleavage to liberate tRNA, rRNA or other functional RNAs from transcription units is called "processing".

4. Translation

Now let us look at translation, the path from mRNA to completed polypeptide. Translation is a more complicated process than transcription. The most important factors needed in a system for translation are the following:

> Ribosomes + mRNA or Polyribosomes
> Amino-acyl-tRNA synthetases
> Amino acids
> ATP
> tRNA
> Various factors for the initiation, propagation and termination of translation.

a. Ribosomes and polyribosomes

Ribosomes are more or less round organelles found in the cytoplasm, plastids and mitochondria, which consist of roughly 60% RNA and 40% protein. The well-studied ribosomoes of pea seedlings possess the shape of a spheroid with axes 250 and 160 A long. Their sedimentation constant is 80 S, somewhat larger than that of the *E. coli* ribosomes (70 S) which have also been well-studied. If a suspension of ribosomes is deprived of Mg^{++}, each dissociates into subunits, one of 60 S and one of 40 S (in the case of *E. coli,* 50 S and 30 S). The RNA of the ribosomes is designated *rRNA* (ribosomal RNA). It can constitute over 90% of the total cellular RNA. However, very little is known with certainty about its function.

Up to now we have mentioned only ribosomes from the cytoplasm of plants. It is interesting that the *ribosomes of the plastids and mitochondria* are extremely similar to bacterial ribosomes. Like those of *E. coli* the ribosomes of these organelles are 70 S ribosomes, each of which can be dissociated into a 50 S and 30 S subunit. This finding lends support to the so-called endosymbiotic hypothesis concerning the origin of plastids and mitochondria. According to this hypothesis both organelles are supposed to have been originally bacteria or blue algae which lived symbiotically within the cell. Later, in the course of evolutionary history, they are thought to have become fully integrated in the form of plastids or mitochondria.

Polyribosomes are associations between mRNA and a variable number of ribosomes. The ribosomes are strung on the mRNA like beads on a thread. Often only half a dozen ribosomes are involved. Electron optical micrographs of polyribosomes often reveal a helical structure. Polyribosomes are the sites of polypeptide synthesis. At least in the living organism it is likely that polypeptide synthesis proceeds primarily on such polyribosomes and more rarely on single ribosomes.

b. Activation and Transfer of Amino Acids to tRNA

Before amino acids can be used in peptide synthesis they must be activated. This activation occurs, as is often the case in the living cell, with the help of the energy source ATP. In a reaction utilizing ATP, the amino acid is transferred to a third kind of RNA molecule — we have already met with mRNA and rRNA — which is called tRNA (transfer RNA). We will presently discuss tRNA in more detail. First, let us summarize the activation and transfer reactions:

(1) Activation:

Amino acid + ATP + Enzyme $\rightleftharpoons$ Amino-acyl-AMP-enzyme + PP

(2) Transfer:

Amino-acyl-AMP-enzyme + tRNA $\rightleftharpoons$ Amino-acyl-tRNA + AMP + Enzyme

Activation consists of the formation of the monoadenylate of the amino acid which remains bound to the enzyme, and the liberation of pyrophosphate. In the second reaction the activated amino acid is transferred to tRNA, and AMP and enzyme are set free. Both reactions are catalyzed by one complex enzyme that has been given the impressive name of amino acyl tRNA synthetase since the ultimate product of its action is amino acyl tRNA.

c. tRNA

The situation becomes more complicated when we realize that there is not simply one tRNA but at least one tRNA for each protein amino acid, and a corresponding synthetase. Let us consider these tRNAs more closely.

They are relatively small molecules containing about 80 nucleotides. Less common purines and pyrimidines occur quite often among the base building blocks but we do not need to bother with their formulae here. In all probability certain regions of the RNA strand are paired with each other, leading to the postulated clover leaf structure. All known tRNA molecules terminate at one end with the nucleotide sequence — CCA. In 1965, Holley and his colleagues succeeded in elucidating the nucleotide sequence of a tRNA specific for the amino acid alanine. Since then the nucleotide sequence of other kinds of tRNA has also become known (Fig. 10).

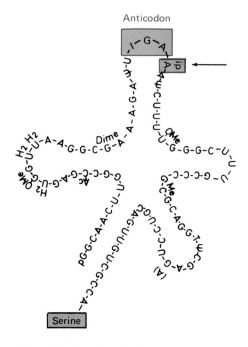

Fig. 10. Structural model of serine tRNA. Many rare bases are conspicuous in the clover leaf structure, e.g. I = inosine or IPA (page 205) directly adjacent to the anticodon (modified from Zachau et al. 1966).

Each tRNA possesses three functionally important regions:
(1) Recognition site for the appropriate amino-acyl-tRNA synthetase
(2) Amino acid attachment site
(3) Template recognition site or anticodon.

By means of the *recognition site for the amino-acyl-tRNA synthetase* the tRNA selects from a mixture of different amino acids the one for whose transfer the tRNA is designed. For example, a tRNA for serine interacts at this site with a synthetase which has bound and activated a serine molecule. What this recognition site looks like in detail and how it functions are still matters of conjecture. In any case, once correctly recognized, the amino acid is transferred to the *amino acid attachment region*. The amino acid attachment site of a tRNA molecule is, in every case, the -CCA end mentioned above.

d. Translation on the ribosome

We shall now pass on to consider the process of translation or polypeptide synthesis proper. First we will discuss how translation proceeds on a single ribosome and then extend the case to polyribosomes.

As already mentioned, the mRNA associates first with ribosomes. This *initiation* of translation is a complex process in which mRNA first binds formyl-methionyl-tRNA to its starter codon (AUG) and simultaneously forms a complex with initiation factors and the smaller ribosomal subunit. Only then does the larger ribosomal subunit join the complex and convert it into a functional ribosome. The propagation (elongation)step now begins and it is assumed that, in the course of it, only one codon of the mRNA on a given ribosome is exposed, in the sense that it can be "read," at any one time. This reading is achieved by an amino acyl tRNA with a complementary *anticodon* recognizing the codon (Fig. 11). Thus, at this point the third functionally important region of the tRNA, the template recognition site or anticodon, comes into play. The amino acyl tRNA is aligned, and that means that the amino acid concerned has been brought into position by means of the tRNA. In the next step, the mRNA is moved one codon in one direction — in principle, the effect is the same whether the mRNA is considered to move in one direction or the ribosome in the other. The result is that a new codon can now be read. Another amino acyl tRNA brings its amino acid into position. The first and second amino acids are enzymatically joined together in peptide linkage, a reaction in which the elongation factors participate. The first tRNA is set free in the cytoplasm where it can be recharged with its amino acid. This mode of alignment and joining of the amino acids is repeated until all of the genetic message has been translated from the mRNA. The stop signal is given by the termination codons. In the *termination* process the polypeptide chain is then released from the last tRNA and the ribosome by means of termination factors. The ribosome dissociates into its subunits which then can be utilised in a new cycle of synthesis.

The sequence of code words is thus translated into a corresponding sequence of amino acids in the growing polypeptide. When the polypeptide has been completed it is released from the ribosome.

e. Translation on the polyribosome

Translation on a polyribosome will now be briefly outlined. On such an assembly line, translation proceeds on each individual ribosome by the mechanism sketched above. At any given moment translation of the mRNA on each of the ribosomes has progressed to varying extents. If we consider the movement of the mRNA to be from left to right (Fig. 12), then a ribosome on the left has almost completed translation and, correspondingly, carries an almost completed polypeptide chain. On the other hand, a ribosome on the right has only just begun to translate, and, consequently, the polypeptide chain attached to it is very short. Per unit time more molecules of polypeptide can be synthesized on such a polyribosome than would be the case if a mRNA were to move its entire

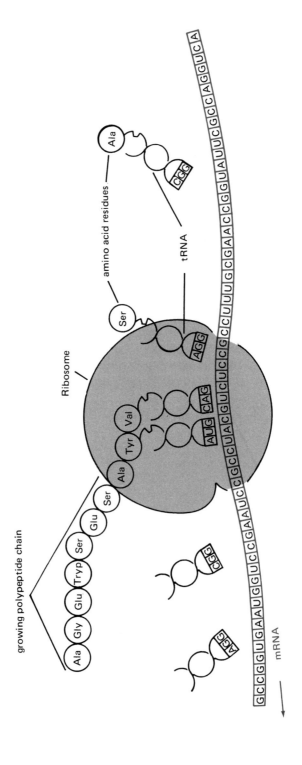

Fig. 11. Translation on the ribosome (modified from Kimball 1970).

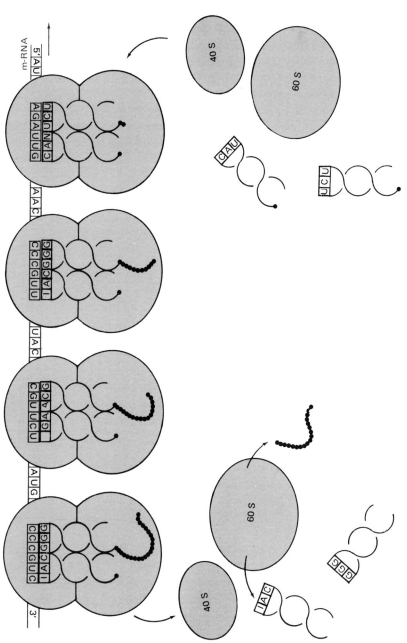

Fig. 12. Translation on the polyribosome (modified from Bennett 1970).

length over one ribosome before becoming attached to the next. Polyribosomes thus represent a very economical synthetic device.

5. Antimetabolites of Transcription and Translation

The biologist often needs to know whether the genetic material participates in a particular developmental process or not. Such participation can only occur via transcription and translation. It is therefore necessary to know whether the process under investigation involves transcription and translation or not. For this purpose substances that cause inhibition at a known step of transcription or translation are useful tools. For if the process under investigation is blocked after administration of specific inhibitors of this kind, then the involvement of transcription and translation is clearly implied.

Substances which block intermediary metabolism are called general antimetabolites. Two kinds of such antimetabolites are of interest here, structural analogs and antibiotics.

a. Structural analogs

Most of these are synthetic substances which are very similar to the natural metabolites. For this reason, they can be used as substitutes for the natural metabolites in metabolism. In the present context the structural analogs of the nucleic acid bases and of amino acids are important (Fig. 13). Both groups can cause inhibition by two mechanisms.

(1) *Incorporation.* The structural analogs can be incorporated into nucleic acids or proteins instead of the natural building blocks. In this way 2-thiouracil and 5-fluoruracil are incorporated into RNA instead of uracil, 5-bromouracil into DNA instead of thymine, and ethionine into protein instead of methionine. These substitutions lead to altered nucleic acids or proteins which cannot carry out their functions or do so inadequately. Thus, one speaks sometimes of "fraudulent" nucleic acids or proteins.

(2) *Competitive Inhibition of enzymes.* In the case of competitive inhibition, the structural analogs compete with the natural substrates for the active site of enzymes. The enzymes of nucleic acid, and protein, synthesis can also be inhibited in this way. Thus, the enzyme thymidilate synthetase is competitively inhibited by 5-fluorodeoxyuridine, a derivative of 5-fluorouracil (Fig. 14). Thymidilate synthetase supplies d-thymidine-5'-phosphate (-thymidilate), a substance which is important to the synthesis of DNA. Inhibition of this enzyme thus brings DNA synthesis to a stop and, with it, all further development. We will discuss DNA synthesis and also thymidilate synthetase in more detail later (p. 165).

Irrespective of the mechanism, both kinds of inhibition interfere with transcription or translation. However, one point must always be borne in mind when antimetabolites are being used: both structural analogs and antibiotics, which we shall discuss presently, can produce side effects which have nothing to do with transcription or translation. This is particularly the case when relatively large quantities are used; they can lead to an unspecific toxicity like any other kind of chemical. Thus, when working with antimetabolites careful control experiments must be carried out. For

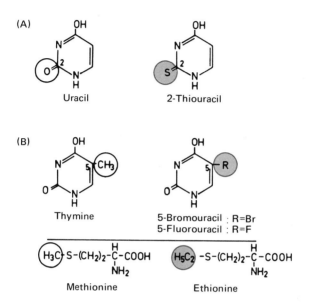

Fig. 13. Structural analogs of nucleic acid bases (A) and amino acids (B).

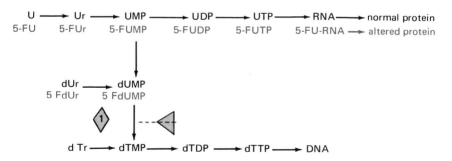

Fig. 14. Effect of 5-fluorouracil (5-FU) and 5-fluorodeoxyuridine (5-FdUMP) on the synthesis of RNA and DNA. 5-FU is incorporated into RNA whereas 5-FdUMP inhibits thymidilate synthetase competitively and blocks DNA synthesis. The conversion of thymidine (dTr) into thymidilate (dTMP) by means of thymidine kinase is only a subsidary routine to thymidilate which can become important, however, under certain conditions such as in developing pollen (page 185). U = uridine triphosphate, Tr = thymidine, TMP = thymidine monophosphate = thymidilate, TDP = thymidine diphosphate, TTP = thymidine triphosphate, 5-F = 5-fluoro, d = deoxy, 1 = thymidine kinase, 2 = thymidilate synthetase.

example, when structural analogs are being used control experiments should demonstrate the following.

(a) Evidence for the incorporation of the analog into the appropriate structure, e.g. RNA, DNA, or protein.

(b) Suspension of inhibition by addition of the corresponding natural building block — an excess of the normal metabolite should displace the analog from its site of action.

(c) Evidence that no unspecific toxic effect is operative. Such an effect can be produced not only by structural analogs but also by antibiotics, whereas the first two types of control apply particularly to experiments with structural analogs.

The last-mentioned requirement is met ideally in the case where only the transcription and translation steps of the process under investigation are inhibited. If possible, other processes which also depend on transcription and translation should not be impaired. In other words, one should try to interfere with the process under investigation as selectively as possible. It goes without saying that the attainment of such *selective inhibition* is not easy and certainly cannot always be achieved. A little later we will discuss several examples in some detail.

b. Antibiotics

These are naturally occurring substances — unlike structural analogs which, as already mentioned, are usually synthesized chemically — which, in quite low concentrations, inhibit processes of growth and development, although they themselves are not enzymes. Contrary to a commonly accepted view, one finds such antibiotics not only in microorganisms, but also in animals and higher plants. Recent research has shown that very many herbs mentioned in the medical writings of antiquity and the herbals of the Middle Ages contained antibiotics.

We are interested here in only a small group of the great number and variety of presently known antibiotics, namely, those that interfere with the heterocatalytic function of DNA. A few examples may be listed: actinomycin C_1 is an inhibitor of transcription, puromycin, chloramphenicol, streptomycin, and actidione are inhibitors of translation. All of them are being used more and more to verify the dependence of a process upon transcription and translation.

Actinomycin C_1 (Fig. 15)

It consists of a chromogenic group which bears two identical, cyclic pentapeptide components. Several actinomycins are known; they differ from each other in their peptide components. As is often the case with physiologically active peptides, uncommon amino acids e.g. sarcosine, are found as structural components. The occurrence of uncommon components and their cyclization to a ring structure protect the physiologically

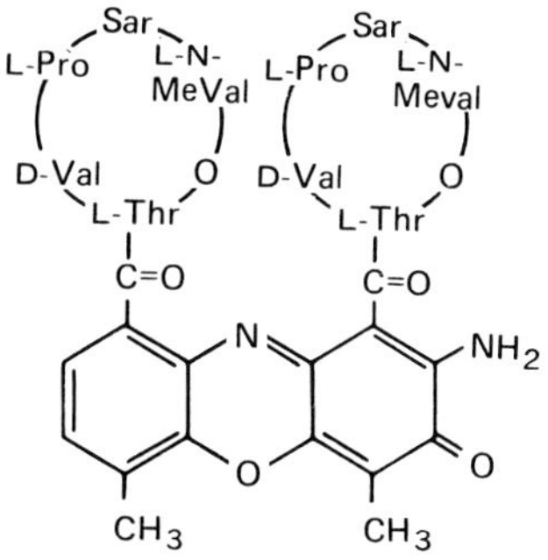

Fig. 15. Actinomycin C_1 = D The amino acids in both peptide rings are indicated by their initial letters. Sar = sarcosine, MeVal = methylvaline.

active peptide from being attacked by proteolytic enzymes of the producing organism. Actinomycin C_1 becomes bound to the nitrogen atom 7 of guanine in DNA. As a result, the DNA is masked and the function of the RNA polymerase is blocked. Thus, mRNA can no longer be formed.

Puromycin (Fig. 16)

This antibiotic is a naturally occurring analog of amino acyl tRNA, especially of tRNA which is charged with phenylalanine or tyrosine. Puromycin becomes linked through its amino group to the growing

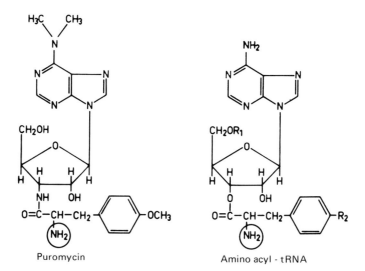

Fig. 16. Puromycin as naturally-occurring structural analog of the tRNA for phenylalanine and tyrosine. Incorporation into the growing polypeptide chain occurs through the encircled amino group. R_1 = polynucleotide chain of the tRNA R_2 = H: phenylalanine; R_2 = OH: tyrosine.

polypeptide chain already present on the ribosome. Since puromycin, unlike tRNA, cannot detach itself from the polypeptide, the growth of the polypeptide chain is stopped. And since puromycin also possesses no anticodon, it cannot pair with mRNA. Consequently, the incomplete polypeptide chain, to one end of which puromycin is attached, is set free from the ribosome.

6. Evidence for mRNA in Higher Plants

Central to the concepts of transcription and translation is the existence of mRNA. For this reason, we shall examine some of the few pieces of evidence for mRNA in higher plants and, in so doing, the knowledge which we have just acquired about antimetabolites will come in handy.

a. mRNA in hypocotyls of the soya bean

Isolated sections from the hypocotyl of the soya bean continue to lengthen after they have been excised. If such hypocotyl sections are treated with 5-fluorouracil, the lengthening proceeds uninterrupted. If, however, actinomycin C_1 is added to the hypocotyls, then the lengthening is inhibited. More detailed analyses of their nucleic acids showed what had happened in the hypocotyls under the influence of the antimetabolites.

First a word concerning the procedure used. Columns are filled with methylalbumin-kieselgur (MAK) and a mixture of nucleic acids is applied to such a column. This mixture contains all of the nucleic acids to be found in the plant, namely, DNA, mRNA, rRNA, and tRNA. After the nucleic acid mixture has seeped into the column, the column is eluted with an NaCl solution of increasing molarity. The ease with which the nucleic acids, adsorbed to the methylalbumin, are eluted by NaCl differs. Some are eluted at quite low molarities of NaCl and so are rapidly displaced from the column. Others are eluted only at higher molarities of NaCl and, accordingly, are displaced later from the column. The nucleic acids are collected from the column in fractions. The amount of nucleic acids in each of the fractions is determined by measurements of the extinction at 260 nm (the extinction maximum of purines and pyrimidines) and, where necessary, of the radioactivity.

If the nucleic acids from the hypocotyl of the soya bean are separated in this way, the following sequence of fractions is obtained (Fig. 17): tRNA, DNA, rRNA, and a further RNA fraction which is presumed to contain mRNA. If the nucleic acids of our sections are analyzed after treatment with 5-fluorouracil, it becomes apparent that the synthesis of tRNA and rRNA is strongly inhibited. In contrast, the synthesis of the supposed mRNA remains unimpaired. This is not the case after treatment with actinomycin C_1. In this instance, as already mentioned, growth is inhibited, and the synthesis of the supposed mRNA is also strongly im-

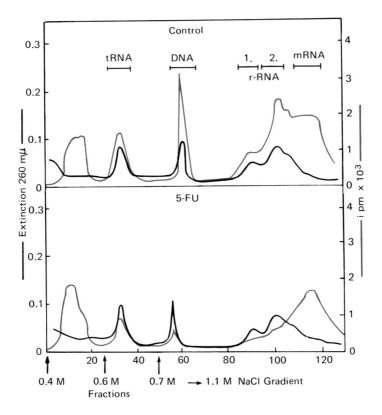

Fig. 17. mRNA as evidence for growth in hypocotyls of the soya bean (modified from Key and Ingle 1966).

paired. This correlation between inhibition of growth and inhibition of mRNA synthesis by actinomycin C_1 shows us that in the RNA fraction in question there is, in fact, mRNA that is responsible for growth.

b. mRNA in the ground nut

We will mention one more piece of evidence for the existence of mRNA, which will enable us, at the same time, to become acquainted with a new technique. This is the evidence for mRNA in the cotyledons of the ground nut *(Arachis hypogaea)*. During germination of such cotyledons a considerable increase in the activity of a whole series of enzymes which are necessary for the mobilization of reserve materials, occurs. It seemed likely that this increase in activity was due, at least in a large part, to the *de novo* synthesis of enzymes. However, such *de novo* synthesis is impossible without mRNA. The technique with which we are concerned here is the *hybridization of nucleic acids*. If a DNA double helix is slowly heated, a sharp increase in the extinction of 260 mμ over a particular temperature range occurs (Fig. 18). One speaks here of a *hyperchromic effect*. It is due to the fact that, with increasing temperature, the two single strands of the double helix are caused to separate. This separation is known as the "melting" of DNA. The two single strands, however, give rise to a higher extinction than the corresponding double helix. Now, after the melting, the DNA solutions may be allowed to cool slowly. On cooling, the separated single strands unite again to form DNA double helices. This reunion is known as renaturation. It is associated with a decrease of the extinction at 260 nm, i.e. with a hypochromic effect. Renaturation takes place according to the principle of base pairing.

Let us extend the technique a little. We have just considered the melting of a given DNA preparation and then its renaturation by slow cooling. One can also melt two preparations of different origin, mix the two preparations together, and then carry out the renaturation. As mentioned, renaturation takes place according to the principle of base pairing. If the two DNA preparations contained complementary base sequences, then they would pair with each other in the regions of complementarity. DNA segments which contained no complementary base sequences would remain unpaired. This pairing of DNA from different origins is called hybridization of DNA. The extent of hybridization is a measure of the relatedness of the organisms from which the DNA preparations were obtained. The more closely related the organisms under consideration are, the more their base sequences will correspond and the amount of hybrid formation will be proportionally higher. In chemosystematics, DNA hybridization is in fact used to devise tests of the degree of relatedness of species at the molecular level.

And now a further variation that finally brings m-RNA into the picture. DNA single strands can be hybridized not only with other DNA single strands but also with single strands of RNA, giving rise to *DNA-RNA hybrids*. It is quite obvious that such a hybridization between DNA and RNA can occur. The fact that the RNA is encoded in the DNA implies the presence of complementary base sequences. DNA-RNA hy-

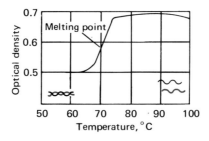

Fig. 18. Melting diagram of DNA from peas (modified from Bonner and Varner 1965).

bridization has been used, among other techniques, to detect mRNA in the cotyledons of the ground nut.

The nucleic acids are first separated on a MAK column as outlined above. The individual RNA fractions are then hybridized with DNA from the ground nut. In order to understand the result of these experiments we must not lose sight of the fact that every cellular RNA is coded for by DNA. Now relatively few DNA segments are needed for the formation of tRNA and the apparently relatively homogeneous rRNA. On the other hand, numerous DNA segments are needed for the coding of mRNA, the intermediary between the large number of genes and the events of the cytoplasm. In the context of our hybridization experiments, this means that, in a given DNA preparation, tRNA and rRNA will find few complementary regions whereas mRNA will find many. Thus, if we find an RNA which hybridizes particularly well with DNA, it is highly probable that this RNA is a mRNA.

In the hybridization experiments with the ground nut, an RNA was, in fact, found, which hybridized particularly readily with DNA (Table 2). This RNA was eluted from the MAK column by the NaCl gradient at the same position as the mRNA from the hypocotyls of the soya bean mentioned above. It may be mentioned that in the case of the separation of bacterial nucleic acids, mRNA is also eluted in the same position. Thus, there is scarcely any doubt in the case of the ground nut that the RNA in question is indeed a mRNA.

Table 2. Hybrid formation between the individual RNA fractions from the ground nut (cotyledons) and DNA from the same source

RNA-Fraction	DNA-RNA-Hybrids %
1. Fraction tRNA	3.9
2. Fraction tRNA	2.4
Light rRNA	3.5
Heavy rRNA	1.7
Supposed mRNA	11.4
After purification by rechromatography	15.4

c. Long-lived mRNA in higher plants

The mobilization of the reserve materials in the cotyledons of the ground nut lead us to a new problem. Over a certain period of time, in this case several days, enzymes are needed, which always have the same function, namely, to mobilize certain reserve materials. It is known from bacteria that the mRNA is very short-lived. A given mRNA molecule can be degraded only a few seconds after it has been synthesized. As we shall learn later, this short life span of bacterial mRNA enables bacteria to adjust rapidly to changing external conditions. Such a rapid adjustment, for instance spore formation under unfavorable conditions, is essential for bacteria. Now in the case of higher plants the same demands will be made on the living system for comparatively long periods of time, e.g. the demand to supply enzymes for the mobilization of reserve materials. In such cases it would be economical to have long-lived mRNA. This long-lived mRNA could be used again and again as matrix for the synthesis of the necessary enzymes. Thus, it would be unnecessary for the transcription machinery to be kept going to provide a continuous supply of new mRNA.

It is well established that in higher organisms, plants as well as animals, long-lived mRNA does in fact occur. One demonstration of this was made in cotton seedlings (*Gossypium hirsutum*). The seedlings were treated with actinomycin C_1, which, as we know, inhibits transcription. Nonetheless the seedlings formed protein and continued to do so for 16 hours after the beginning of actinomycin treatment. Protein synthesis requires the existence of mRNA. It can, in fact, be shown that mRNA is present in the seedlings 16 hours after the cessation of *de novo* synthesis of mRNA. Thus, ultracentrifugation experiments showed that polyribosomes are still present in seedlings that have been treated for 16 hours with actinomycin (Fig. 19). Polyribosomes are, of course, associations of several ribosomes and mRNA.

Let us note, then, that mRNA can also be detected in higher plants and this finding helps to support the concept of molecular genetics. This mRNA is, in part, short-lived and, in part, long-lived too.

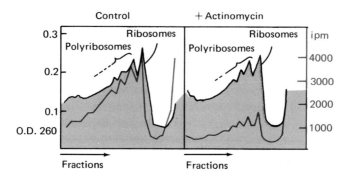

Fig. 19. Evidence for long-lived mRNA in seedlings of the cotton plant. The seedlings were incubated for 12 hours in P^{32} (control) or P^{32} + actinomycin. The ribosomes and polyribosomes were then isolated and separated in a sucrose density gradient. The high peak of the curve shown represents the monomeric ribosomes and to the left of it are the associations of the oligo- to polyribosomes. The ribosome-polyribosome profile after actinomycin treatment is not drastically different from that of the control, although RNA synthesis is drastically inhibited (red curves) (modified after Dure and Waters 1965).

7. Transcription and Translation in a Cell-free System

The final demonstration that a biochemical reaction proceeds as supposed should be provided, if possible, by experiments in a *cell-free system*. Although such a cell-free system does not contain any intact cells, in many cases it does contain certain organelles, for example, chloroplasts or ribosomes. In addition, all necessary enzymes and cofactors are present. Compared with the living cell, cell-free systems that still contain organelles are very simple and are suitable for the elucidation of still unanswered questions.

Transcription and translation can also be carried out in a cell-free system. As an example, let us consider experiments in which transcription and translation are occurring successively in a cell-free system rather than only one process or the other. Leguminosae carry certain reserve globulins in their cotyledons. If we isolate the DNA from a leguminosa, namely pea seedlings, then this DNA must contain the genetic information for the synthesis of these reserve globulins. For the sake of brevity let us call these proteins pea globulins. We now add the DNA to a cell-free system which contains all factors necessary for the synthesis of, first, mRNA and, second, protein. Among other components our cell-free system must contain RNA polymerase (for the synthesis of mRNA) and ribosomes (as organelles of translation). The bacterium *E. coli* can be used as a source of these two components since the RNA polymerase and the ribosomes of *E. coli* work particularly well in a cell-free system. Furthermore, our system contains, in addition to the usual protein amino acids, an amino acid labeled with C^{14}, leucine C^{14}.

After a certain time of incubation the cell-free system was examined to ascertain whether protein had been made (Table 3). This proved to be

Table 3. Transcription and translation in a combined cell-free system from pea seedlings and *E. coli*. Chromatin or DNA came from the pea seedlings. The results of the experiments also demonstrate that the histones contained in the chromatin of the buds of shoots repress the synthesis of pea globulin (cf. page 188).

Matrix (from pea)	Incorporation of Leucine C^{14}(ipm)		% pea globulin in total soluble protein after subtraction of the blank of 0.13
	Total soluble protein	Pea globulin	
Chromatin from the buds of shoots	41200	54	0
Chromatin from cotyledons	8650	623	7.07
DNA from the buds of shoots	15200	60	0.27
DNA from cotyledons	5600	22	0.26

the case since radioactivity was found in total protein and this was due to the incorporation of leucine C^{14} in newly formed protein. Indeed, a small fraction of this new protein was pea globulin. Thus, in the cell-free system, mRNA for pea globulin was first formed on the added DNA matrix and this was translated on the ribosomes into pea globulin. This experiment shows us that transcription and translation in higher plants can be carried out in a cell-free system—even if with lower efficiency. Furthermore, the cooperation of factors from quite different organisms (DNA from peas + RNA polymerase and ribosomes from *E. coli*) demonstrates to us the universal validity of molecular biological data, which may often be found.

8. One Gene-One Polypeptide

In the 1940s Beadle and Tatum advanced the *one gene-one enzyme hypothesis,* mainly on the basis of their own experiments with mutants of the ascomycetous fungus, *Neurospora crassa.* According to this hypothesis, one gene induces the synthesis of one particular enzyme and engages in character formation via this enzyme.

We must make this statement more precise. We have learned in the preceding sections that one gene induces the synthesis of one polypeptide. A single polypeptide may itself be an enzyme protein. However, in many cases the properly functional enzyme protein arises only after several polypeptides of the same or a different kind have combined. Finally, it should not be forgotten that polypeptides are not only constituents of enzyme proteins but also of structural proteins. Hence, let us substitute the formulation *one gene-one polypeptide* for the statement one gene-one enzyme: each gene engages in character formation via the induction of a gene-specific polypeptide.

a. The structure of proteins

Before we consider the relation one gene-one polypeptide in more detail, we must become acquainted with certain principles relating to the structure of proteins. There is a whole hierarchy of structural principles, which are referred to as primary, secondary, tertiary, and quaternary structure (Fig. 20). The term *primary structure* refers to the sequence of the individual amino acids in the polypeptide chain. As already outlined, this amino acid sequence is laid down in transcription and translation. The bonds which determine the primary structure are the peptide bonds between the different amino acid residues.

Under natural conditions each individual polypeptide chain can adopt a particular three-dimensional structure, a native conformation. This structure is maintained primarily by hydrogen bonds between the oxygen and the nitrogen atoms of the peptide linkages. This kind of three-dimensional structure of a polypeptide chain which depends on bonds between

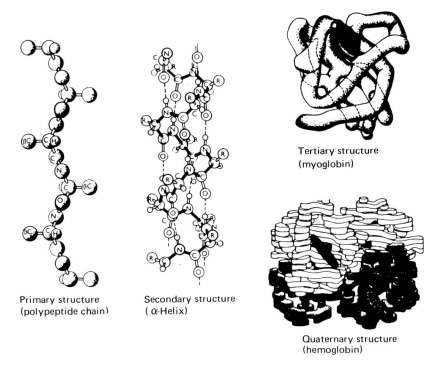

Primary structure
(polypeptide chain)

Secondary structure
(α-Helix)

Tertiary structure
(myoglobin)

Quaternary structure
(hemoglobin)

Fig. 20. Primary, secondary, tertiary, and quaternary structure of proteins. In the secondary structure, the small unlettered circles represent hydrogen atoms, R = sidechains of the amino acids. In the tertiary structure the black disc represents the hemin group. In the quaternary structure the two α-chains of hemoglobin are white, the two β-chains black. One of them is easy to distinguish from the white background of the α-chains (modified from Sund as presented in Wieland and Pfleiderer 1969).

the peptide groupings is described as *secondary*. A well-known secondary structure is the α-helix analyzed by Pauling.

Until now we have spoken only of the bonding between the peptide linkages within a given polypeptide chain. Now the side chains of such a polypeptide chain can also make contact with each other. The kinds of bonds which are involved are, in the first place, hydrogen bonds again, then hydrophobic bonds and, finally, disulphide bridges, which are formed between two SH-groups (Fig. 21). A three-dimensional structure that arises in this way from interactions between the side chains of a polypeptide is called *tertiary*. It is of a higher order than the secondary structure, i.e. a secondary structure such as a α-helix can be folded into a particular tertiary conformation.

Now we must contemplate a still more complex situation, one in which we do not limit ourselves to a single polypeptide chain as in the case of primary, secondary and tertiary structure but bring two or more

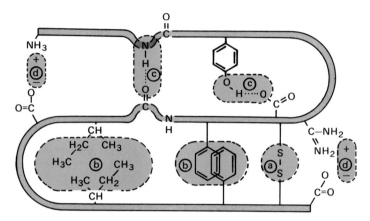

Fig. 21. Types of bonding that give rise to the secondary and tertiary structure of proteins. a = disulphide bridge, b = hydrophobic bonds, c = hydrogen bonds, d = electrostatic bonds. The electrostatic bonds are often included under "hydrogen bonds" (after Lynen 1969).

polypeptide chains into play. The side chains of several different polypeptides can associate with each other just as the side chains within a given polypeptide. The kinds of bonds are the same as those involved in the tertiary structure: hydrogen bonds, hydrophobic bonds and disulphide bridges. By definition, peptide bonds are excluded. A structure which arises from the association of several polypeptide chains as a result of interactions between their side chains — not as a result of peptide bonds — is known as *quaternary*.

b. Isoenzymes

Armed with these new insights into primary, secondary, tertiary, and quaternary structure of proteins, let us now return to the question of the relation one gene-one polypeptide. In doing so we will not concern ourselves with the enumeration of instances of one gene-one enzyme relationships. (There is a variety of obvious examples in higher plants.) Instead we want to verify whether in fact one gene does induce one polypeptide, which can then, possibly, combine with other polypeptides of the same or a slightly different kind to form a quaternary structure. To do this we must study *isoenzymes*.

Isoenzymes are enzymes having the same function but differing somewhat in structure. The existence of isoenzymes, isozymes, or also multiple forms of enzymes has been known for some time. However, it was not until zone electrophoresis on suitable supports provided a method of separation that it could be shown that isozymes are of almost universal distribution.

In maize there is an esterase that shows maximal activity at a pH value of 7.5. It is known as the pH 7.5 esterase. The quaternary structure

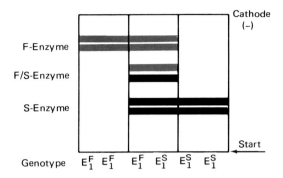

Fig. 22. Isoenzymes of the pH 7.5 esterase of maize. Separation by zone electrophoresis. The F/S enzyme is a "hybrid enzyme."

of the enzyme consists of two polypeptide chains that are derived from the different alleles of the gene locus E. For the moment, let us concern ourselves with only two of these alleles, E^F_1 and E^S_1

Plants which carry the homozygote $E^F_1 E^F_1$ show only a single, fast-moving band in zone electrophoresis (Fig. 22). Each component of this enzyme band consists of two identical polypeptides, namely two F-polypeptides. Plants that contain the other homozygous allele ($E^S_1 E^S_1$ - plants) give rise to a slowly moving enzyme band in zone electrophoresis. Each molecule from this enzyme zone consists of two S-polypeptides.

Now let us prepare a heterozygote which contains both alleles (E^F_1 E^S_1. In this heterozygote two different kinds of polypeptides, F-polypeptide and S-polypeptide, combine with each other. There are three possible ways of combining the two polypeptide chains, F-polypeptide and S-polypeptide, to give the functional esterase: FF, FS, and SS.

Thus three different bands of esterase activity should be found on zone electrophoresic separation, two of which should show the same mobility as that of the two homozygotes. This is the case. The third zone shows an intermediate mobility. This happens because this zone consists of enzyme molecules whose quaternary structure is built-up from one F- and one S-polypeptide. The observed intermediate mobility is the resultant of the dissimilar mobilities of these two polypeptides.

These experiments on the pH 7.5 esterase of maize provide evidence that the one gene-one polypeptide relation also holds in higher plants. The polypeptide chains derived from the genes can then combine to form quaternary structures of a higher order.

And now a word about the function of the isoenzymes. It has been established that the individual isoenzymes adapt to the physiological circumstances in the different tissues. For example, after infection new isoenzymes of peroxidase can be formed, which appear to assume a protective function. Perhaps, though, another aspect is more important:

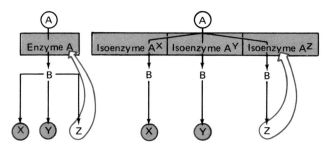

Fig. 23. Fine regulation by feed-back inhibition of isoenzymes.

isoenzymes can play an important role in the *fine regulation of branched, biosynthetic pathways* (Fig. 23).

In such branched systems, different synthetic pathways require the delivery of a common intermediate by a certain enzyme (enzyme A in this example). It can happen that the organism already has a sufficient supply of substance Z. As we shall see later, feedback mechanisms ensure that no further supply of the common intermediate is delivered. That can be brought about by inhibition of enzyme synthesis or, also, by inhibition of enzyme activity. In our example such inhibition would block the further production of substances X and Y. For in these cases, too, the enzyme A stands at the beginning of their respective synthetic pathways.

The way out of this dilemma is that there is not one enzyme A but *three isoenzymes* A that all supply the substance B. Each of the isoenzymes, and, thus, also substance B, stands at the beginning of a branch in the biosynthetic system. In the case of feedback inhibition, only that isoenzyme will be affected that stands at the beginning of the respective branch. Thus, for example, feedback inhibition from Z will be exerted only on isoenzyme A^Z. The other two isoenzymes, A^X and A^Y, will not be affected, and so substances X and Y can still be produced. We shall become acquainted with examples of the quite frequent need for such fine regulation by isoenzymes when we discuss the metabolism of phenylpropane and the biosynthesis of the amino acids.

Photosynthesis

In the preceding chapters, we have seen how genes supply enzymes and structural proteins. Enzymes are the most important catalysts of the living cell. If we now turn to the processes in living organisms that are controlled by enzymes, then we shall want to begin with photosynthesis. This is the process in which, ultimately, all terrestial life originates. In photosynthesis light energy is converted into chemical energy. By means of this chemical energy, the CO_2 in the atmosphere and water is incorporated into organic compounds. The conversion of a foreign substance into a body constituent is known as assimilation. Thus, sometimes one speaks of CO_2 assimilation instead of photosynthesis. However, since photosynthesis is not the only means by which CO_2 can be assimilated, i.e. bound into organic acceptor molecules, and since in this case the utilization of the light energy is the distinctive factor, the more exact description *photosynthesis* has prevailed.

Photosynthesis is not only qualitatively an exceedingly important process. Every year approximately 200–500 billion tons of carbon are transformed by photosynthesis. Thus, photosynthesis is also quantitatively a decisive process. Indeed, the carbon cycle, which begins with the fixation of CO_2 during photosynthesis, is quantitatively the most important chemical process and the second most important of all processes on the earth. In quantitative terms it is exceeded only by the water cycle.

A. Division of Photosynthesis into Primary and Secondary Processes

In photosynthesis the CO_2 of the air and H_2O are converted into carbohydrates. Carbohydrates are compounds which contain, in addition to carbon, the elements hydrogen and oxygen in the same ratio as that in which they occur in water. Thus, the simplest carbohydrate has the formula CH_2O, as is shown by the equation:

$$CO_2 + H_2O \rightarrow (CH_2O) + O_2$$

In fact, there is a compound that corresponds to the formula of the simplest carbohydrate, namely formaldehyde. At one time it was assumed that formaldehyde was the first carbohydrate to be synthesized in the course of photosynthesis. However, this assumption has been shown to be incorrect.

Nevertheless, let us consider for a moment which of the two starting materials, CO_2 and H_2O, furnishes the atoms of our simplest carbohydrate. Without doubt, the carbon is derived from CO_2 and the hydrogen from H_2O. The question is, though, whether the oxygen of the carbohydrate is derived from water or from the CO_2. Conversely, one can ask which of the starting materials supplies the oxygen which, according to the overall equation, is liberated. This question was approached experimentally by using a system containing heavy water, H_2O^{18}. O_2^{18} was found to be liberated and thus it was proved that this oxygen comes from water. Conversely, the carbohydrate formed contained no heavy oxygen. Thus, its oxygen comes from CO_2:

$$CO_2 + H_2O^{18} \rightarrow (CH_2O) + O_2^{18}$$

According to this equation O_2^{18} is liberated. This implies that we must revise the overall equation: instead of H_2O or H_2O^{18} we must write $2H_2O$. We thus arrive at the following overall equation:

$$CO_2 + H_2O^{18} + H_2O^{18} \rightarrow (CH_2O) + O_2^{18} + H_2O$$

We must also take into account the fact that photosynthesis leads eventually to important products, the hexoses, that is carbohydrates containing six carbon atoms. This leads to a further revision of our overall equation:

$$6\,CO_2 + 12\,H_2O \xrightarrow{\;675\ kcal\;} C_6H_{12}O_6 + 6\,H_2O + 6O_2$$

From this overall equation we can draw two important conclusions (Fig. 24):

(1) In the course of photosynthesis water must be cleaved. All of the processes that are related to the cleavage of water-photolyse, are called the *primary processes of photosynthesis*. To these belong not only *photolysis* but also the *noncyclic electron transport,* which is tightly coupled to it, and *cyclic electron transport.*

(2) CO_2 must be reduced to carbohydrate by means of the hydrogen set free by photolysis. This reduction takes place only after the CO_2 has been bound to an organic acceptor. The binding and the reduction of the CO_2 are known as the *secondary processes of photosynthesis.*

B. Primary Processes of Photosynthesis

1. Electron Transport Chains

Let us first discuss briefly the electron transport chains that we shall encounter in the primary processes of photosynthesis and also later, in a discussion of respiration (Fig. 25).

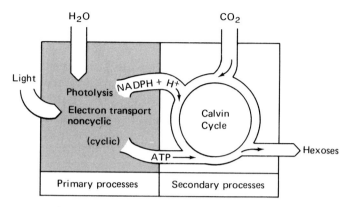

Fig. 24. Scheme of the primary and secondary processes of photosynthesis.

At the beginning of such a chain there is an electron donor, a substance of high "electron pressure." An electron is transferred from the electron donor to an electron acceptor, a substance with a lower electron pressure and higher "electron affinity" than the donor. This first acceptor can then transmit the electron to a second acceptor. It then functions as

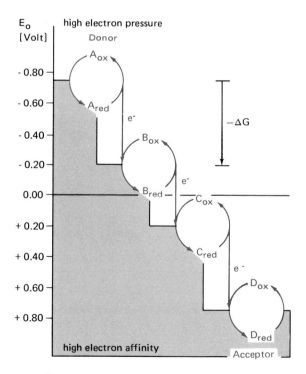

Fig. 25. Scheme of an electron transport chain.

donor with respect to the second acceptor. This kind of transmission of the electron can then be repeated.

Electron pressure and electron affinity are descriptions of the electric potential of a system. We can replace them with the expressions negative and positive redox potential. A substance with a high electron pressure has a high negative redox potential and a substance with a high electron affinity has a high positive redox potential. Thus, our electron transport chain is a series of redox systems of increasing redox potential arranged in tandem. The electron is transmitted along this chain. It falls downhill, as it were, in steps, where the individual steps are redox systems of increasing positive potential. In falling from one step to the next, the electron attains a lower and lower energy level. One could say that at each transition to the next redox system the electron loses a portion of the energy it originally possessed. The energy so liberated can be stored in a chemically bound form as ATP. Conversely, an electron can also be transported uphill. In this case energy must be expended, e.g. ATP must be consumed.

2. Redox Systems in the Primary Process of Photosynthesis

a. Chlorophylls

Only a small portion of the sun's electromagnetic radiation that falls on the earth corresponds to visible light. In turn, only well-defined spectral regions of the visible light are utilized for photosynthesis. This becomes clear when one sets up an action spectrum or an efficiency spectrum of light in photosynthesis. This is done by letting light of constant intensity but of different wavelengths fall on a green plant and examining the effect of the light on the extent of photosynthesis. The result of this is that the action spectrum shows a maximum in both the red and the blue regions of the spectrum. In other words, the middle part of the spectrum — that corresponding to the wavelengths of green light — is not utilized for photosynthesis. Alone the green pigmentation of the photosynthetic plants shows us that green light is remitted. Thus, the action spectrum of photosynthesis leads us to the supposition that green pigments, i.e. pigments that transmit green light but absorb light of other wavelengths, such as blue and red, could be important in photosynthesis. The chlorophylls are green pigments of this kind. Indeed, chlorophyll a, for example, shows an absorption spectrum that coincides almost exactly with the action spectrum of light in photosynthesis (Fig. 26). This coincidence between the absorption spectrum of chlorophyll and the action spectrum of light suggests that the chlorophylls could be the photosynthetically active pigments. This supposition has been confirmed.

A whole series of different chlorophylls is known, called chlorophyll a, b, c, etc. The basic structure of all of these chlorophylls is a porphyrin system. This porphyrin system is formed from 4 pyrrole rings, which are linked together by methylene groups to form a ring system. A sequence of

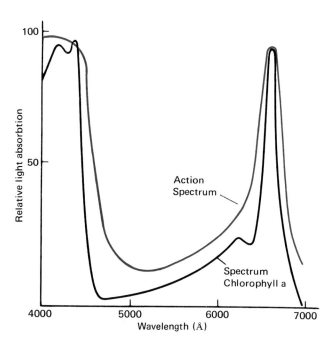

Fig. 26. Absorption spectrum of chlorophyll a and action spectrum of particular wavelengths of light in photosynthesis (after Lehninger 1969).

conjugated double bonds that is responsible for the color of the molecule runs through this ring system. In the center of the porphyrin system is a polyvalent metal, which in the case of vitamin B_{12} is cobalt, in hemoglobin divalent iron and in the case of the chlorophylls, magnesium. The divalent magnesium, $Mg++$, is complexed with the nitrogen atoms of the four pyrrole rings. This porphyrin skeleton bears substituents which are characteristic of the compound. The characteristic substituent of the chlorophylls is an alcohol with 20 carbon atoms, phytol, which is bound in ester linkage to the pyrrole ring IV. According to its biogenesis phytol is classified as a terpenoid. It is responsible for the lipoid solubility of the chlorophylls. The individual chlorophylls differ from each other with respect to the remaining substituents. We will simply note that the frequently occurring chlorophyll b bears an aldehyde function in the position of the pyrrole ring II at which chlorophyll a bears a methyl group (Fig. 27).

The absorption of light quanta from the blue and red regions of the spectrum causes a transition of the chlorophyll molecule to the excited state and, simultaneously, an energy-rich electron is ejected from the chlorophyll molecule. Thus, the chlorophyll molecule becomes ionized. The electron can be accepted by certain acceptors (Fig. 28).

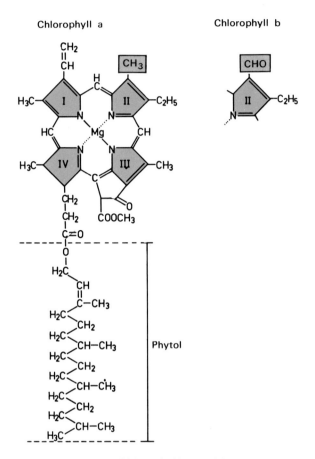

Fig. 27. Chlorophylls a and b.

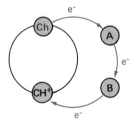

Fig. 28. Ionization of chlorophyll. The ejected electron can return to the chlorophyll molecule via several redox systems. A cycle of this kind is found in the cyclic electron transport of photosynthesis.

b. Cytochromes

Cytochromes are formally closely related to the chlorophylls. They also possess the porphyrin skeleton which, in their case, complexes iron, not Mg^{++} (Fig. 29). A whole series of other biologically important substances such as the peroxidases, catalases, and the red blood pigment hemin also contain these same structural components (iron + porphyrin). All of these substances, including the cytochromes, are known collectively as cell hemins. In the living cell the iron-containing porphyrin system is linked to protein. Thus, the cell hemins occur as proteids.

A series of different cytochromes is known, which can be further subdivided into the groups a, b, and c. Cytochromes of the b and c types are primarily of importance in photosynthesis.

All cytochromes are redox systems for the reason that the central iron atom can oscillate between the divalent and trivalent states. Electron emission causes it to shift from the divalent to the trivalent state and electron uptake causes a shift in the opposite direction.

c. Plastoquinone 45

Another important redox system which is engaged in electron transport in photosynthesis is plastoquinone. Its chemical structure shows similarity to the vitamins of the K series (Fig. 29). Like them, it is characterized by a benzoquinoid nucleus. In the case of plastoquinone this nucleus is substituted with two methyl groups and a side chain of nine 5-carbon units. This side chain shows terpenoid character. Since the important plastoquinone in photosynthesis bears a side chain of 45 C atoms, it is known as plastoquinone 45.

Plastoquinone can function as a redox system because by acceptance of 2H it can be converted to its hydroquinone. This conversion is reversible. This can be formulated differently by substituting $2H^+ + 2e^-$ for 2H ($2H = 2H^+ + 2e^-$). Hence, a molecule of plastoquinone can accept two electrons and a molecule of hydroquinone can give up two electrons. In each case the transport of both electrons is coupled with a translocation of two H^+ ions. This is known as a 2 electron transition.

In the case of the chlorophylls, the cytochromes, and a few other redox systems, which we shall discuss, only one electron was exchanged per moelcule. Thus, in these cases there was a 1 electron transition.

d. Flavoproteins

Yellow (hence the name) prosthetic groups, *flavin adenine dinucleotide (FAD)* and the somewhat less common *flavin mononucleotide* (FMN) are linked to protein in flavoproteins (Fig. 29). The nomenclature is not quite accurate since 6,7-dimethyl-isoalloxazine, which is responsible for the color, is not linked to ribose, as in the nucleotides, but to the corresponding sugar alcohol ribitol. In both FAD and FMN the 6,7-dimethyl-

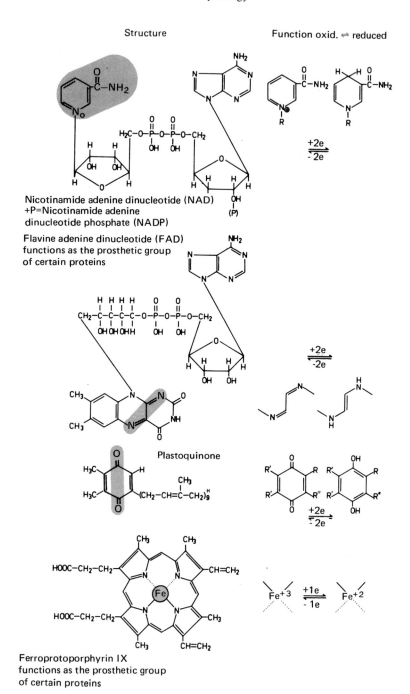

Structure

Function oxid. ⇌ reduced

Nicotinamide adenine dinucleotide (NAD)
+P=Nicotinamide adenine
dinucleotide phosphate (NADP)

Flavine adenine dinucleotide (FAD)
functions as the prosthetic group
of certain proteins

Plastoquinone

Ferroprotoporphyrin IX
functions as the prosthetic group
of certain proteins

Fig. 29. Redox systems that take part in electron transport in the primary processes of photosynthesis (modified from Goldsby 1968).

isoalloxazine component forms the redox system. By uptake of 2H it can be reversibly reduced. Thus, in this case too, there is a 2 electron transition. This is because $2e^- + 2H^+$ can be written for 2H, as was shown for plastoquinone 45. The ferredoxin-$NADP^+$ reductase of higher plants bears FAD.

e. $NAD^+ NADP^+$ (Pyridine Nucleotides)

Nicotinamide adenine dinucleotide (NAD^+) and nicotinamide adenine dinucleotide phosphate ($NADP^+$) are structurally dinucleotides (Fig. 29). The two substances differ from each other by one phosphate residue, which is present in $NADP^+$ and is attached to the $2'$ hydroxyl group of one of the ribose residues. The actual redox system is the nicotinamide which can be reversibly reduced, again by a 2 electron transition. Flavoproteins are very often coupled with NAD^+ or $NADP^+$. In photosynthesis reduced FAD can transmit 2H ($= 2e^- + 2H^+$) to $NADP^+$, leading to $NADPH + H^+$.

f. Ferredoxin

Ferredoxin is an iron-containing protein, which contains two atoms of iron per protein molecule. The redox character of ferredoxin is due to the valence shift of iron between the divalent and trivalent states.

g. Plastocyanin

Plastocyanin is a copper-containing protein. The reversible transition of copper between the monovalent and divalent states is responsible for its reducing and oxidizing character respectively.

h. FRS

Ferredoxin reducing substance is as yet an unidentified redox system.

i. Y

Y is another as yet unidentified redox system.

3. Pigment Systems I and II of Photosynthesis

In discussing the chlorophylls we have already dealt with the most important pigments involved in photosynthesis. A number of other pigments whose exact function is not known is associated with them. For example, various carotinoids come into this category. All of these pigments are combined in two pigment systems. These pigment systems, in turn, are responsible for the so-called 1 and 2 light reactions of photosynthesis. The pigment system I is responsible for the first light reaction and pigment system II for the second light reaction.

a. The Emerson Effect

The first piece of evidence that two pigment systems, and also two light reactions, participate in photosynthesis was obtained by Emerson in experiments with algae. Similar experiments can be carried out with higher plants. If algae are exposed to light of wavelength $> 680m\mu$, for example, a certain rate of photosynthesis is obtained. In the same way, exposure to light of wavelength $< 680m\mu$ leads to a certain effect on photosynthesis. The two values can be summed to give the overall value. If algae are now exposed to light of both wavelengths simultaneously, not in separate experiments as just described, an effect on photsynthesis is observed that exceeds the sum of the values obtained in the separate experiments. This indicates that the two pigment systems cooperate; only a synergism of this kind can explain the observed increase of the rate of photosynthesis over the sum of that found in the two separate experiments (Fig. 30).

b. The two pigment systems of photosynthesis

Both pigment systems or collectives have this in common that, in both of them, many chlorophyll molecules are combined with accessory pigments to form a unit. Of pigment I it is known with certainty and of pigment II with some probability that only one of the many chlorophyll molecules can be excited by a quantum of incident light.

Pigment system I. In pigment system I only one of the very many chlorophyll molecules which are present can be excited, i.e. can transmit an energy-rich electron. Evidence for this was obtained from experiments with strong and weak light flashes. Let us suppose that an intense flash of incident light is given which should be capable of exciting practically all of the chlorophyll molecules present in the pigment collective. The effect

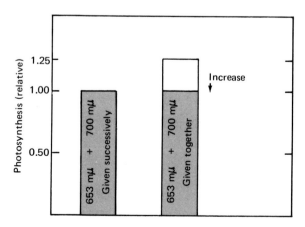

Fig. 30. The Emerson effect (modified from Goldsby 1968).

on photosynthesis, as measured by the amount of O_2 liberated, was, however, very much less than would have been expected for the excitation of all of the chlorophyll molecules present. It could be calculated that only one of approximately 500 chlorophyll molecules is excited by a light quantum. Further support for this finding was obtained from experiments with weak light flashes. Thus, the O_2 yield after a weak flash of light was just as high as that after an intense light flash. We have just established that *only one* of approximately 500 chlorophyll molecules is *activated*. After a weak flash of light, i.e. the supply of only a few light quanta, the chance that just this active chlorophyll will be hit is extremely small. As mentioned, the O_2 yield was, however, as high as after an intense flash of incident light. It is difficult to explain this discrepancy other than by postulating the transmission of the incident light quantum to the active chlorophyll.

This brings us to the imaginary concept of the so-called "trapping center" (Fig. 31). One visualizes that the chlorophyll molecules combined in a pigment system form, as it were, a pit, in which the incident light quanta are trapped. The light quanta that happened to fall into the pit are then transmitted from one chlorophyll molecule to the next until, ultimately, a light quantum reaches the active chlorophyll. This active chlorophyll is then excited and gives up an energy-rich electron. In pigment system I, chlorophyll a_I is the active chlorophyll. It is also known as P-700, since it has an absorption maximum at 700 mμ. It is a chlorophyll a molecule which has special properties, presumably as a result of binding to protein. Thus, it is not a chemically different kind of chlorophyll molecule but a chlorophyll a molecule bound in a different way.

Pigment system II. Much less is known about pigment system II than about pigment system I. The active chlorophyll in this case has not been unambiguously identified. It is assumed that the principle of the trapping center applies also in the case of pigment system II. The active

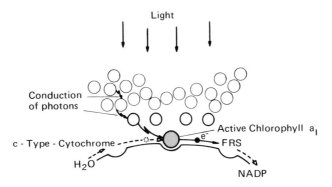

Fig. 31. Model of the "trapping center" (modified from Kok as presented in Wilkins 1969).

chlorophyll is known as chlorophyll a$_{II}$. It shows an absorption maximum at 680 mμ, which leads to another name, P-680. Here again we are concerned with a chlorophyll a molecule, the special properties of which are presumably due to its binding to protein. Let us recapitulate the important point that in both pigment systems which are important for photosynthesis chlorophyll a is the electron donor. The role of chlorophyll b is still a matter of controversy; however, it is unlikely to play an indispensable one in photosynthesis. This is because entire systematic units of algae and blue algae, and mutants of the most diverse higher plants are known that possess no chlorophyll b and yet are able to carry out photosynthesis quite normally.

4. Primary Processes of Photosynthesis

We have become acquainted with the redox systems that participate in the primary processes of photosynthesis. We have also learned that two pigment collectives are involved in the primary processes and in both cases chlorophyll a, present in a distinctive form, is the active component. We are now left with the problem of explaining the interplay of these components in the photolysis of water and the subsequent electron transport, both noncyclic and cyclic. To do this we shall draw up an energy diagram in which we arrange the individual components in the right order and according to their redox potential (Fig. 32). This sounds

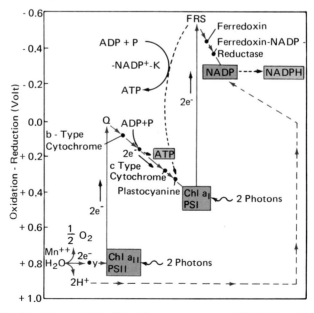

Fig. 32. Electron transport in the primary processes of photosynthesis. Red — noncyclic, black dotted line = cyclic (modified from Levine 1969).

easier than it turned out to be in practice. Many well-known scientists, including Arnon, De Wit, Duysens, Gaffron, Kessler, Kok, Pirson, Trebst, and San Pietro have contributed to the solution of this problem but many questions are still unanswered.

a. First light reaction

Let us begin with the first light reaction. For each light quantum absorbed, one molecule of chlorophyll a_I is excited. As a result, its redox potential is so reduced (from $+0.46$ to -0.44 V) and thus its electron pressure so elevated that it is capable of reducing FRS by giving up an electron. The electron then falls downhill from the FRS redox system via ferredoxin and flavoprotein to the terminal acceptor $NADP^+$. In order to reduce flavoprotein and $NADP^+$ two electrons and two protons ($2H^+$) are needed (we may recall the concept of a 2 electron transition). Now there is no great difficulty: instead of one, we simply start with two quanta of incident light, two molecules of chlorophyll a_I are excited, etc. The two H^+ ions, however, are derived from the photolysis of water. We shall return to this point presently. For the moment it is sufficient to note that we have reached one end of the electron transport chain. At this end are $NADPH + H^+$.

b. Second light reaction

As mentioned, there are still uncertainties about this reaction. The following sequence of events may be assumed: each light quantum excites a chlorophyll a_{II} molecule. Let us consider two molecules of chlorophyll a_{II} because we are concerned with 2 electron transitions in the sequence. Thus, two molecules of chlorophyll a_{II} give up two electrons after appropriate excitation. The electron flux then passes to the substance Q, probably plastoquinone. The reduction of plastoquinone takes place in a 2 electron transition.

The next stages are not known in detail. Probably, the electrons are first transmitted to a b type cytochrome and from it via a still unknown redox system to a c type cytochrome, plastocyanin, and, finally, chlorophyll a_I. Thus, we have made the connection with the first light reaction. In the latter two molecules of chlorophyll a_I had given up two electrons. Here now, in this last reaction, two electrons are returned.

At one of the steps between plastoquinone and the c type cytochrome the decrease in the free energy is utilized to convert ADP into ATP. Thereby physical energy is converted into chemical energy which the plant can use. This is known as *photophosphorylation*.

c. Photolysis

We have still not discussed the actual photolysis. Here, too, fundamental questions have not been answered. One of the current concepts is as follows. The chlorophyll a_{II}, oxidized as a result of giving up electrons, meets its electron deficit by drawing electrons from a hypothetical redox system Y. Here again we must think in terms of 2 electron transitions. Thus, we must consider two ionized chlorophyll a_{II} molecules that draw two electrons from Y. Y, in turn, obtains electrons from the photolysis of water. The details of the photolysis are unknown. It is known, for example, that Mn^{++} is necessary but it is not known what function Mn^{++} has. Photolysis can be summarized as follows:

$$2\,H_2O \rightarrow 2\,H^+ + 2\,OH^-$$
$$2\,OH^- \rightarrow 2\,(OH) + 2\,e^-$$
$$2\,(OH) \rightarrow 1\,H_2O + \tfrac{1}{2}\,O_2$$

Sum: $\quad 2\,H_2O \rightarrow 2\,H^+ + 2\,e^- + 1\,H_2O + \tfrac{1}{2}\,O_2$ or
$\qquad\quad 1\,H_2O \rightarrow 2\,H^+ + 2\,e^- + \tfrac{1}{2}\,O_2$

The photolysis of water has brought us to the other end of the electron transport chain and we can now summarize the sequence of events. The electrons which accumulate from the photolysis of water are conducted via the electron transport chain to $NADP^+$. By means of two electrons and two H^+ ions, which are also derived from photolysis, $NADP^+$ is reduced to $NADPH + H^+$. Two motors maintain the transport: the excitation of the chlorophylls a_I and a_{II} in the first and second light reactions. During the downhill transport of electrons between chlorophyll a_{II} and a_I ATP is formed.

Thus, electrons flow from one end of the chain to the other, from the electron donor, water, to the terminal electron acceptor, $NADP^+$. In this case one speaks of a *noncyclic electron transport*. In addition to oxygen, which is derived from the photolysis of water, its products are $NADPH + H^+$ and ATP.

There is another possibility. A released electron can return to the original electron donor after passing through several other redox systems. In photosynthesis a *cyclic electron transport* like this can follow the first light reaction. Electrons that have been set free from chlorophyll a_I by light quanta are not ransmitted to the FRS redox system and then to ferredoxin, but are allowed to return to chlorophyll a_I via an unknown redox system. ATP is also formed in this cyclic electron transport. The second light reaction, and thus the photolysis of water, are not engaged in this process. $NADPH + H^+$ is not formed; neither, or course, is oxygen. Thus, the product of cyclic electron transport is exclusively ATP. It is still a matter of controversy to what extent this cyclic electron transport is of importance for higher plants.

5. Quantum Yield of Photosynthesis

The question here is in regard to the number of light quanta which are needed to form a molecule of O_2. From our scheme for noncyclic electron transport we infer that four light quanta are needed for the conduction of two electrons over the chain of redox systems and, thus, for the liberation of ½ O_2 which is coupled with it—two for each electron. This implies that 8 light quanta are required for the liberation of a molecule of O_2. Most experimental data pertaining to the determination of the quantum requirement are in good agreement with our calculation.

C. Secondary Processes of Photosynthesis

Of the products of noncyclic electron transport, O_2, ATP, and NADPH + H^+, O_2 will not interest us further. On the other hand, ATP and NADPH are the substances that link the primary processes with the secondary processes. They do so in that they arise in the primary processes and can then be utilized in the secondary processes for the fixation and reduction of CO_2.

1. The CO_2 Acceptor

To which substance is CO_2 bound? Which substance is the CO_2 acceptor? These questions, along with many others associated with the secondary processes, were answered by Calvin. Extracts of cultures of the green alga *Scenedesmus*, which had been photosynthetically active for a prolonged period of time, were examined by means of two dimensional paper chromatography. They contained, among other compounds, the derivative of a pentose, ribulose-1,5-diphosphate and 3-phosphoglyceric acid. If the algae were then deprived of the CO_2 needed for photosynthesis, there was an immediate and rapid increase in the amount of ribulose-1,5-diphosphate and a decrease in 3-phosphoglyceric acid (Fig. 33). Thus, it could be concluded that ribulose-1,5-diphosphate is the CO_2

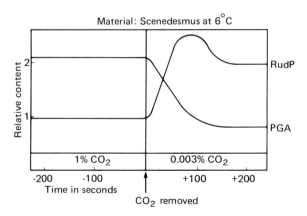

Fig. 33. Experiment to determine the CO_2 acceptor. Removal of CO_2 leads to an increase in the concentration of ribulose-1,5-diphosphate (RudP) and a decrease in that of 3-phosphogylceric acid (PGA) (from Baron 1967).

acceptor. If no more CO_2 is available ribulose-1,5-diphosphate is no longer utilized and accumulates. Furthermore, it seemed likely that the product derived from ribulose-1,5-diphosphate and CO_2 was converted to phosphoglyceric acid in subsequent reactions. These initial inferences were confirmed by subsequent experiments. For example, it was found that the addition of ribulose-1,5-diphosphate promotes the binding of CO_2 and the formation of phosphoglyceric acid.

2. The Connection With the Primary Processes

In the primary processes of photosynthesis ATP and NADPH + H$^+$ are formed (and, in addition, O_2 is liberated, a fact which is of no interest in the present context). In which chemical reaction or reactions of the secondary processes are these two substances utilized? This question could be approached experimentally in that the reaction concerned must be indirectly light-dependent. Indirect in the sense that light directly excites the chlorophylls in the first and second light reactions which have been discussed.

If plants that have been exposed to light for a certain length of time are placed in the dark, the content of most of the substances which are of interest here falls. An exception is 3-phosphoglyceric acid which increases significantly (Fig. 34). Thus, the utilization of 3-phosphoglyceric acid is, apparently, light-dependent. 3-Phosphoglyceric acid disappears in a light-dependent reaction but into which product is it converted in this reaction? Here, autoradiography was a help. Calvin supplied algal suspensions (*Chlorella* or *Scenedesmus*) with $C^{14}O_2$. After very short intervals of only a few seconds and then after longer and longer time intervals the algae were killed and extracted, for example, in hot 80% ethanol. The extracts were separated by means of two dimensional paper chromatography and subjected to autoradiography. In this way it was shown that even after

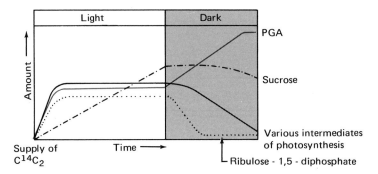

Fig. 34. Experiment to determine which reaction is indirectly light-dependent in the secondary processes of photosynthesis. When it becomes dark the amount of 3-phosphoglyceric acid (PGA) increases, in contrast to that of the other components (modified from Baron 1967).

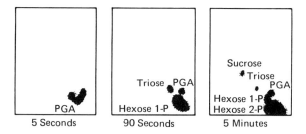

Fig. 35. Experiment to ascertain the path of carbon in the secondary processes of photosynthesis. $C^{14}O_2$ was supplied to algal suspensions. The algae were then killed at specified times and extracts from them were separated in two-dimensional paper chromatography. PGA = 3-phosphoglyceric acid, triose = 3-phosphoglycerinaldehyde + dihydroxyacetone phosphate. The blackened spots on the copies of the chromatograms shown indicate that C^{14} is present in the respective compounds (modified after Baron 1967).

very short periods of photosynthesis 3-phosphoglyceraldehyde was radioactively labeled, as well as 3-phosphoglyceric acid (Fig. 35). Further investigations confirmed that 3-phosphoglyceraldehyde is the product of the light-dependent reaction.

We can thus state definitely that in the secondary processes of photosynthesis 3-phosphoglyceric acid is converted into 3-phosphoglyceraldehyde. For this reaction ATP and NADPH + H$^+$ formed in the primary processes, are utilized. Thus, the reaction is indirectly light-dependent, and, by means of it the primary and secondary processes of photosynthesis are linked.

3. The Calvin Cycle

We have already become familiar with the first step in the fixation and reduction of CO_2. The subsequent reaction steps were also discovered by Calvin's group using the technique outlined above in connection with 3-phosphoglyceraldehyde. Let us consider the complete sequence of reactions, including the first step once again (Fig. 36).

CO_2, present in the form of HCO_3-, is fixed in the acceptor, ribulose-1,5-diphosphate, by means of the enzyme carboxydismutase. An intermediate with 6 C atoms is formed, the identity of which is still unknown. This substance is unstable. It decomposes into two molecules of 3-phosphoglyceric acid. The latter is then reduced to 3-phosphoglyceraldehyde by means of the ATP and NADPH + H$^+$ formed in the primary processes. 3-Phosphoglyceraldehyde exists in equilibrium with its isomer, dihydroxy acetone phosphate. The equilibrium is controlled by the enzyme triose phosphate isomerase. 3-Phosphoglyceraldehyde and dihydroxy acetone phosphate are referred to collectively as triose phosphate.

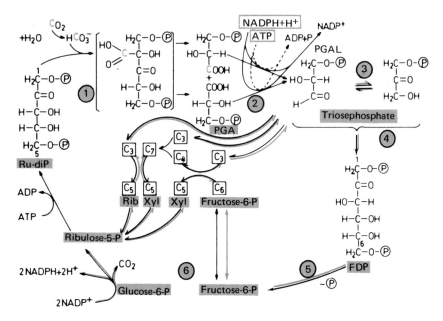

Fig. 36. The Calvin cycle (black lines) and pentose phosphate cycle (red lines). PGA = 3-Phosphoglyceric acid, PGAL = 3-phosphoglyceraldehyde, Rib. = ribose-5-phosphate, Xyl = xylulose-5-phosphate, Ru-diP = ribulose-1,5-diphosphate, C_4 = erythrose-4-phosphate, FDP = fructose-1,6-diphosphate. A few of the enzymes participating are encoded, 1 = carboxydismutase, 2 = triose phosphate dehydrogenase, 3 = triose phosphate isomerase, 4 = aldolase, 5 = phosphatase, 6 = phosphoglucoisomerase. Details of the conversion of glucose-6-P into ribulose-5-P are given in Fig. 43. It should be pointed out that the pentose phosphate cycle presents only here and there a true reversal of the Calvin cycle. In many instances the mechanisms and enzymes are different.

At triose phosphate the pathway branches: in one direction two molecules of triose phosphate (one each of 3-phosphoglyceraldehyde and dihydroxy acetone phosphate) combine to form a molecule of fructose-1,6-diphosphate. The reaction mechanism is a so-called aldol condensation, and, accordingly, the enzyme controlling the step is called aldolase. A phosphatase can then cleave a phosphate residue from fructose-1,6-diphosphate. Fructose-6-phosphate is formed which can be converted into other sugars. In this way, the CO_2 of the air is utilized for the synthesis of carbohydrates.

In the other direction, triose phosphates participate in the regeneration of the CO_2 acceptor ribulose-1,5-diphosphate. A balance sheet presents the following picture. In a complicated sequence of interconversions, which we will not consider in detail here, 3 molecules of triose phosphate and one molecule of fructose-6-phosphate, thus, a total of 15 C atoms, are put in. Three molecules of ribulose-5-phosphate, i.e. 15 C atoms again, are put out. Ribulose-5-phosphate is then converted into

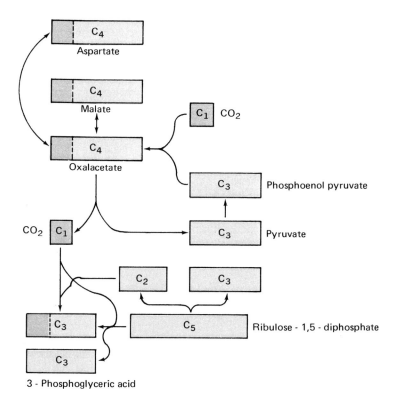

Fig. 36(A). Outline of the C_4 dicarboxylic acid pathway.

ribulose-1,5-phosphate by means of ATP. With that, the CO_2 acceptor is regenerated. A part of the reaction sequence outlined above can run in the opposite direction and in doing so glucose-6-phosphate is oxidized to 6-phosphogluconic acid. This reversal is important in so far as it provides a means of supplying pentoses in the form of their phosphates. For this reason it is known as the *pentose phosphate* cycle (page 62).

Let us now return to the experiments of Calvin. Starting from and returning to ribulose-1,5-diphosphate we find a cycle which provides a supply of (a) carbohydrates, e.g. fructose-6-phosphate, and (b) the CO_2 acceptor, ribulose-1,5-diphosphate. This cycle is known as the *Calvin cycle*.

4. The C_4 Dicarboxylic Acid Pathway

Until recently it was assumed that the Calvin cycle was the only universally feasible pathway of CO_2 fixation, even though the data for maize and sugar cane were not in complete agreement with this view. In 1966, Hatch and Slack demonstrated that, in fact, there is another pathway of CO_2 fixation in higher plants. It was first demonstrated in gramineae, and then later in other monocotyledons and in dicotyledons too.

Many plants that exhibit this new pathway of photosynthesis display an anatomic peculiarity: the photosynthetically active cells that contain chlorophyll are arranged radially around the vascular tissue.

In the meantime it has been possible to form well-founded ideas as to the course of the reaction. We shall learn later that succulents especially can fix CO_2 into C_3 bodies, leading ultimately to the formation of malate (page 275). In all probability the CO_2 is fixed in phosphoenol pyruvate to form, initially, oxalacetate. Oxalacetate can then be converted to malate, or even to aspartate. In the case of the "Hatch and Slack pathway" of CO_2 fixation, too, phosphoenol pyruvate is carboxylated to oxalacetate, which can then be converted to malate and aspartate. One of these C_4 dicarboxylic acids, probably oxalacetate, can, however, transfer the newly fixed CO_2 to another acceptor. This other acceptor is either a C_2 unit, which is furnished by degradation of ribulose-1,5-diphosphate, or a C_5 unit, perhaps ribulose-1,5-diphosphate itself. If the CO_2 is transferred to the C_5 unit, then the product is 3-phosphoglyceric acid. If the CO_2 is transferred to the C_5 unit a C_6 body is first formed, which, in turn, breaks down into two units of 3-phosphoglyceric acid. Thus, 3-phosphoglyceric acid is formed — and the link is made with the known reactions of the Calvin cycle (cf. Fig. 36).

The difference between this route and the Calvin cycle lies in the fact that the CO_2 is fixed, not into ribulose-1,5-diphosphate, but into phosphoenol pyruvate with the intermediary formation of C_4 dicarboxylic acids. Hence the name, the *C_4 dicarboxylic acid pathway*. One of the C_4 dicarboxylic acids transmits the CO_2 further with the formation, ultimately, of 3-phosphoglyceric acid.

The ecophysiological significance of the C_4 dicarboxylic acid pathway is still a matter of discussion. It is striking, though, that it is also found in a large number of halophytes. This has led to the assumption that the C_4 dicarboxylic acids formed via the Hatch-Slack pathway might play a role in osmoregulation in these species. However, this is only one of several possibilities.

D. The Chloroplast: Site of Photosynthesis

Sites of photosynthesis in higher plants are the chloroplasts, as a rule lens-shaped organelles with a rather large diameter of 5–10 mμ. Under a powerful light microscope an internal structure can be detected: the disc-shaped grana are embedded in a ground substance, the stroma. The grana look intensely green owing to their high chlorophyll contact.

The electron microscope permitted a further differentiation. The entire chloroplast is interlaced with a system of self-contained membranes which are called thylacoids ("sack-like") after the proposal of Menke. The grana consist of thylacoids laid in piles, one on top of the other. Individual thylacoids of such piles also pass through the region outside the grana

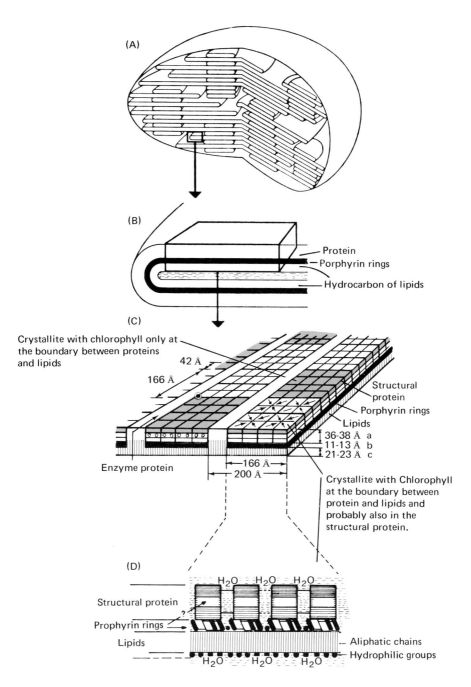

Fig. 37. Model of the structure of the chloroplast (modified from Kreutz 1966 and 1969).

which is called stroma. Nowadays the following nomenclature is, in general, used: chloroplasts consist of membrane systems, the thylacoids, which are embedded in a ground substance, the matrix. Whether chlorophyll is found in the thylacoids outside the grana is still a controversial point (Fig. 37).

Up to this point the opinions of the individual scientists who investigate the structure of the chloroplast by means of the electron microscope, are essentially in agreement. However, their opinions regarding the ultrastructure of thylacoids are quite divided. Here an interpretation is given which is maintained by Kreutz (Fig. 37). According to him, the membranes consist of an outer protein-containing layer and an inner lipid layer. The protein-containing layer is built of two different "crystallites"; one type consists only of structural protein, the other probably contains chlorophyll in addition to structural proteins. In addition, chlorophyll is found in the region between the protein-containing layer and lipids: its porphyrin ring is oriented towards the proteins, its phytol chains towards the lipids.

According to Kreutz enzymes are located between the crystallites. Mühlethaler and other scientists have been able to show that other enzyme systems, the carboxydismutase and an ATPase (an enzyme that controls the equilibrium $ATP + HOH \rightleftharpoons ADP + P$ and thus might play a role in photophosphorylation) are located on the outside of the thylacoid membranes.

According to all the data presently available the primary processes of photosynthesis take place in the membrane structures and the secondary processes at the boundaries of the membranes and in the matrix. Inspite of the hypothetical nature of all statements concerning the relation between ultrastructure and function of chloroplasts two central principles can be sifted out:

(1) *Surface enlargement.* All metabolically active structures strive for this and it is achieved here by the arrangement of the membrane systems as thylacoids.

(2) *Compartmentalization.* A compartmentalization, i.e. a separation of reaction zones, makes it possible for different processes to proceed with highest efficiency and in close proximity without mutual disturbance. Such a compartmentalization is attained by the localization of the primary processes in the thylacoids and of the secondary processes in the adjoining matrix.

Carbohydrates

The first products to be furnished by the Calvin cycle belong to the carbohydrates. They can be converted, in a series of reactions, to other carbohydrates, some of which are polymers. The whole group is of central importance for plants (and also for animals). This is because carbohydrates

(1) serve as a reservoir of energy and

(2) constitute the starting material for the synthesis of all other organic substances to be found in plants and animals

Carbohydrates are substances which, per carbon atom, contain hydrogen and oxygen in the same ratio as that in which they occur in water. Thus, for each carbon atom there are two H atoms and one O atom. By far the most important carbohydrates are known as "sugars." Sugars bear either an aldehyde or a keto function, giving rise to aldoses, e.g. glucose, or ketoses, e.g. fructose, respectively. Sugars can occur singly or be linked. Simple sugars are known as monosaccharides. Linked sugars result from the joining together of simple sugars by means of glycosidic bonds. We will discuss the formation of glycosides in more detail below. Depending on the number of sugars that are linked to each other by a glycosidic bond one distinguishes between disaccharides, oligosaccharides, and polysaccharides.

A. Monosaccharides

Monosaccharides can be grouped according to the number of their C atoms into trioses (3C), tetroses (4C), pentoses (5C), and hexoses (6C). Sugars with a higher number of C atoms, such as heptoses with 7C, are rare but are important as intermediates in metabolism. The formulae of a few monosaccharides are compiled in Figure 38. We will become more familiar with some of them by outlining a few of the alterations to which the monosaccharides are subject in the metabolism of plants.

1. Phosphorylation (Kinases)

Sugars very often participate in metabolism in a phosphorylated form. We have just become acquainted with an illustration of this in the Calvin

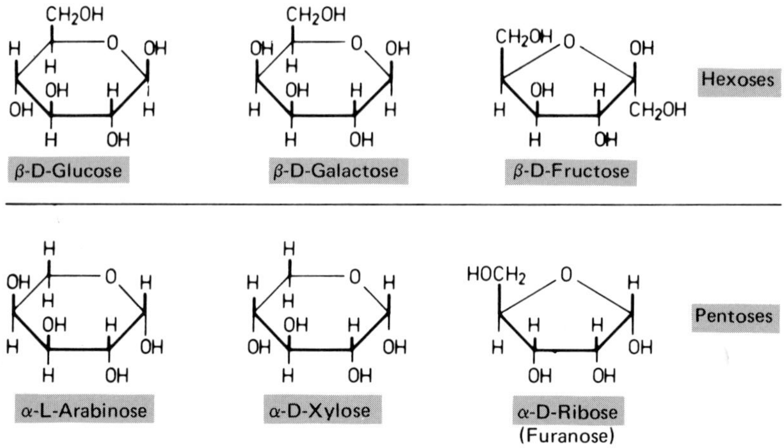

CH₂OH / β-D-Glucose, CH₂OH / β-D-Galactose, CH₂OH / β-D-Fructose (Hexoses)

α-L-Arabinose, α-D-Xylose, α-D-Ribose (Furanose) (Pentoses)

Fig. 38. A few monosaccharides (hexoses and pentoses).

cycle. Phosphorylation is brought about by means of ATP. The enzymes which can transfer a phosphate residue from ATP to a sugar are known as kinases. The sugar phosphate bonds are relatively energy-rich. Thus, the sugar is activated for further metabolic reactions. An example of a kinase is hexokinase which catalyzes the conversion of glucose into glucose-6-phosphate. In this reaction the equilibrium lies on the side of glucose-6-phosphate (Fig. 39).

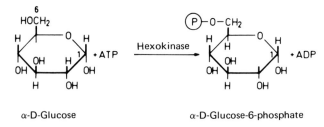

α-D-Glucose   α-D-Glucose-6-phosphate

Fig. 39. Function of hexokinase.

2. Intramolecular Migration of Phosphate (Mutases)

The end effect of mutases is to shift phosphate residues within a sugar molecule. In actual fact, several reactions are involved in which phosphorylated mutases take part. For example, phosphoglucomutase converts glucose-6-phosphate into glucose-1-phosphate by a phosphate shift. The existence of this enzyme in plants has not yet been proved. However, a number of reactions, which have been observed, presuppose its existence.

3. Sugar Nucleotides (UDPG)

An important active form of glucose in higher plants is uridine diphosphate glucose (UDPG). UDPG is often simply called "active glucose". It is obtained from glucose-1-phosphate and uridine triphosphate with the accompanying release of pyrophosphate (Fig. 40). UDPG is undoubtedly one of the most important sugar nucleotides. However, a large number of other sugar nucleotides, e.g. those of ADP, GDP, CDP, and TDP, are known which are utilized in particular reactions.

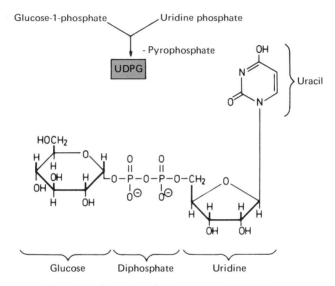

Fig. 40. Formation and structure of UDPG.

4. Inversion of an OH Group (Epimerases)

Epimerization means a change in the steric relationship at a C atom, which is brought about by the inversion of an attached hydroxyl group, the carbon skeleton itself remaining unchanged. The enzymes that bring about this reaction are known as epimerases (Fig. 41).

An example is the 4-epimerase which converts UDPG into UDP galactose by inversion of the hydroxyl group at C atom Nr.4. In subsequent reactions this galactose can be exchanged for glucose and be set free in the form of galactose-1-phosphate. Another example is the ribulose-3-epimerase. This enzyme controls the equilibrium between the phosphates of ribulose and xylulose by an inversion of the hydroxyl group at C atom 3.

5. Control of the Equilibrium Between Aldoses and Ketoses (Isomerases)

In the case of isomerization, too, the carbon skeleton remains unchanged. We have just learned that monosaccharides can occur as

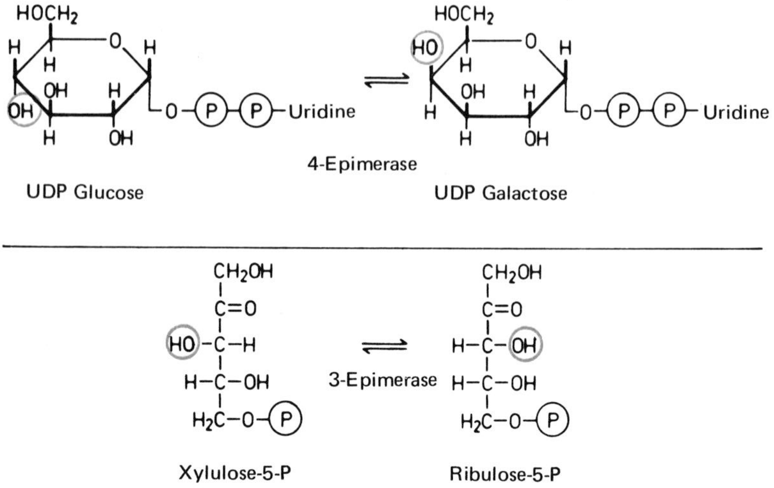

Fig. 41. Function of epimerases.

aldoses and as ketoses. The transition between the two is regulated by the isomerases, an example of which is phosphoglucoisomerase. This enzyme controls the equilibrium between glucose-6-phosphate and fructose-6-phosphate. A second example is phosphoriboisomerase which reversibly converts ribose-5-phosphate into ribulose-5-phosphate. It should be noted that both epimerases and isomerases prefer the sugar phosphates as sustrate.

6. Oxidative Degradation of 1 C Atom (Hexose-pentose Transition)

a. Elimination of C 6 (Fig. 42)

If a sugar is oxidized at its HO-C^6H_2 group, a uronic acid is obtained. Thus, glucuronic acid arises from glucose. Uronic acids can be decarboxylated. As an example we shall mention the formation of UDP xylose. UDPG is converted into UDP glucuronic acid by means of NAD^+. By decarboxylation of the latter UDP-xylose is formed. In this way, a hexose is converted into a pentose. Among other things, UDP-xylose can be utilized in the synthesis of high polymer pentose derivatives, the xylans. We must mention one further use of UDP-glucuronic acid. A 4-epimerase can convert the compound into UDP-galacturonic acid by inversion of the HO group at C atom 4. Galacturonic acid is important as the building block of the pectin substances. In higher plants it appears to arise principally in this way.

b. Elimination of C 1 (Fig. 43)

If the aldehyde function of a sugar is oxidized instead of its HO-C^6H_2 group, a lactone is first obtained which can be converted to an "onic" acid on hydrolysis. Thus, glucose-6-phosphate is first dehydrogenated to gluconolactone-6-phosphate. The lactone ring can then be cleaved hydrolytically to form gluconic acid-6-phosphate. A second dehydrogenation of this compound gives rise to an intermediate, whose constitution is still unknown and which leads to ribulose-5-phosphate upon decarboxylation. Here again CO_2 can be eliminated subsequent to the oxidation, i.e. the conversion of hexoses to pentoses occurs in this reaction sequence too.

This transition takes place with the formation of $NADPH + H^+$ which can be used by the cell for synthetic purposes. This is not the main reason why the conversion of glucose-6-phosphate into ribulose-5-phosphate interests us, but because it forms the beginning of the pentose phosphate cycle.

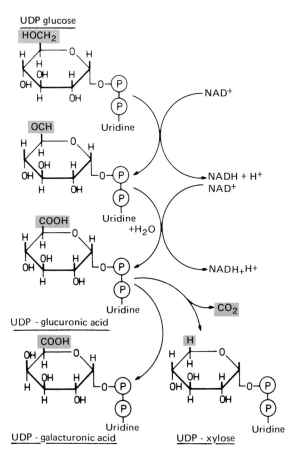

Fig. 42. Oxidative degradation of 1C atom: Formation of UDP xylose, UDP glucuronic acid and UDP galacturonic acid.

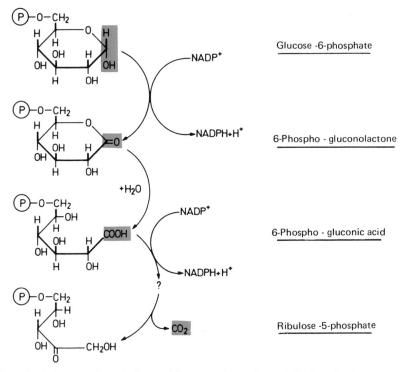

Fig. 43. Oxidative degradation of 1C atom: Formation of ribulose-5-phosphate = beginning of the pentose phosphate cycle.

7. The Pentose Phosphate Cycle (Fig. 36)

This is a pathway for the direct breakdown of glucose along which pentose phosphates are formed. The sequence of reactions requires no great feat of memory, provided that we ignore certain details which are important to the biochemist. The cycle begins with the conversion of glucose-6-phosphate to ribulose-5-phosphate as just discussed. From ribulose-5-phosphate onwards we are in familiar territory: we simply need to consider a portion of the Calvin cycle running backwards. The first products to be derived from ribulose-5-phosphate are ribose-5-phosphate and xylulose-5-phosphate, two other pentose phosphates. The name is thus appropriate. Finally, the pentose phosphates can be converted into triose phosphate and fructose-6-phosphate. Now we must complete the cycle. This is done by 6-phosphoglucoisomerase, which has already been mentioned, converting fructose-6-phosphate into glucose-6-phosphate.

The pentose phosphate cycle is only a subsidiary pathway for the breakdown of glucose. In higher plants its importance lies primarily in its supplying $NADPH + H^+$. This is because the formation of pentoses from glucose-6-phosphate seems to be less critical in photosynthetically active green plants. They have sufficient triose phosphate at their disposal from the secondary processes of photosynthesis. This is the source from which pentose phosphates can be formed in the subsequent steps of the Calvin cycle.

B. Oligosaccharides and Polysaccharides

1. Glycosides

Oligosaccharides and polysaccharides belong to the group of the glycosides. These are a chemically very heterogeneous collection of substances, which have in common the property that the molecules are linked together and contain at least one sugar unit. Some of them consist entirely of sugar units. Each sugar is linked through a glycosidic bond to either a nonsugar component, an aglycone, or with another sugar. What kinds of linkages are these glycosidic bonds?

a. Hemiacetal

We shall first become acquainted with the expression hemiacetal. Hemiacetals are formed by the addition of an alcoholic component to a carbonyl function C=O (Fig. 44).

b. Internal Hemiacetal

An alcoholic component needed to participate in the formation of a hemiacetal need not be a foreign molecule; it may be a part of the same molecule as the carbonyl function. In such a case one speaks of an internal hemiacetal. Sugars can form internal hemiacetals, which are then present in a ring structure (Fig. 44).

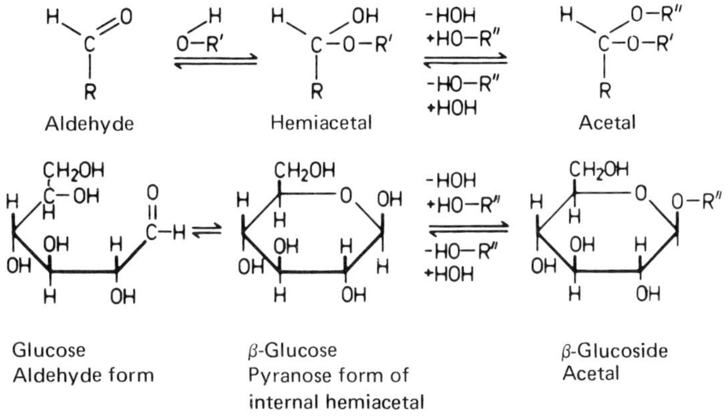

Fig. 44. Hemiacetal, internal hemiacetal, and acetal.

c. Acetal

A hemiacetal, even an internal hemiacetal, can react with an additional alcohol molecule. We must emphasize the word react because this is not a simple addition; it entails bond formation with the splitting out of water. The product is an acetal. If a sugar takes part in acetal formation, a glycoside results. The bond between the alcoholic component and the sugar is known as a glycosidic bond (Fig. 44).

d. Types of Glycosides

As already mentioned, glycosides constitute a group of chemically very heterogeneous substances. For the sake of cataloguing, at least two large groups can be distinguished, O-glycosides and N-glycosides.

O-Glycosides. The alcoholic component reacts with the sugar through a HO group. In addition to true alcohols, organic acids, phenols, and sugars can function as alcoholic components. Indeed, the most diverse phenols occur very often as the aglycone in plants. If, however, a sugar serves as the alcoholic component so that one sugar is linked with another through a glycosidic bond, then oligo- or polysaccharides are formed.

N-Glycosides. The alcoholic component reacts with the sugar through a NH-group. N-Glycosidic linkages are found in the nucleosides and, thus, in the nucleotides and polynucleotides.

e. Glycosidases

Glycosidic bonds can be cleaved hydrolytically by enzymes with the liberation of sugars. These enzymes are known as glycosidases. They can be classified in different groups. Thus, β-glycosidases attack β-glycosidic bonds. The reverse reaction, the synthesis of glycosides from sugars by means of glycosidases, can also occur. However, other enzymes that utilize sugar nucleotides or phosphates are usually more important for synthesis.

2. Oligosaccharides

a. Disaccharides

In the plant kingdom disaccharides are widely distributed either as such or linked to an aglycone. Let us consider the formulae of a few of them (Fig. 45). Maltose, cellobiose, and gentiobiose all consist of two glucose molecules. However, the hydroxyl groups which participate in the glycosidic linkage are attached either to different C atoms (1–4 or 1–6 linkage) or to the same C atom but in a different steric position (α- or β-position of the hydroxyl on the C atom). By far the most important disaccharides in higher plants is sucrose which is built up from glucose and fructose. The glycosidic hydroxyls of both sugars, those on the C atom 1 of glucose and C atom 2 of fructose, take part in the glycosidic bond. A consequence of this involvement of both glycosidic hydroxyls is that sucrose is a "nonreducing" sugar. This is in contrast to the other disaccharides named above in which the glycosidic hydroxyl of one of the two sugars is still free.

Sucrose can be synthesized from (a) UDPG and fructose and (b) UDPG and fructose-6-phosphate. In the latter case the product is sucrose phosphate which can be cleaved by a phosphatase to sucrose and phosphate. The enzymes catalyzing the synthesis are called sucrose synthetase (1) and sucrose phosphate synthetase (2), depending on the product.

(a) $\text{UDPG} + \text{fructose} \xrightarrow{1} \text{UDP} + \text{sucrose}$

(b) $\text{UDPG} + \text{fructose-6-P} \xrightarrow{2} \text{UDP} + \text{sucrose-P}$

$\downarrow$ phosphatase

sucrose + phosphate

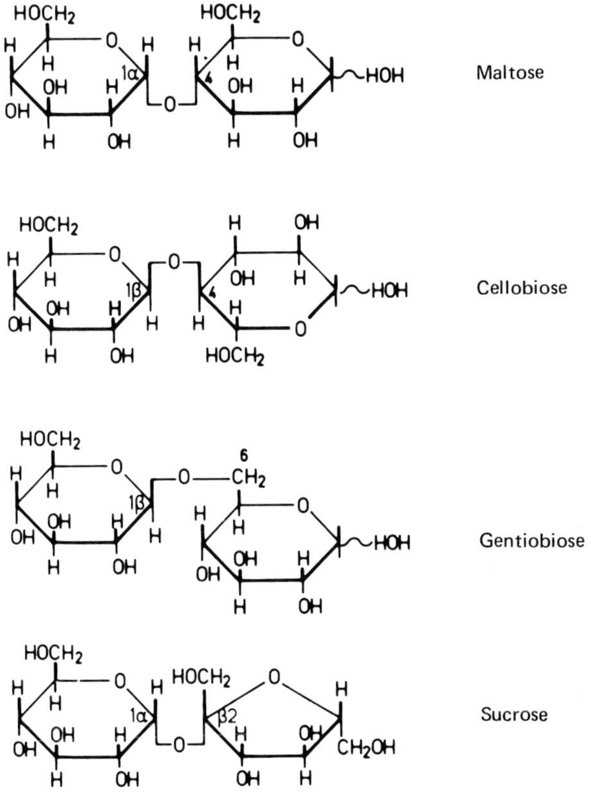

Maltose

Cellobiose

Gentiobiose

Sucrose

Fig. 45. A few disaccharides.

Sucrose can be split hydrolytically by a glycosidase known as saccharase or invertase. Invertase, because as a result if its action the plane of polarization of polarized light is "inverted": sucrose is dextrorotatory (+). In the hydrolysate the highly laevorotatory (−) fructose outweights the more weakly dextrorotatory glucose so that overall a laevorotation results. More chemically expressed, invertase is a β-fructofuranosidase since it is in the form of a β-fructofuranose that fructose takes part in the glycosidic bond of sucrose and the enzyme cleaves this bond on the side of fructose.

b. Trisaccharides

Of the trisaccharides only raffinose will be mentioned here, it being the most frequently occurring oligosaccharide in higher plants after sucrose. Its structure, expressed only in terms of the sugars and their mode of linkage is:

> galactose1−6glucose1−2fructose

That is nothing other than

> galactose1−6sucrose

It is not difficult to visualize how plants might synthesize raffinose: from UDP-galactose + sucrose, by analogy with the formation of sucrose from UDPG + fructose (or fructose-6-P). In fact, it was possible to isolate an enzyme from the broad bean (*Vicia faba*) which catalyzes the synthesis of raffinose in this way:

$$\text{UDP-galactose} + \text{sucrose} \xrightarrow{\text{enzyme}} \text{UDP} + \underline{\text{galactose1−6sucrose}}$$
$$\text{raffinose}$$

c. Tetrasaccharides

As an example of a tetrasaccharide we shall choose stachyose which occurs very frequently in the Leguminosae. Its structure is:

> galactose1−6galactose1−6glucose1−2fructose
> or galactose1−6galactose1−6sucrose
> or galactose1−6raffinose

In view of what we have just learned about the syntheses of sucrose and raffinose, it would seem probable that plants form stachyose from UDP-galactose + raffinose. But with plants one can never be sure, and according to findings of Tanner and Kandler with the bean (*Phaseolus vulgaris*) the galactose donor is not UDP-galactose but a glycoside of the sugar alcohol myoinositol and galactose called galactinol:

$$\underline{\overset{\text{galactinol}}{\text{myoinositol 1−1galactose}}} + \text{raffinose}$$
$$\downarrow \text{enzyme}$$
$$\text{myoinositol} + \underline{\text{galactinose 1−6raffinose}}$$
$$\text{stachyose}$$

3. Polysaccharides

We shall begin with compounds whose biosynthesis requires sucrose as primer just like that of the oligosaccharides mentioned above. Subsequently, we shall discuss other polysaccharides whose biosynthesis does not start with sucrose. However, even with these substances we shall come across well-known principles again, such as the transfer of a sugar from a sugar nucleotide to a particular primer.

a. Fructosans

Fructosans consist predominantly, but not exclusively, of fructose. They are synthesized according to a principle similar to that which we saw, for example, in the case of raffinose. In that case a galactose unit was attached to a primer sucrose molecule. Examination of the structure of the fructosans shows that a large number of fructose units are attached to the fructose component of a sucrose molecule. We shall see presently that the sucrose molecule acts as primer for the biosynthesis. The fructose units can be linked to each other in different ways which makes it necessary to distinguish between the inulin type and the phlein type.

The inulin type (Fig. 46). The linkages of the inulin type run from C 1 to C 2. If side chains are present they are linked to the main chain by a 6-2 linkage. Inulin itself is the prototype.

The phlein type (Fig. 46). The fructose units are linked to each other through 6-2 linkages. Side chains are linked to the main chain through 1-2 linkages.

The most important compound is inulin, which occurs as reserve material, particularly in Compositae. It consists of a chain of 32–34 fructose units which are linked to each other through β-glycosidic 1-2 linkages. Fructose is present in the form of a fructofuranose. At one end of the molecule there is a sucrose unit. The biosynthesis of inulin proceeds by the attachment of fructose units, one after the other, to the sucrose molecule. The fructose donor is presumably a sugar nucleotide, probably UDP-fructose. The presence of UDP-fructose in the inulin-containing bulbs of dahlias is consistent with this supposition. Thus, our

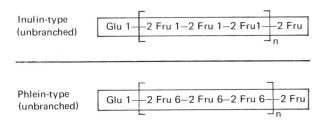

Fig. 46. Fructosans: Inulin-type (including inulin itself) and phlein-type. Only unbranched chains are shown.

schematic description of the polyfructosan structure (sucrose plus several to many fructose units) corresponds to the mode of biosynthesis.

b. Starch

Starch is the most important reserve carbohydrate of plants. Two components of starch can be distinguished according to their structure, amylose, and amylopectin. The building block of both of them is α-glucose.

Amylose (Fig. 47). Amylose consists of many α-glucose units which are linked in series by 1-4 glycosidic bonds. The number of glucose units can vary from about 200 to about 1,000. The amylose chains adopt a helical structure. I_2 can be bound in the turns of the helices. The resulting inclusion compound is blue-black. This iodine-starch reaction is used as a test for starch.

Amylopectin (Fig. 48). Amylopectin also consists of chains of α-glucose units linked 1-4 glycosidically with each other. However, in contrast to amylose, amylopectin is branched. The side chains are attached through α-glycosidic 1-6 linkages. Amylopectin, which lacks rather long helical glucose chains, gives only a pink coloration with iodine.

How is starch, i.e. how are amylose and amylopectin, formed in the plant? We can already foresee that two kinds of enzymes will be necessary: those that can form α-glycosidic 1-4 links and those that can form α-glycosidic

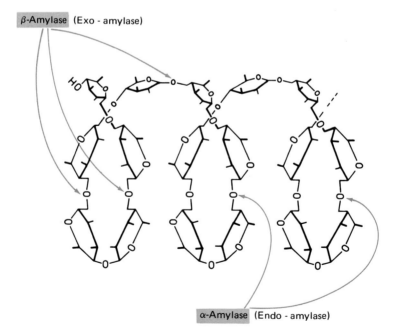

Fig. 47. Structure of amylose and sites of attack of α- and β- amylases (modified from Karlson 1970).

1-6 linkages. The expected enzyme systems have indeed been found.
ADPG-and UDPG-Starch Transglucosylases. These enzymes form α-glycosidic 1-4 bonds. They use either ADPG or UDPG as glucose donor. The glucose is transferred to a primer. The smallest possible glucose acceptor is maltose. Thus, the mechanism of biosynthesis is the same, in principle, as that for the carbohydrates discussed above: additional sugar units are transferred to a primer molecule.

$$\text{ADPG (UDPG)} + \underset{\text{acceptor}}{\underline{\text{G1}-\text{4G}}} \xrightarrow{\text{starch transglucosylase}} \text{G1}-\text{4G1}-\text{4G} + \text{ADP (UDP)}$$

One of the questions which has still not been answered with certainty is whether ADPG or UDPG is preferred as glucose donor in plants. The transglucosylases turn over ADPG much more rapidly than UDPG but the concentration of the latter in plants is up to 10 times higher.
P-Enzyme. This is a starch phosphorylase, which can also catalyze the formation of 1-4 linkages, at least, in a test tube. Here again a primer, which must contain at least three glucose units, is necessary. It is still a matter of controversy whether the P enzyme plays a role in the biosynthesis of starch in plants. It is possible that it can only attach glucose units to the growing amylopectin, and not to amylose.

$$\text{G1}-\text{P} + \underset{\text{acceptor}}{\underline{\text{G1}-\text{4G1}-\text{4G}}} \xrightarrow{\text{P-enzyme}} \text{P} + \text{G1}-\text{4G1}-\text{4G1}-\text{4G}$$

Q-Enzyme. This enzyme forms 1-6 linkages such as occur in amylopectin. It can attach glucose chains to other glucose chains through 1-6 linkages. In this way, the 1-6 branched amylopectin systems arise.

$$\text{G1}-\text{4G1}-\text{4G}-\text{enzyme} + \text{G1}-\text{4G1}-\text{4G1}-\text{4G1}-\text{4G1}-\text{4G}-$$
$$\longrightarrow \text{G1}-\text{4G1}-\text{4G1}$$
$$|$$
$$6$$
$$\text{G1}-\text{4G1}-\text{4G1}-\text{4G1}-\text{4G}- + \text{enzyme}$$

As already mentioned, starch is the most important reserve carbohydrate of plants. Such reserve materials must be mobilized when they are needed. This, in turn, requires that they be broken down into smaller units which can be transported to the sites where they are needed. Starch is degraded by means of phosphorylases and hydrolases.
The P-enzyme already mentioned is one of the *phosphorylases.* It cleaves the α-glycosidic 1-4 linkages with the incorporation of phosphate. The cleavage product is glucose-1-phosphate which can very readily be utilized further in metabolism.

The *hydrolases* are enzymes that break bonds with the incorporation of water and that, under certain conditions, can catalyze bond formation. A large number of hydrolases are needed to degrade starch to glucose. They belong to the subgroup of the glycosidases.

α-Amylase. This enzyme splits α-glycosidic 1-4 linkages. These linkages must be 6 to 7 glucose units removed from the end of the chain. Thus, the site of attack of the enzyme lies within the starch molecule, i.e. α-amylase is an endoamylase (Fig. 47). 1-6 Linkages are not hydrolyzed; neither do they inhibit the α-amylase; it simply "by-passes" them.

β-Amylase. This enzyme also cleaves α-glycosidic 1-4 linkages. In contrast to α-amylase, however, β-amylase works from the chain ends, splitting off a maltose unit each time (Fig. 47). For this reason β-amylase is also known as exoamylase. 1–6 Linkages are not cleaved and also are not by-passed. Thus, if amylopectin is treated with β-amylase the branched nucleus remains intact. The products of incomplete hydrolysis are known as limit dextrins.

Isoamylase. We have not yet mentioned enzymes that are capable of splitting α-glycosidic 1-6 linkages. These enzymes are the isoamylases. They have been detected many times in higher plants.

Maltase. Up to now, we have encountered enzymes that can cleave α-glycosidic 1-4 and 1-6 linkages. The smallest cleavage products thus obtained are isomaltose (a disaccharide containing two glucose units in α-glycosidic 1-6 linkage, which corresponds to the branch points in amylopectin) and maltose. Of these, maltose is much the more predominant. It can be broken down into two glucose molecules by cleavage of its α-glycosidic 1-4 linkage. The enzyme responsible for this is called after its substrate, maltase.

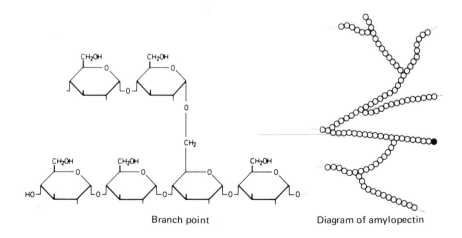

Branch point   Diagram of amylopectin

Fig. 48. Structure of amylopectin (from Karlson 1970).

c. Cellulose

Cellulose is composed of β-glucose units, linked to each other through 1-4 linkages. The number of such glucose molecules linked in series varies. It is likely, though, that a macromolecule comprises 3,000–10,000 glucose units. In contrast to the helices of amylose, cellulose molecules exist in an extended form. Several cellulose chains arranged in parallel form crystalline regions over parts of their chain length, which are called micelles or elementary fibrils (Fig. 49). Less ordered cellulose chains lie between and around micelles and form paracrystalline regions. Up to 100 cellulose chains can participate in the formation of a micelle. The diameter of a micelle is about 5 mμ. Several micelles or elementary fibrils can assemble to form microfibrils whose diameter can range up to 30 mμ. On the periphery of such microfibrils crosslinkages to noncellulosic material can be formed. The microfibrils are embedded in a ground substance, a matrix, in the cell wall and in this matrix other carbohydrate polymers of noncellulosic character are found. Examples of these are the xylans, which consist predominantly of xylose, the mannans, which are

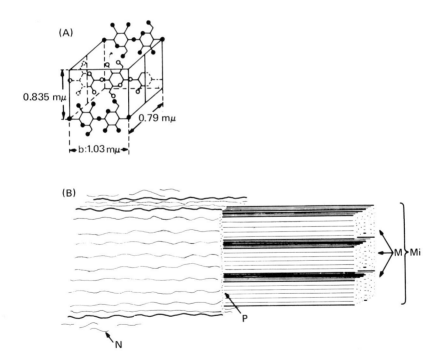

Fig. 49. Model of cellulose structure. (A) Deposition of cellulose chains. (B) Construction of a cellulose microfibril. Up to 100 cellulose chains are deposited side by side as in (A) to form a micelle or elementary fibril (M). Several micelles form a microfibril (Mi) which contains paracrystalline cellulose regions (P). Non-cellulosic carbohydrates = N (modified from Clowes 1968).

composed principally of mannose, other groups of carbohydrates, which are difficult to define and are known collectively as hemicelluloses, and pectin substances. These materials can also combine to form more or less crystalline regions. Finally, it has been demonstrated that the cell wall also contains proteins. One of these is extensin which we shall discuss later (p. 238).

Earlier investigations of the biosynthesis of cellulose were carried out on *Azobacter xylinum*. This bacterium possesses a cellulose membrane. Cell-free systems of *Azobacter* use UDPG as glucose donor for the synthesis of cellulose. That does not necessarily imply that higher plants synthesize their cellulose in the same way. Admittedly, it was at first assumed that UDPG is the glucose donor in higher plants. However, further experiments make it seem very likely that GDPG is the glucose donor in higher plants. The corresponding transfer reactions from GDPG to particular acceptor molecules have been demonstrated many times in a cell-free system.

Cellulose is a very robust substance. However, it must be capable of being degraded rapidly, otherwise a cellulose film would cover the surface of the earth in a short time. This degradation is carried out by different microorganisms. With the help of *cellulases,* they break down the cellulose chains to cellobiose. By means of the appropriate enzymes, *cellobiases,* cellobiose can then be broken down to the fundamental building block β-glucose.

In higher plants cellulases are quite rare. They occur principally in seedlings where they have the function of facilitating the bursting of the seed coat by degrading cellulose.

d. Pectin substances

Pectin substances are macromolecules that consist essentially of galac-

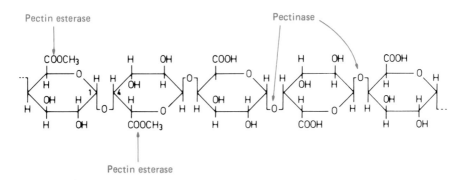

Fig. 50. Diagram of a pectin chain showing sites of attack of pectin esterases and pectinases.

turonic acid units linked by α-glycosidic 1-4 linkages. In each case several hundred units are involved.

It is necessary to distinguish between pectin acids, pectins, and protopectins (Fig. 50). The *pectin acids* consist of galacturonic acid chains. Thus, they are polygalacturonic acids. In the *pectins* some of the carboxyl groups of the galacturonic acid units are methylated. Thus, pectins are partial methyl esters of pectin acids. Finally, *protopectins* are insoluble pectin substances of varying composition. Probably carboxyl groups of the pectins which are still free are crosslinked by divalent metal ions, e.g. Ca^{++} and Mg^{++}, and also by phosphoric acid. Protopectins are found particularly frequently in the middle lamellae. The synthesis of the pectin substances starts from glucuronic acid. UDP-Glucuronic acid is converted initially into UDP-galacturonic acid by a 4-epimerase. UDP-Galacturonic acid then serves as galacturonic acid donor in the synthesis of the pectin substances. The methyl groups which are found in the pectins are only incorporated after the main polygalacturonic acid framework has been completed. Here, as in many other instances in biology, the donor of the methyl group is *S*-adenosyl methionine (cf. Fig. 108).

Some data are also available on the degradation of the pectin substances. *Pectin esterases* cleave the methyl groups and *pectinases* break the α-glycosidic 1-4 bond between the galacturonic acid building blocks. Both kinds of enzymes are frequently found in microorganisms. In particular, phytopathogenic bacteria and molds use them to attack the wall substance of plant cells so that they can then penetrate the cells. In higher plants, on the other hand, enzymes that degrade pectin have been found in seedlings. Here again it is likely that they assist in the bursting of the seed coat.

Biological Oxidation

We have already noted several times the involvement of the most important energy reservoir of the cell, ATP, in reactions without mentioning in detail where this ATP actually comes from. As a matter of fact we have mentioned one source of ATP: ATP is formed in the primary processes of photosynthesis. This kind of ATP formation, which is linked to the light reactions of photosynthesis, is called photophosphorylation (cf. page 47). The living organism is, however, equipped with other means of producing ATP and they are connected with biological oxidation.

We shall now concern ourselves with biological oxidation and this is a topic we can deal with quite briefly. The principles of biological oxidation are the same for all organisms, even if differences of detail such as those of the respiratory chain exist between plants and animals.

Biological oxidation proceeds in four stages if we introduce a carbohydrate, such as the ubiquitous glucose, into the process (Fig. 51):

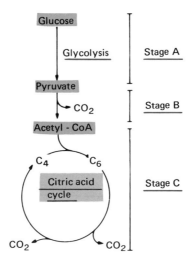

Fig. 51. The stages of biological oxidation (endooxidation in the respiratory chain is not shown).

74

A. *Glycolysis.* Degradation of glucose to pyruvate.

B. *Oxidative Decarboxylation of Pyruvate.* A relatively short step during which a 2-C body arises from the 3-C body pyruvate with the elimination of CO_2.

C. *The Citric Acid Cycle.* Degradation of the 2-C body furnished by stage B to CO_2. It is also known as the Krebs cycle after Krebs who, together with other scientists, elucidated this pathway; a third name is tricarboxylic acid cycle which it owes to the participation of acids with three carboxyl functions.

D. *Terminal Oxidation in the Respiratory Chain.* The hydrogen which is extracted from the substrates in stages A to C is finally combined with oxygen to form water. For this to occur a hydrogen transport takes place along a chain of redox systems, i.e. via an electron transport chain. Energy set free by this transport is utilized for the formation of ATP.

A. Glycolysis

At least the formulae appearing in the first steps of glycolysis should not cause any great difficulty. They are familiar to us from the interconversions of the monosaccharides and the Calvin cycle. But let there be no mistake, glycolysis is in no way a reversal of the Calvin cycle. Indeed, the enzymes concerned are only in part the same as those of the Calvin cycle. And, finally, the events take place in different locations: the Calvin cycle occurs only in the chloroplasts whereas glycolysis takes place in the cytoplasm itself. Let us now consider the fate of a glucose molecule subject to glycolysis (Fig. 52). In reactions with which we are already familiar it is converted to glucose-6-phosphate which is converted to fructose-6-phosphate by isomerization. Phosphofructokinase attaches an additional phosphate unit so that fructose-1,6-diphosphate is formed. In a reaction with which we are already familiar aldolase converts fructose-1,6-diphosphate to triose phosphate which is a mixture of 3-phosphoglyceraldehyde and dihydroxyacetone phosphate. The equilibrium of the aldolase reaction lies on the side of the hexose and that for the triose phosphates favors dihydroxyacetone phosphate. However, the equilibrium is displaced in the direction of glycolytic degradation by removal of 3-phosphoglyceraldehyde from the system (Fig. 53). This removal is brought about by phosphotriose dehydrogenase, a HS-enzyme, i.e. an enzyme with a functionally important HS group. The enzyme adds to the carbonyl function of the 3-phosphoglyceraldehyde and NAD^+ takes up 2H from the addition product, thus giving rise to an energy-rich thiol ester bond. The enzyme is now displaced by phosphate at this thiol ester bond. In this way phosphotriose dehydrogenase catalyzes the formation of 1,3-diphosphoglyceric acid in which an "energy-rich" phosphate has become bound to C1. This is an important step because in the next reaction this energy-rich phosphate is cleaved to form ATP. The second

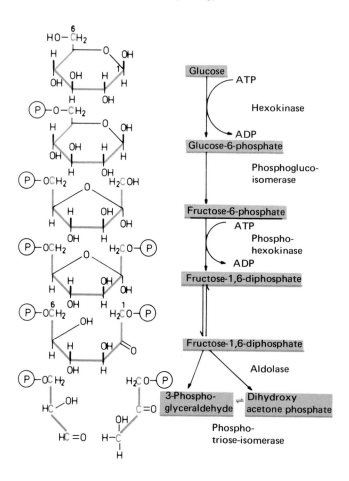

Fig. 52. Glycolysis I: from glucose to triose phosphate.

product of this reaction is 3-phosphoglyceric acid. Phosphoglyceromutase then catalyzes the conversion of 3-phosphoglyceric acid into 2-phosphoglyceric acid. Removal of the elements of water from the latter leads to 2-phosphoenolpyruvate, a compound with an energy-rich phosphate bond (the enzyme responsible for this conversion is called enolase). In the next step, and with the help of pyruvate kinase, the energy of this bond is utilized to form ATP. Concomitantly, the enol form of pyruvate, which exists in equilibrium with its keto form, is set free.

    That completes what needs to be said about glycolysis. What is true of biological oxidation as a whole is also valid for its constituent process glycolysis: it serves to supply certain intermediates that can be used for synthesis and energy. In the present discussion we are concerned only with the supply of energy:

2 ATP are consumed in the phosphorylation of glucose: −2 ATP

2 ATP are furnished by pyruvate kinase for each glucose introduced

This phosphate is derived from the ATP utilized in the initial phosphorylation: + 2 ATP

Up to this point the ATP account is balanced. However,

2 additional ATP are furnished per glucose molecule by the transition

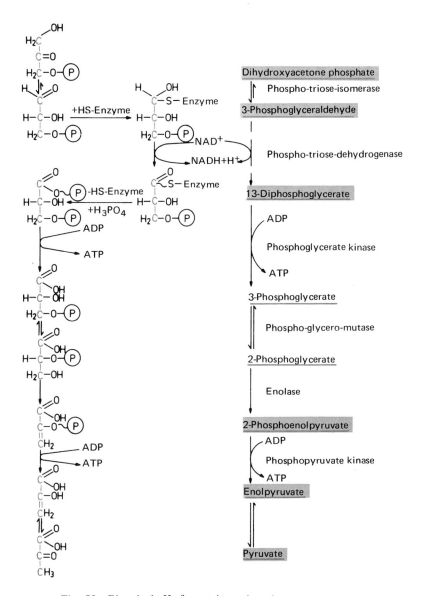

Fig. 53. Glycolysis II: from triose phosphate to pyruvate.

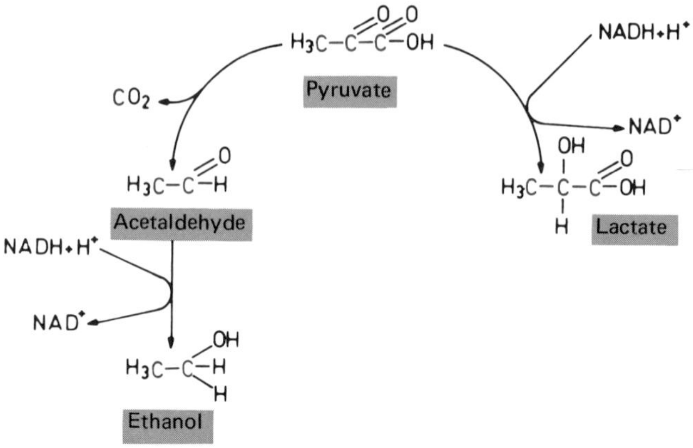

Fig. 54. Scheme of alcoholic and lactic acid fermentation.

from 1,3-diphosphoglyceric acid to 3-phosphoglyceric acid: + 2 ATP
Since this ATP formation occurs directly on the substrate it is referred to
as *substrate level phosphorylation*. It supplies 2 ATP per molecule of glucose
introduced.

One more point: the phosphotriose dehydrogenase supplies two
NADH + H$^+$ per molecule of glucose. The fate of this reduced NAD$^+$ is
varied. One possibility is for it to be utilized in the respiratory chain for
the formation of more ATP. More about that later. The NADH + H$^+$ furnished by glycolysis can, however, also be used in fermentation processes.

Fermentations

Let us consider very briefly two kinds of fermentations, alcoholic fermentation and lactic acid fermentation (Fig. 54).

Alcoholic fermentation. The pyruvate which is furnished by glycolysis
is decarboxylated to acetaldehyde and this is further reduced to ethanol
by the NADH + H$^+$ mentioned above.

Lactic acid fermentation. The pyruvate supplied by glycolysis is
reduced directly to lactate by means of NADH + H$^+$ again. Thus, NADH
+ H$^+$ is consumed in both kinds of fermentation. Thereby the possibility
of a further gain of energy as would have been the case had the reduced
coenzyme been fed into the respiratory chain is lost. Thus, the energy
gain in the case of both kinds of fermentation is limited to the two ATP
molecules, which are derived from the substrate level phosphorylation.

B. Oxidative Decarboxylation of Pyruvate, Formation of Active Acetate

The second stage of biological oxidation can be summarized as follows
(Fig. 55): Here, for the first time, we encounter a multienzyme complex.

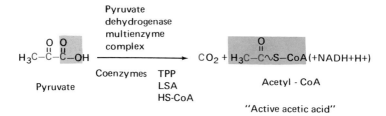

Fig. 55. Schematic representation of the formation of active acetic acid. The participation of NAD^+ (cf. Fig. 57) is merely hinted at.

This is an association of several enzymes that usually catalyze a sequence of closely linked reactions. The kind of reactions involved will become clear by examining the function of the coenzymes assigned to each of the apoenzymes (Figs. 56, 57).

TPP: Thiamine pyrophosphate. The decarboxylation of pyruvate to acetaldehyde takes place on TPP. The acetaldehyde bound to TPP is called "active aldehyde". From its function TPP derives the name cocarboxylase.

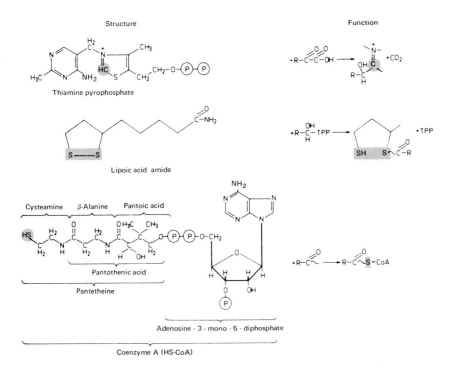

Fig. 56. Coenzymes which take part in the formation of active acetate (cf. Fig. 57 for the function of flavoprotein and NAD^+).

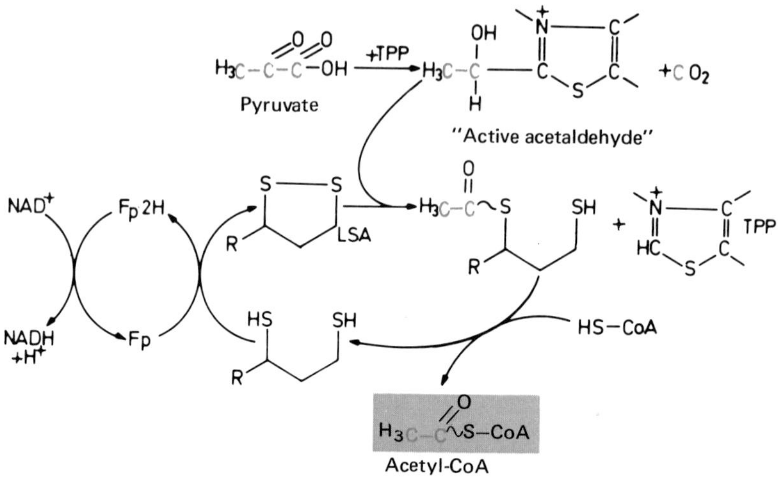

Fig. 57. Formation of active acetate.

LAA: Lipoic acid amide. Acetaldehyde is dehydrogenated to the acetyl group on LAA. It is brought about by the transfer of acetaldehyde from TPP to LAA with accompanying reductive cleavage of the disulphide bridge. The newly formed bond is "energy-rich." TPP is liberated.

HS-CoA: Coenzyme A. HS-CoA accepts the acetyl residue with its sulphydryl (HS) group. The acetyl residue linked to CoA through an energy-rich thiol ester bond is known as acetyl CoA or active acetate. This active acetate is a key substance in metabolism, to which we shall often have need to return. We are indebted to Lynen for a major part of our knowledge of this nodal point of metabolism.

We must now discuss the regeneration of LAA. The liberation of reduced LAA, which bears two HS-groups, occurs simultaneously with the formation of active acetate. Removal of 2H leads to the reformation of the oxidized form of LAA with an intact disulphide ring. Initially the hydrogen atoms are accepted by a flavoprotein, which transfers them to NAD^+. The NADH + H^+ so formed can then enter the respiratory chain where it is utilized for ATP formation. We have discussed the decarboxylation of pyruvate in some detail. As justification for this it should be pointed out that thiamine is identical with vitamin B_1. Thus, thiamine pyrophosphate serves as a good example of the kinds of function vitamins can exercise: they can be coenzymes or constituents of coenzymes. In addition, we shall note other cases of decarboxylation which proceed, in part, according to the same mechanism. Two instances are alcoholic fermentation and the decarboxylation of α-ketoglutarate in the citric acid cycle which we come to now.

C. Citric Acid Cycle

Our subject in this chapter is the total degradation of glucose to CO_2 and H_2O and the production of ATP which this degradation allows. Up to this point degradation has led to a 2-C body, active acetate. These last two C atoms must now be eliminated in the form of CO_2 and that takes place in the citric acid cycle (Fig. 58).

Active acetate is linked to a 4-C body, oxalacetate, by the "condensing enzyme" in a kind of aldol condensation. The product is citrate, which exists in equilibrium with *cis*-aconitate and isocitrate, the equilibrium being controlled by aconitase.

Isocitrate dehydrogenases transfer hydrogen from the secondary hydroxyl group of isocitrate to either NAD^+ or $NADP^+$. $NADH + H^+$ then

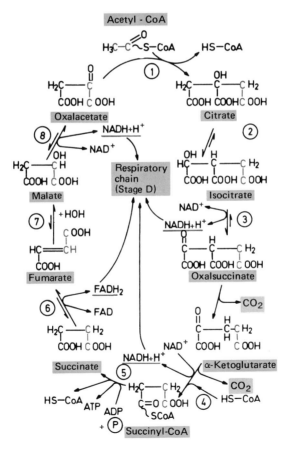

Fig. 58. The citric acid cycle. The enzymes involved are encircled. 1 = condensing enzyme, 2 = Aconitase, 3 = Isocitrate dehydrogenase, 4 = α-Ketoglutarate dehydrogenase, 5 = Succinyl-CoA-synthetase, 6 = Succinate dehydrogenase, 7 = Fumarate dehydrogenase, 8 = Malate dehydrogenase.

enters the respiratory chain and NADPH + H$^+$ can be utilized for synthesis. The product of the dehydrogenation is a labile substance, oxalsuccinate, which is decarboxylated to α-ketoglutarate by isocitrate dehydrogenases too. With that, one of the two C atoms introduced into the cycle as active acetate has been lost as CO_2.

α-Ketoglutarate now becomes the substrate of a multienzyme complex called α-ketoglutarate dehydrogenase. The mechanism of action of this multienzyme complex is similar to the one involved in the oxidative decarboxylation of pyruvate to active acetate: the coenzymes are TPP, LAA, FAD, and NAD$^+$, and, finally, HS-CoA. The products are CO_2 (and with this the second C atom has been eliminated), NADH + H$^+$, and succinyl CoA, the CoA derivative of succinic acid.

Succinyl CoA synthetase (at the mention of this name we must not forget that enzymes catalyze reactions in the forward and in the backward direction) breaks down succinyl CoA into succinate and HS-CoA. In so doing, the energy of the thiol ester bond is used for the formation of ATP. At least that is so with the succinyl CoA synthetase from spinach. In animals, GTP is formed at this point, which then transfers its terminal phosphate residue to ADP, thus leading indirectly to ATP.

A flavoprotein with FAD as its prosthetic group, succinate dehydrogenase, now dehydrogenates succinate to fumarate. Succinate dehydrogenase is particularly interesting in that it is a component of the respiratory chain, as we shall see presently. One further point may be noted here: succinate dehydrogenase is inhibited by malonate, a structural analog of succinate (Fig. 59). This is an outstanding example of competitive inhibition (page 192).

In the next step nature shows us a trick which is to be found time and again: water is first added across a double bond and then hydrogen is abstracted from the addition product. In the present case fumarate hydratase catalyzes the addition of water to the double bond of fumarate to give malate which is then dehydrogenated to oxalacetate by malate dehydrogenase. NAD$^+$ serves as the hydrogen acceptor.

With the formation of oxalacetate the cycle is completed. Let us draw up a balance sheet: in the citric acid cycle the two C atoms introduced in the form of active acetate are formally eliminated as CO_2. Per molecule of active acetate one ATP is formed. Further, hydrogen becomes bound to coenzymes: 3 NADH + H$^+$ and 1 FAD $\cdot$ H$_2$ are furnished per molecule of active acetate introduced.

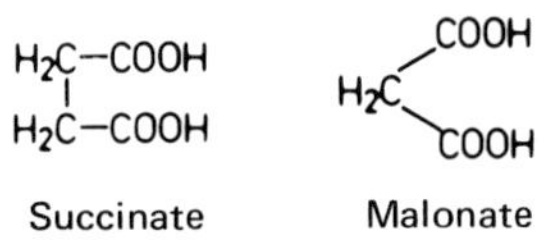

Fig. 59. Succinate and malonate.

D. The Respiratory Chain

We are interested now in the fate of the reduced coenzymes or, more precisely, of the hydrogen with which the coenzymes are charged. This hydrogen is introduced into a chain of redox systems. Electron transport chains of this kind are not new to us. We already know of them in connection with the primary processes of photosynthesis. Here again, just as in a few transport steps in that case, we must visualize 2 H as $2 H^+ + 2 e-$.

The electron transport chain into which the hydrogen of the reduced coenzymes of the citric acid cycle is channelled is called the respiratory chain. In this chain hydrogen or electrons are conducted downhill from redox systems of high to redox systems of low electron pressure. The energy set free in the process is utilized for the formation of ATP. At the end of the chain hydrogen is oxidized to water.

What are the components of the respiratory chain? At the outset it must be said that the respiratory chain in animals, bacteria, and many lower plants is different from that in higher plants. Some of the details are still unexplained. Nonetheless the following redox systems are components of the respiratory chain of higher plants:

NAD^+

Flavoproteins

Ubiquinone (Fig. 104)

3 Cytochromes b (Cytochrome b complex, typical of higher plants)

2 Cytochromes c (c_{549} and c_{547}. The naming of individual cytochromes of the a-, b-, c- groups is based on one of the characteristic absorption maxima that appear in the reduced state.)

2 Cytochromes a (a and a_3 = cytochrome oxidase complex)

The serial arrangement of the redox systems in the respiratory chain is shown in Fig. 60. The scheme is quite tentative. $NADH + H^+$ is supplied by the reactions of the citric acid cycle, described above, and hydrogen or, rather, the electrons derived from it are transported from left to right. At the right end of the chain is the cytochrome oxidase complex, consisting of cytochrome a and cytochrome a_3. In animals these two components are very tightly coupled. This seems not to be the case in plants. However, up to now attempts to isolate a and a_3 separately have been no more successful in plants than in animals. Cytochrome oxidase a_3 makes direct contact with oxygen and is therefore known, together with a few other enzymes, (peroxidases, catalases, and phenolases), as a "direct" oxidase. The name cytochrome oxidase comes from the fact that the complex accepts electrons from cytochrome c, its immediate predecessor in the series, and, thus, oxidizes it. The electrons come from the hydrogen which was introduced initially ($2H = 2H^+ + 2e^-$). By means of them cytochrome oxidase reduces oxygen to O^{--} which then combines with $2H^+$ to form H_2O.

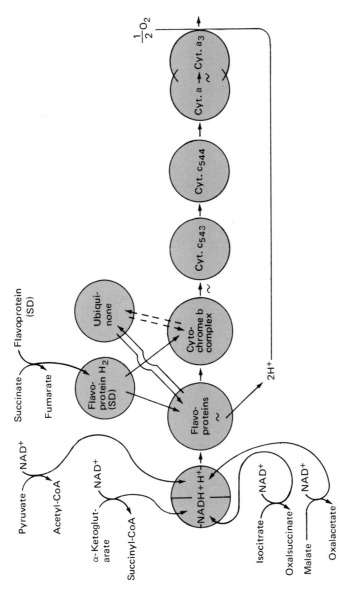

Fig. 60. The respiratory chain of higher plants. Ubiquinone appears to serve as an "electron reservoir." ∼ = probable site of ATP formation. SD = succinate dehydrogenase. It used to be assumed that, with the exception of the reaction catalyzed by SD, the hydrogen acceptor in dehydrogenation reactions was NAD^+ and that the hydrogen then entered the respiratory chain in the form of $NADH+H^+$. In reality the situation is more complicated since the lipoic acid oxidizing flavoproteid of the pyruvate dehydrogenase and the α-ketoglutarate dehydrogenase complexes—in both cases the same flavoproteid is involved—can establish direct contact with the flavoproteins of the respiratory chain just like succinate dehydrogenase. ∼ associated with encircled flavoproteins means that ATP can be formed as a result of transitions between the various flavoproteins, except those involving SD.

Reduced flavoprotein is also supplied by the citric acid cycle. It transfers its electrons to the cytochrome b complex. They are then transported to the right as outlined above.

Three ATPs are formed for each NADH + H^+ introduced into the respiratory chain and 2 ATPs per reduced flavoprotein. That is understandable because the reduced flavoprotein enters the respiratory chain somewhat later than NADH + H^+. The site of ATP formation on the respiratory chain is not known exactly for higher plants.

ATP formation in the respiratory chain is known as *oxidative phosphorylation*. We have already become acquainted with two other types of ATP formation, photophosphorylation and substrate level phosphorylation. The terminal oxidation of the respiratory chain concludes biological oxidation. We have mentioned that the whole process has a dual function, namely, the production of energy and the supplying of intermediates for synthesis. We shall now draw up an overall balance sheet for biological oxidation with regard to ATP production. In doing so we shall refer to glucose as starting material but it should be quite clear that other substances can also be used as substrates in the degradative pathway. The ATP yield for each of the stages of degradation is given in Table 4. According to that, a total of 38 molecules of ATP are derived from one molecule of glucose (Fig. 61).

That completes our discussion of energy production by biological oxidation. Nothing has yet been said about the starting materials for synthesis which accrue in the course of biological oxidation. We shall deal

Table 4. Yield of ATP in the individual stages of biological oxidation calculated for the degradation of one molecule of glucose

Stage		in the terminal oxidation
A. Glycolysis		
3-phosphoglyceraldehyde →		
1,3-diphosphoglyceric acid:	$2\,NADH+H^+ →$	6 ATP
1,3-diphosphoglyceric acid		
→ 3-phosphoglyceric acid:		2 ATP
		(substrate chain phosphorylation)
B. Formation of active acetate	$2\,NADH+H^+ →$	6 ATP
C. Citric acid cycle		
Isocitrate → α-Ketoglutarate	$2\,NADH+H^+ →$	6 ATP
α-Ketoglutarate → Succinyl-CoA:	$2\,NADH+H^+ →$	6 ATP
Succinyl-CoA → Succinate		2 ATP
		(substrate chain phosphorylation)
Succinate → Fumarate	$2\,FAD-H_2→$	4 ATP
Malate → Oxalacetate	$2\,NADH+H^+ →$	6 ATP
	Sum	38 ATP

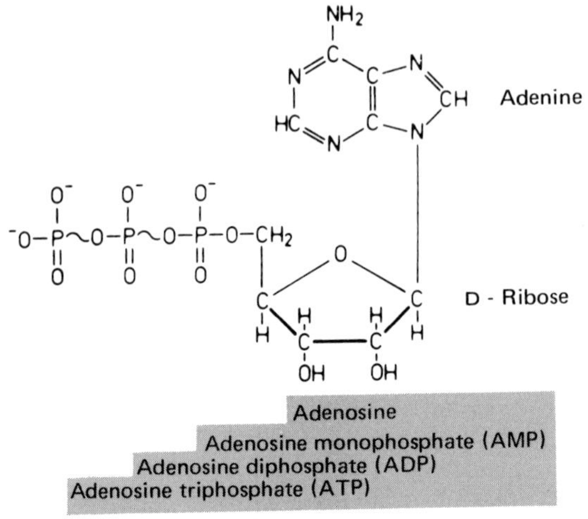

Fig. 61. The structures of AMP, ADP, and ATP (from Lehninger 1969).

with them one at a time as we discuss the metabolism of particular classes
of substances in the coming chapters.

E. Mitochondria as Power Plants

As already mentioned, the enzymes of glycolysis are located in the
cytoplasm. One the other hand, the enzymes and cofactors for the formation of active acetate, the citric acid cycle and the respiratory chain, including the systems of oxidative phosphorylation, are found in the
mitochondria. The enzymes for the β-oxidation of fatty acids (page 95)
are also stationed there. Individual components are also known to occur in
the cytoplasm but the complete systems are, nonetheless, limited to the
mitochondria.

The form of the mitochondria can vary very much according to the
type of cell and species. Often they are elongated "ellipsoids" of 1 μ
diameter. Their number also varies from about 10 to 200,000. Whereas
they are present in aerobic cells they are absent from anaerobic cells.

Two membranes form the inner and outer surfaces of the
mitochondria. Both appear to show the kind of cross-section discussed for
the unit membrane: protein layer-lipid layer-protein layer (Fig. 81). The
outer membrane covers the surface of the mitochondrium tightly but the
inner membrane is deeply folded and can form, for example, cone-like
projections in the interior of the mitochondrium, the *mitochondrial cristae*.
The space enclosed within the inner membrane is filled with a
cytoplasmic ground substance, the matrix.

There are various hypotheses as to where the systems described
above are localized. One of them is illustrated in Fig. 63. Even though the

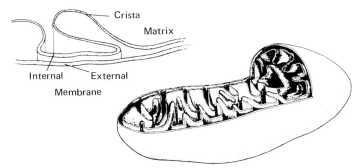

Fig. 62. Construction of a mitochondrium (after Lehninger (1969).

details are unsettled it has been established that the components of the respiratory chain lie in close spatial association with those of oxidative phosphorylation and that both systems are localized in or on the inner mitochondrial membrane. Thus, for example, spherical bodies, which show ATPase activity have been detected electron microscopically on that surface of the inner membrane that is oriented towards the matrix. Probably they are involved in ATP synthesis *in vivo* and thus are components of oxidative phosphorylation. On the other hand, the components of the citrate cycle and those of the β-oxidation pathway of the fatty acids (p.

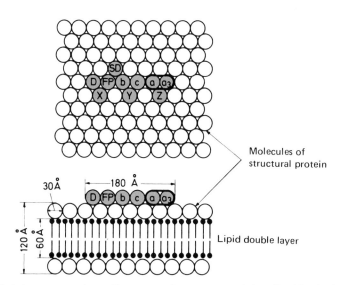

Fig. 63. Model representation of how a respiratory group is localized in the inner mitochondrial membrane. Above, side view; below, cross-section. D = NAD$^+$, FP = flavoprotein, SD = flavoprotein of succinate dehydrogenase, b = cytochrome B complex, c = cytochrome c_{543} + C_{544}, a+a$_3$ = cytochrome oxidase complex. X, Y, Z = links to oxidative phosphorylation (modified from Lehninger 1969).

95) are found in the matrix. Just as with the chloroplasts two construction principles are apparent in the mitochondria:

(1) Surface enlargement by invaginations of the inner membrane.

(2) Compartmentalization by localization of particular systems in the membranes or in the matrix.

Fats

One of the first groups of substances that is derived from the intermediates of biological oxidation is the fats. Chemically the fats are esters formed between the trihydric alcohol glycerol and fatty acids (Fig. 64). All three hydroxyl groups of glycerol can be esterified. The resulting triglycerides are known as neutral fats. It may also be that one or two of the hydroxyl groups are not esterified with fatty acids. They may remain free or bear other substituents, e.g. galactose. In this chapter we shall be concerned only with the neutral fats, which, for the sake of convenience, we shall refer to as "fats." Fats usually have a very heterogeneous composition. That means that, as a rule, the three hydroxyl groups of glycerol are esterified with three different fatty acids. So we need first to look into the chemical constitution of the fatty acids.

A. Chemical constitution of the Fatty Acids

The fatty acids, which are important constituents of the fats, are chain-like molecules containing an even number of C atoms. This fact alone suggests that the fatty acids may be synthesized from 2-C units, a supposition which proves to be correct. A few of the more important fatty acids are shown in Figure 65. In the fats of plants the predominant saturated fatty acids are palmitic and stearic acids, and the predominant unsaturated fatty acids are oleic, linoleic, and linolenic acids. There is a point to be noted about these unsaturated fatty acids. The constitution of the

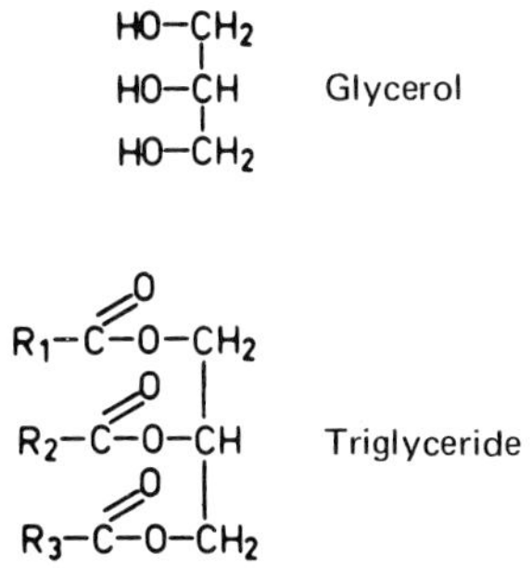

$$HO-CH_2$$
$$HO-CH \quad \text{Glycerol}$$
$$HO-CH_2$$

$$R_1-C\overset{O}{\diagup}-O-CH_2$$
$$R_2-C\overset{O}{\diagup}-O-CH \quad \text{Triglyceride}$$
$$R_3-C\overset{O}{\diagup}-O-CH_2$$

Fig. 64. Structure of a neutral fat.

Plant Physiology

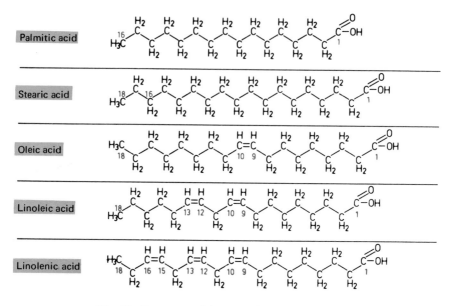

Fig. 65. Structure of the most important fatty acids.

three acids mentioned can be easily derived if one bears two facts in mind:

(1) The double bonds are not conjugated but are always separated by a CH_2 group.

(2) The first double bond lies between C atoms 9 and 10, whereby the C atom of the carboxyl group is taken as No. 1. Thus, the first and only double bond in oleic acid lies between C 9 and 10, linoleic acid has an additional double bond between C 12 and 13, and linolenic acid, finally, has a third double bond between C 15 and 16.

B. Biosynthesis of the Fatty Acids

The starting materials are acetyl CoA and malonyl CoA. Successive molecules of malonyl CoA are added to a primer molecule of acetyl CoA with accompanying decarboxylation. Malonate consists of three C atoms. After decarboxylation a 2 C body remains. Thus, in effect, the biosynthesis of the fatty acids consists in arranging 2 C units in series, as we had already deduced from the constitution of the fatty acids. In discussing the biosynthesis in detail the following constituent processes have to be recognized:

1. Formation of Malonyl CoA (Fig. 66)

This can occur in two ways, one of which is by the fixation of CO_2 in acetyl CoA. The coenzyme which is effective here is biotin charged with

CO_2, "active CO_2." This seems to be the main pathway adopted by those parts of the higher plants which are above ground. In the roots, on the other hand, another pathway is widely used: a peroxidase oxidizes oxalacetate to CO_2 and malonate. Malonate is then converted to its CoA ester, malonyl CoA.

2. Fatty Acid Synthesis Proper (Fig. 66)

It can be subdivided into initiation reaction, chain elongation, and termination reaction.

a. Initiation reaction

In the initiation reaction acetyl CoA transfers its acetyl group to one of the HS-groups of a multienzyme complex. All of the enzymes necess-

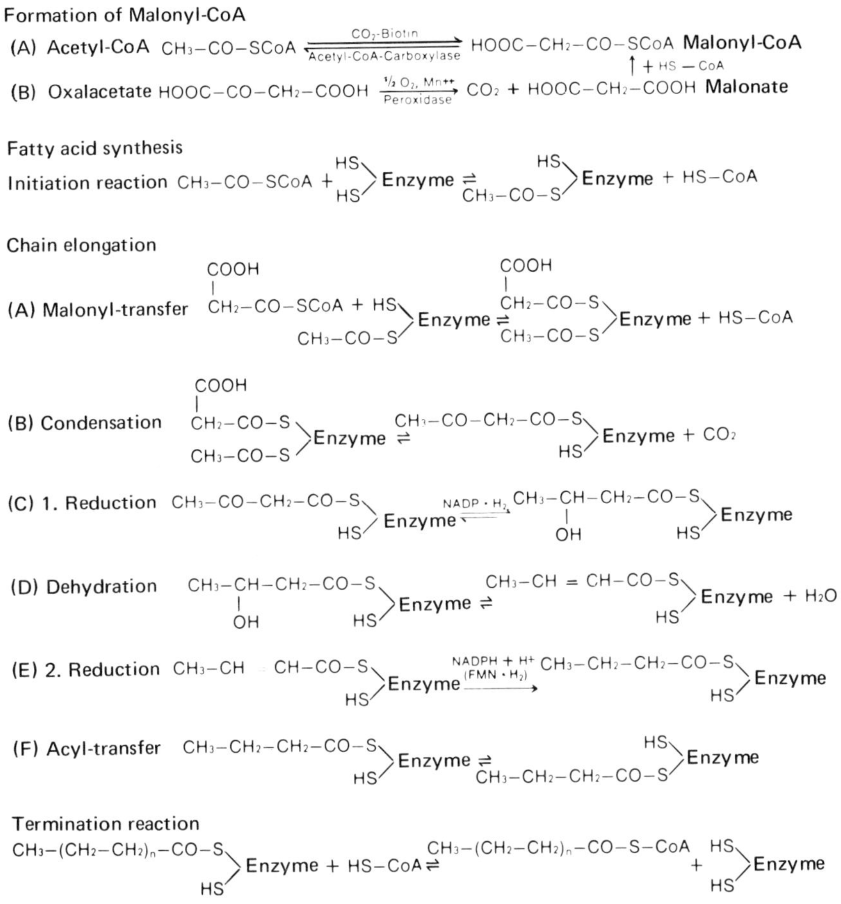

Fig. 66. Biosynthesis of the fatty acids (from Hess 1968).

ary for the synthesis of the fatty acids are integrated in this multienzyme complex. Thus, one speaks of a fatty acid synthetase.

b. Chain elongation

Chain elongation begins with a malonyl transfer. A malonyl group is transferred from malonyl CoA to a second HS-group of the multienzyme complex. In the next step, the condensation, the acetyl group becomes linked to the malonate residue. At the same time, the free carboxyl group of the malonate residue undergoes decarboxylation. The equilibrium in this condensation reaction lies completely on the side of chain elongation. The condensation reaction leads to a chain of 4 C atoms. This unit is converted to a saturated fatty acid residue by three successive reactions: a reduction, a dehydration, and a second reduction. If the chain is to be lengthened further—and the most important fatty acids are 16 or 18 C atoms long—then an acyl transfer first occurs: the fatty acid residue is transferred back to the HS-group to which the acetyl residue was transferred in the initiation reaction. Subsequently, the cycle begins again with a malonyl transfer, condensation etc.

c. Termination reaction

When the definitive chain length has been attained a termination reaction occurs instead of the acyl transfer. The acyl group is not transferred to the other HS-group of the multienzyme complex but the HS-group of CoA. The coenzyme A derivative of the fatty acid can then be utilized for the synthesis of the fats, as we shall discuss presently.

But first a few supplementary remarks about the *multienzyme complex of fatty acid synthesis.* It has been photographed in the electron microscope, and as expected, several subunits could be seen. When the complex isolated from bacteria, yeast, higher plants, and animals was disassembled, one protein component was found which did not possess enzyme character. This protein has a pantetheine side chain which is linked to the protein through a phosphate group. It is very likely that the acyl residue is bound to the HS-group of pantetheine during fatty acid synthesis. By means of this "arm" it is then passed on from one enzyme of the complex to the next. This protein with its pantetheine arm which is probably located at the center of the multienzyme complex, is therefore called acyl carrier protein (Fig. 67).

Up to now we have spoken only of saturated fatty acids. As a matter of fact it is the *unsaturated fatty acids,* which are essential for man. Unfortunately, their synthesis in higher plants has not been completely elucidated. For example, the origin of the first double bond which appears in oleic acid between C 9 and 10 is still a matter of controversy. It is assumed that this first double bond is introduced at the point when a chain length of 10 or 12 C atoms has been attained and that chain elongation then continues until the final chain length is reached. Oleic acid with its 18 C

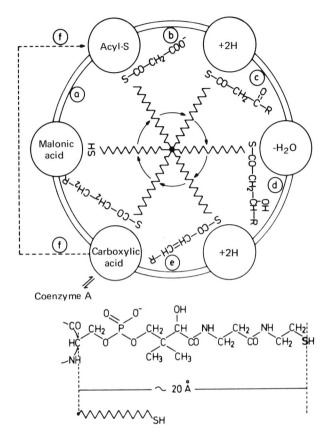

Fig. 67. Model of fatty acid synthetase. In the center of the multienzyme complex is the acyl carrier protein with its pantetheine arm which is shown again in detail below. The arm wheels around from one enzyme of the complex to the next in the direction of the arrow. In the process the individual reactions of chain elongation take place, reactions which are marked with the same letters as in Figure 66 (modified from Lynen 1969).

atoms and one double bond might arise in this way. On the other hand, there are findings which suggest that a dehydrogenation of stearic to oleic acid might occur. Although the mechanism of the biosynthesis of oleic acid is still not clear the derivation of linoleic and linolenic acids has been established: they arise by dehydrogenation of oleic acid in the appropriate positions.

C. Biosynthesis of the Neutral Fats

The route from the fatty acids to the fats is quite simple (Fig. 68): dihydroxyacetone phosphate is hydrogenated to glycerol phosphate by means of $NADH + H^+$. The acyl residues of the fatty acid CoA derivatives

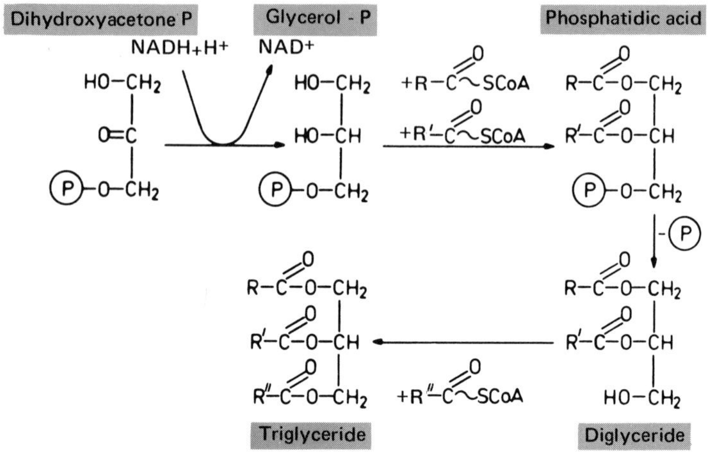

Dihydroxyacetone P   Glycerol - P   Phosphatidic acid

NADH+H+   NAD+

$$HO-CH_2$$

$$O=C$$

$$\textcircled{P}-O-CH_2$$

$$HO-CH_2$$

$$HO-CH$$

$$\textcircled{P}-O-CH_2$$

$$+R-\overset{O}{C}\sim SCoA$$

$$+R'-\overset{O}{C}\sim SCoA$$

$$R-\overset{O}{C}-O-CH_2$$

$$R'-\overset{O}{C}-O-CH$$

$$\textcircled{P}-O-CH_2$$

$$\downarrow -\textcircled{P}$$

$$R-\overset{O}{C}-O-CH_2$$

$$R'-\overset{O}{C}-O-CH$$

$$R''-\overset{O}{C}-O-CH_2 \quad +R''-\overset{O}{C}\sim SCoA$$

Triglyceride

$$R-\overset{O}{C}-O-CH_2$$

$$R'-\overset{O}{C}-O-CH$$

$$HO-CH_2$$

Diglyceride

Fig. 68. The biosynthesis of the neutral fats.

are then transferred to the free hydroxyl groups of glycerol phosphate. The products are phosphatidic acids from which the phosphate residue is cleaved by a phosphatase. The third hydroxyl group, thus set free, can then also be esterified, and the neutral fat is formed. Summarizing its derivation from the intermediates of biological oxidation, we can say that the fatty acid components of the fats come from acetyl CoA and malonyl CoA and the glycerol component from dihydroxyacetone phosphate.

D. Degradation of the Fats

Fats occur widely as reserve materials in higher plants which can be stored, for example, in seeds. The products of their degradation can be again subjected to biological oxidation and so be utilized for ATP production. They can also be used for synthesis, as in the glyoxylate cycle for the synthesis of glucose. The degradation of fats begins with the break-down of the neutral fats into glycerol and fatty acids. This is brought about by lipases, hydrolytic enzymes belonging to the group of esterases (Fig. 69).

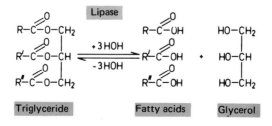

Fig. 69. Hydrolysis of the neutral fats by lipase.

The glycerol produced can, of course, be converted to triose phosphate which, in turn, can either be degraded glycolytically or used for the synthesis of hexoses. The fatty acids liberated can be subjected to either β-oxidation or α-oxidation. The names of the two degradative paths signify that an oxygen function is introduced at either the β-carbon atom or at the α-carbon.

1. β-Oxidation (Fig. 70)

β-Oxidation is usually the more important degradative path. The fatty acids arising from hydrolysis of the fats are first converted to their CoA esters with the consumption of ATP. They are thus "activated." As a result they are dehydrogenated between the α- and β-C atoms. The hydrogen acceptor is FAD. After addition of water to the double bond so generated a second dehydrogenation occurs. This time NAD$^+$ is the hydrogen acceptor. The β-C atom now bears a carbonyl function. In the subsequent "thioclastic" cleavage acetyl CoA is severed from the fatty acid residue which is simultaneously converted into its CoA thiol ester. In this way, 2 C have been set free as acetyl CoA. The fatty acid residue can be subjected again and again to the same degradative cycle. If the degradation was carried out on a fatty acid with an even number of C atoms, as is usually the case, then the degradation proceeds to the last molecule of acetyl CoA. The acetyl CoA can then be degraded in the citric acid cycle or be used for synthesis.

Let us check the energy balance of the β-oxidation. As we can deduce from the reaction sequence, 1 FADH$_2$ and 1 NADH + H$^+$ are formed per 2C fragment. 1 FADH$_2$ furnishes 2 ATP and 1 NADH + H$^+$ 3 ATP in oxidative phosphorylation. Thus, 5 ATP are derived from each 2 C unit.

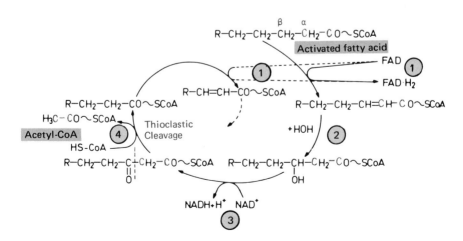

Fig. 70. β-Oxidation of fatty acids.

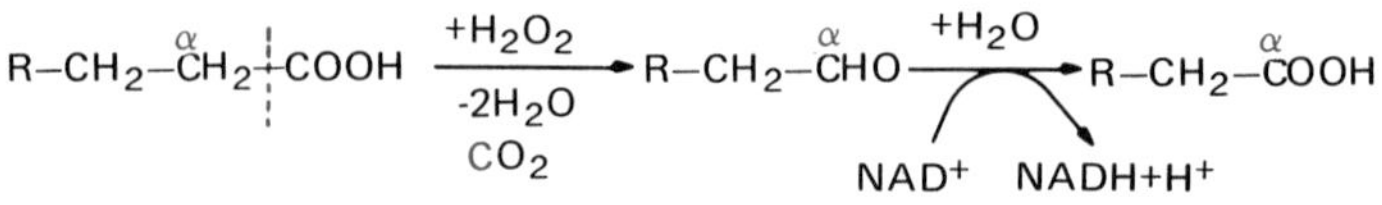

$$R-CH_2-\overset{\alpha}{C}H_2\!\mid\!COOH \xrightarrow[\substack{-2H_2O \\ CO_2}]{+H_2O_2} R-CH_2-\overset{\alpha}{C}HO \xrightarrow[\substack{NAD^+ \quad NADH+H^+}]{+H_2O} R-CH_2-\overset{\alpha}{C}OOH$$

Fig. 71. α-Oxidation of fatty acids.

Superficial inspection of the β-oxidation might suggest that it is a reversal of biosynthesis. This is not so. First, malonyl CoA does not play a role, secondly the thioclastic cleavage is not a reversal of the condensation and, finally, the hydrogen acceptors are different coenzymes from those which act as hydrogen donors in synthesis, where indeed $NADPH + H^+$ supplied the hydrogen.

2. α-Oxidation (Fig. 71)

In higher plants there is another route to the degradation of the fatty acids which is called α-oxidation. This does not occur in animals. The substrates are long chain fatty acids with 13 to 18 C atoms. Shorter fatty acids are not degraded. α-Oxidation proceeds in two steps. In the first, the fatty acid is converted to an aldehyde one C atom shorter by means of a fatty acid peroxidase, the reaction being accompanied by decarboxylation. In the second, the hydrate of this aldehyde is dehydrogenated. NAD^+ is the coenzyme of the dehydrogenase responsible.

The significance of α-oxidation is disputed. The energy gain is less than by β-oxidation. The degradation of 1 C indeed generates one molecule of $NADH + H^+$ and thus — via oxidative phosphorylation — also 3 molecules of ATP. But no acetyl CoA is formed, which could be used in the citric acid cycle for the production of more ATP. It is possible that α-oxidation is a means to produce acids with an odd number of C atoms. For example, if an 18 C fatty acid is degraded by α-oxidation to a 13 C fatty acid and this is then subjected to β-oxidation, the 3 C body, propionic acid, would ultimately be obtained (Fig. 72).

E. The Glyoxylate Cycle

Seeds are often rich in fats. At germination these fat reserves are used, to some extent, as energy sources, e.g. via oxidative degradation as outlined above. However, the fats can also be converted to carbohydrates. In this case they are first broken down by lipases into glycerol and fatty acids. It has already been mentioned that glycerol can be converted to carbohydrate via the triose phosphates. However, the fatty acids can also be

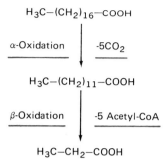

Fig. 72. A possible route to the formation of propionic acid.

converted into carbohydrates and for this the glyoxylate cycle is of central importance.

The glyoxylate cycle may be considered as a variant of the citric acid cycle (Fig. 73). Isocitrate is broken down into succinate and glyoxylate by isocitratase=isocitrate lyase. Succinate is converted to malate in reactions known from the citric acid cycle. However, glyoxylate, too, can give rise to malate by being combined with acetyl CoA by malate synthetase. In germinating seeds this acetyl CoA comes from the β-oxidation of the fatty acids. The key enzymes of the glyoxylate cycle, isocitratase and malate synthetase, have been demonstrated many times in germinating seeds.

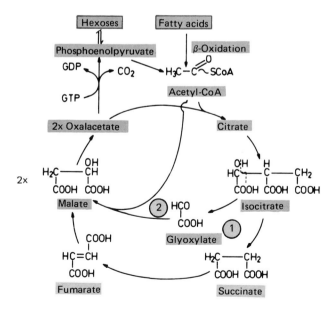

Fig. 73. The glyoxylate cycle. The key enzymes are encircled: I = Isocitratase = Isocitrate lyase, 2 = malate synthetase.

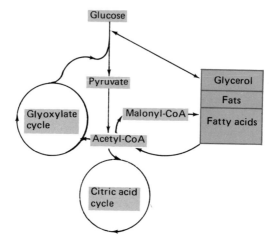

Fig. 74. Position of fats in metabolism.

The result, then, is an augmented supply of malate and, in turn, of oxalacetate. Part of the oxalacetate is condensed with acetyl CoA to form citrate and so keep the glyoxylate cycle going. Another portion of the oxalacetate can also be phosphorylated and decarboxylated to give phosphoenolpyruvate. This in turn can give rise to hexoses via triose phosphate.

We have said that the glyoxylate cycle is *formally* a variant of the citric acid cycle. This statement needs further underpinning: Beevers was able to show that both of the key enzymes, isocitratase and malate synthetase, are localized in special organelles, the "glyoxysomes," which can be distinguished from the mitochondria. Here again we come across the phenomenon of compartmentalization—this time between such closely related processes as the citric acid cycle and the glyoxylate cycle.

Place of the fats in total metabolism (Fig. 74). In conclusion let us summarize the position of the biosynthesis of fats in total metabolism: their glycerol component originates from dihydroxyacetone phosphate and their fatty acids are formed from acetate and malonate. Thus, the biosynthesis of the fats involves the incorporation of certain intermediates of biological oxidation. Conversely, intermediates of biological oxidation are furnished when they are degraded—provided one ignores α-oxidation of fatty acids.

Terpenoids

Acetyl CoA is the starting material for the biosynthesis of the fatty acids. It is used for the synthesis of the fatty acids by the acetate-malonate pathway. With the terpenoids we become acquainted with a second, large group of natural products whose biosynthesis starts from acetyl CoA. The terpenoids are furnished via the acetate-mevalonate pathway.

A. Chemical Constitution

As a glance at their structural formulae shows terpenoids are obviously constructed from 5 C building blocks. Recognition of this fact was summarized by Ruzicka in the "Isoprene Rule." Indeed, it was first assumed that the known chemical compound isoprene is the naturally occurring 5 C building block. It is now known that it is the so-called "active isoprene," isopentenyl pyrophosphate (IPP), and not isoprene itself, that plays a role in the biosynthesis of the terpenoids. The alternative name isoprenoids for the group of the terpenoids is derived from isoprene or active isoprene.

The terpenoids are arranged in subgroups according to the number of 5 C units (Fig. 75). The hemiterpenes consist of only one 5 C unit. Discounting substances so greatly modified that their terpenoid character is scarcely apparent, the hemiterpenes occur principally as constituents of the so-called mixed terpenes. One speaks of mixed terpenes in regard to substances that are constructed from a terpenoid and a nonterpenoid component. For example, hemiterpenes are found as side chains of certain quinones.

Monoterpenes consist of two 5 C units. They may be open chain or ring structures. The cyclic monoterpenes are further divided into monocyclic and bicyclic systems.

The same kinds of structural variations that are found among the monoterpenes (open chain or cyclic, one or several ring systems) are also found among the sesquiterpenes, diterpenes, etc. Only the polyterpenes are, without exception, long, open chains of 5 C units linked in series. When one realizes that the terpenoids can also be subject to additional modifications, such as oxidation or reduction, the addition or subtraction of C atoms, it becomes clear that the number of different terpenoids is

5-C-Units	Group	Several examples; the formulae of some are given in the text.
1 X 5-C	Hemiterpenes	"Prenyl" residue in quinones and coumarins
2 X 5-C	Monoterpenes	Open chain: Citral, Geraniol, Linalool. Monocyclic: Limonene, Menthol, Thymol, Menthone, Carvone, Cineole, Phellandrene. Bicyclic: Camphor, α-and β-Pinene
3 X 5-C	Sesquiterpenes	Open chain: Farnesol Cyclic: β-Cadinene
4 X 5-C	Diterpenes	Open chain: Phytol Cyclic: resin acids, Gibberellins
6 X 5-C = 2 X 15-C	Triterpenes	Open chain: Squalene Cyclic: Triterpene alcohols and acids, Steroids, Gossypol, Cucurbitacine
8 X 5-C =2 X 20-C	Tetraterpenes	Carotenoids (Carotenes, Xanthophylls)
n X 5-C	Polyterpenes	Rubber, Guttapercha, Balata

Fig. 75. Scheme of the groups of the terpenoids (modified from Hess 1968).

enormous. The formulae of only a few of them can be presented in the following discussion of the individual terpenoid groups.

B. Secondary Plant Substances

The terpenoids, the phenols and the alkaloids constitute the three most important groups of the "secondary" plant substances. This is a very unfortunate designation since it is all too easy to equate "secondary" with "of secondary importance" or, even, "unimportant." In actual fact it is not known for many of the substances so classified in what way they could be of use to the plants that produce them. In many cases they do indeed seem to be "waste products" of metabolism which are withdrawn from metabolism by being transferred to vacuoles or stored in the bark without any demonstrable benefit to the plant other than an easing of the demands on its synthetic pathways. That does not mean that a use for such compounds may not be found in the future. There are a number of substances that were assumed to be waste products of metabolism but that recent research has shown to be degraded by the plant. This is true of certain terpenoids, phenols, and alkaloids too. In view of the biochemical dexterity of the plant it would be surprising if this degradation were not used, for example, for the production of ATP.

Many secondary plant substances are, however, essential to the plants. Among these are the phytohormones (the indole derivatives, gibberellins, phytokinins, and abscisins) the purine and pyrimidine bases of the nucleic acids, the porphyrins—think of chlorophyll and the cell hemins—the photochrome system, different kinds of coenzymes and the structural polymer lignin, to mention only a few examples. For man, too, these substances are not "secondary." Pharmacy and technology thoroughly exploit the reservoir of the "secondary" plant substances. Thus, we cannot dismiss the secondary plant substances as being generally of no consequence, either for the plant or for man. How, then, shall we interpret "secondary"? It would certainly be best if this expression were to disappear entirely from the literature. This is too much to expect at the moment although it has been replaced to a large extent by "natural products." Let us then attempt a definition: secondary plant substances are substances which are derived biosynthetically from the metabolism of carbohydrates, fats and amino acids (Fig. 133). Thus, it is their biosynthesis, and not their importance, which is "secondary." Let us bear in mind that this definition is useful, mainly for didactic reasons. It allows the profusion of secondary substances to be discussed separately from the carbohydrates, fats and amino acids, including the proteins, when necessary.

C. Volatile Oils

The volatile oils must not be confused with the "fatty oils." The latter are fats existing in the fluid state. On the other hand, the former are highly volatile, as their name suggests, and they belong to the terpenoids, especially the mono- and sesquiterpenes, or the phenols. Various additional substances may be present.

The synthesis of the volatile oils often occurs in special glandular cells or epithelia. It is known that volatile oils can be formed and secreted by glandular hairs of the leaf surface. In several cases, such as the peppermint (*Mentha piperita*), the secretion process can be observed with both the light and electron microscope. Small "oil vacuoles" are first formed in the cytoplasm whose content is discharged into the space between the cell wall on the one hand and the cuticle on the other as a result of an apparent loosening of the cell wall. The accumulation of volatile oil beneath the cuticle is easily visible in the light microscope. The surface of the cuticle is first enlarged, either by stretching or by growth, so that still more volatile oil can accumulate. Then ultimately the cuticle ruptures and the volatile oils are set free. The secretion process, which has been sketched here for the case of a glandular hair, can take many different forms, depending on the kind of glandular system.

We have just mentioned secretion and before we go any further we must be quite clear what we mean by it. In plants a distinction is made between secretions and excretions which can be defined more or less as follows. *Secretions* are substances, by means of which the secreting cell interrelates with its immediate and more distant environment. *Excretions* are substances whose discharge is not known to mediate such a direct interaction with the environment. Volatile oils that are discharged from glandular structures specially designed for the purpose in flowers, the osmophores, and which serve to attract pollinating insects are, for example, secretions. The volatile oils of peppermint leaves are also secretions. This is because geese, for example, and very many insects are thus deterred from molesting the peppermint. And it should also not be forgotten that we drink peppermint tea from time to time. Thus, the interrelation with the environment exists. Volatile oils that are discharged without having any clearly recognizable function would be excretions. As is often the case with such pretty definitions in biology, there are exceptions and cases where the two overlap. Indeed, it seems quite doubtful whether the clear distinction between secretions and excretions mentioned above can be maintained. It is likely that in every case an intensive search will lead to the discovery of an interrelation with the immediate or more distant environment.

The function of the volatile oils is varied. Volatile oils can inhibit the germination of seeds and the growth of plants, and thus function as weapons in the battle against undesired competition. Certain volatile oils inhibit the growth of bacteria and fungi in the laboratory. Perhaps this inhibitory effect also plays a role in nature. It is in any case certain that it is precisely this effect that has made certain plants therapeutically interesting as "herbs." One has only to think of the peppermint, the camomile, and the eucalyptus. Volatile oils also protect the organisms which produce them from being devoured by mammals, birds, insects, and snails. Or the reverse can happen: volatile oils can attract insects, as in pollination, which is undoubtedly a most important function. However, in other cases it is still unknown what uses plants can make of the volatile oils they produce.

D. Biosynthesis (general)

We shall now discuss the biosynthesis of the terpenoids, an area of research which is linked particularly with the names of Bloch, Lynen, Cornforth, and Popjak. We shall first consider general principles, from which the individual subgroups can then be approached (Fig. 76). Subsequently, we shall discuss the biosynthesis of a few of these subgroups in more detail.

The biosynthesis begins with acetyl CoA which is combined with a second unit of acetyl CoA to give acetoacetyl CoA. A third acetyl CoA molecule is then added and the 6 C body so produced is hydrogenated to mevalonic acid by means of NADPH + H$^+$ with the liberation of coenzyme A. Mevalonic acid, which was discovered as a growth factor for microorganisms, is an important intermediate. The active isoprene, isopentenyl pyrophosphate, is derived from it by decarboxylation, dehydration, and phosphorylation with ATP.

Isopentenyl pyrophosphate (IPP) exists in equilibrium with its isomer, dimethylallyl pyrophosphate. The latter is the fuse without which terpenoid biosynthesis cannot be set in motion. This is because it is only

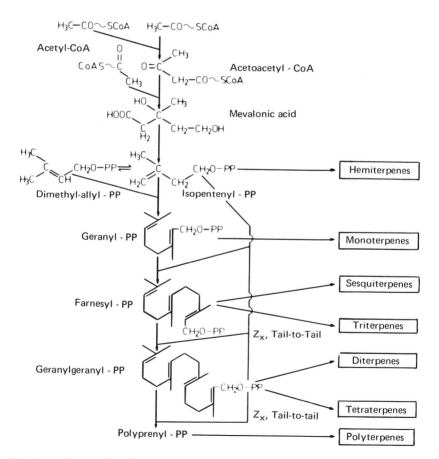

Fig. 76. Scheme of the biosynthesis of terpenoids. Only the most important intermediates are shown (from Hess 1968).

with dimethylallyl pyrophosphate that IPP can unite to form the open chain monoterpene geranyl pyrophosphate, with the liberation of pyrophosphate. Other open chain and cyclic monoterpenes can be formed from geranyl pyrophosphate.

If a further unit of IPP is now added to geranyl pyrophosphate, farnesyl pyrophosphate, a sesquiterpene, is obtained. The addition occurs "head to tail": the CH_2 group, the "head" of IPP, adds to the pyrophosphate end, the "tail," of geranyl pyrophosphate. The resulting farnesyl pyrophosphate can be joined "tail to tail" to form an open chain triterpene. The latter, in turn, serves as the starting material for the synthesis of the cyclic triterpenes which include the steroids, compounds that are essential for all living organisms.
synthesis of the cyclic triterpenes which include the steroids, compounds that are essential for all living organisms.

Let us consider further the consequences of head-tail additions. If an additional molecule of IPP is added head-to-tail to farnesyl pyrophosphate geranylgeranyl pyrophosphate, a diterpene, is obtained. The series of events outlined above can now be repeated at a higher level of complexity: geranylgeranyl pyrophosphate can either be converted to other diterpenes or two molecules of geranylgeranyl pyrophosphate can be joined "tail-to-tail" to give 40 C bodies. In this way tetraterpenes, i.e. carotenoids, are obtained. Further head-to-tail additions of IPP lead, finally, to the polyterpenes rubber, gutta-percha, and balata.

E. Biosynthesis (particular)

1. Monoterpenes

A glance at the structural formulae of a few of the monoterpenes is instructive (Fig. 77). Hints as to possible biosynthetic routes can be gleaned, particularly in cases where several structurally related monoterpenes occur in one plant. Isotope experiments have shown clearly that labeled precursors can be converted to geranyl pyrophosphate and then to cyclic monoterpenes. Such experiments have shown repeatedly that a high proportion of input C^{14} glucose is incorporated into monoterpenes, whereas only very small amounts of input C^{14} mevalonate are similarly incorporated. At first sight that is quite unexpected but can be explained by a compartmentalization of the glandular cells: the glandular cells can take up mevalonate only with great difficulty. Other precursors such as glucose, for example, can be taken up easily, however, and then can be converted into mevalonate and monoterpenes by means of the gland's own enzyme systems.

Once cyclic monoterpenes have been formed, they can be converted by minor modifications into other cyclic monoterpenes. As an example, let us consider the peppermint, which we have already mentioned.

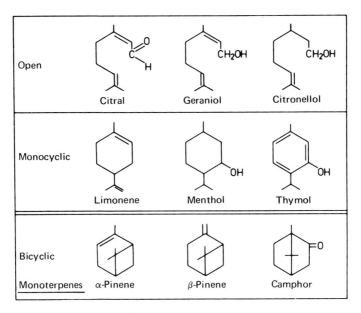

Fig. 77. Several monoterpenes.

Geranyl pyrophosphate is converted to piperitenone via a series of unknown intermediates and this is then transformed into pulegone, menthone, and menthol in three hydrogenation steps (Fig. 78). The hydrogenation of pugelone to menthol via menthone can be carried out in a cell-free system. Additional monoterpenes, which can be integrated into the same biogenetic pattern as those mentioned, are also to be found in the peppermint.

2. Sesquiterpenes

Open chain sesquiterpenes are relatively rare. Farnesol is an example and is an important component of the scent of lily of the valley and lime blossom. It is probably formed from farnesyl pyrophosphate by cleavage of the pyrophosphate group. We shall not give any further consideration to the cyclic sesquiterpenes either except to mention the phytohormone abscisic acid (page 210) as an example.

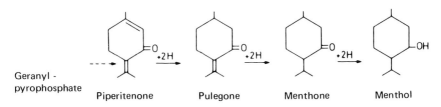

Fig. 78. The biosynthesis of several monoterpenes of the peppermint.

3. Triterpenes

The fact that farnesyl pyrophosphate is converted into triterpenes by tail-to-tail addition (Fig. 79) is of crucial importance. "Tail" refers to the pyrophosphate end. Correctly speaking, it is not two molecules of farnesyl pyrophosphate that react together but one molecule of farnesyl pyrophosphate with one molecule of nerolidyl pyrophosphate, an isomer of farnesyl pyrophosphate. The addition takes place reductively. The product is a symmetrical 30 C entity, sequalene. It is widely distributed in the plant and animal kingdoms, even though steady state concentrations may often be very small. This is what one would expect, since squalene is the

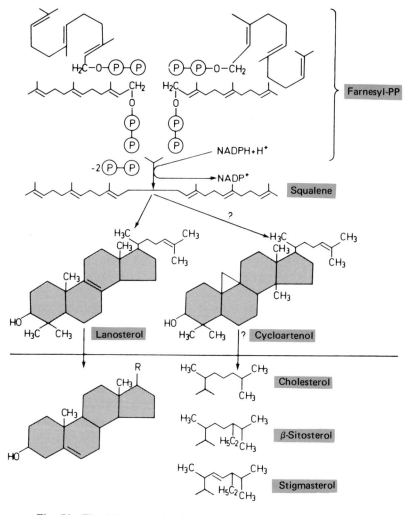

Fig. 79. The biosynthesis of triterpenes, particularly of the sterols.

starting material for the biosynthesis of the cyclic triterpenes, especially of the steroids.

The following groups of substances belong to the steroids:

1. Sterols
2. Bile acids
3. Steroid hormones (e.g. sex hormones and hormones of the adrenal cortex)
4. The vitamins of the D group
5. Steroid saponins
6. Heart glycosides
7. Steroid alkaloids

All of these substances have the skeletal structure of sterane or cyclopentoperhydrophenanthrene, which is then subjected to modifications which vary from group to group (Fig. 80).

Only in animal organisms do *steroid hormones* and *bile acids* have clearly defined functions. It is true that several steroid hormones have been discovered in plants as well as in animals, among them the insect moulting hormones, the ecdysones. Ecdysones have been detected, for example, in the speckled fern *Polypodium vulgare* and in concentrations much higher than those found in insects. In this case it is possible that they serve to protect the plant against predators. Thus, so-called β-ecdysone, when orally administered, has the effect of deterring the cotton stainer *Dysdercus fasciatus* from eating. However, in most cases the function of the steroid hormone in the plant is unknown. Admittedly certain effects of steroid hormones on growth and development are known; for example, the formation of flowers in the chinese aster *Callistephus sinensis* and the duckweed *Lemna minor* can be triggered by oestradiol. It is still a completely open question, though, whether steroid hormones function as flowering hormones under natural conditions, inspite of a few more suggestive pointers in this direction. Of the 5 remaining groups of steroids we shall select the *sterols,* the *heart glycosides,* and the *steroid alkaloids* for discussion. The sterols are examples of steroids that are structurally important, and the heart glycosides and the steroid alkaloids are steroids that are pharmacologically important.

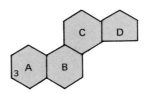

Fig. 80. Sterane = Cyclopentanoperhydrophenanthrene.

a. Sterols

The sterols bear a hydroxyl group on C atom 3, which gives them their name. Of the animal sterols, the zoosterols, cholesterol is the most important. Cholesterol also occurs in plants. However, the most important sterols in higher plants are β-sitosterol and stigmasterol (Fig. 79). They differ from each other with respect to a double bond in the side chain which is present in stigmasterol. The C skeleton of both substances consists of 29 atoms, two of which are attached as a branched chain to the side chain at C 24. This branched chain is absent from cholesterol, which has only 27 C atoms.

In animal organisms the biosynthesis of cholesterol proceeds via the intermediate lanosterol (Fig. 79). The biosynthesis of the phytosterols, on the other hand, is not completely understood. Here, too, apparently the route via lanosterol can be adopted. However, several pieces of evidence make it seem likely that another route via cycloartenol, which is very similar to lanosterol, is practicable. The 2 C branched chain of β-sitosterol and stigmasterol is not derived from the acetate pool, as one might have supposed. Rather, first one C is incorporated and then the second. The supplier of the 1 C units is the amino acid methionine. As usual, it participates in the reaction in the form of S-adenosyl methionine.

The unit membrane. the structure of the cell and its organelles is based, to a large extent, on the existence of membrane systems. Consider the plasmalemma, the tonoplasts, the nuclear membrane, the membrane system of plastids and mitochondria, the dictyosomes and the endoplasmic reticulum. It is assumed that these membrane structures might all be based on a common structural principle, namely, the so-called unit membrane. There are several models of the molecular structure of the unit membrane, one of which is shown in Fig. 81: two protein layers enclose two lipoid layers. The lipoids are so oriented that their polar ends are directed towards the protein and their lipophilic ends towards the other lipoid layer.

In animals, the lipids in such unit membranes are almost exclusively cholesterol and phospholipids. In plants they are phytosterols,

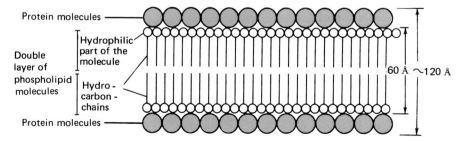

Fig. 81. Construction of a model of the unit membrane (from Lehninger 1969).

phospholipids (e.g. lecithin), glycolipids (especially galactolipids), and sulpholipids.

b. Heart glycosides

In 1785, the English physician Withering used the red foxglove *(Digitalis purpurea)* as a remedy for heart diseases for the first time in Europe. Since then the genus *Digitalis* has become an indispensable accessory to the physician. The efficacious substances of the plants are called heart glycosides after their field of application. They consist of aglycones, which are linked with a varying number of sugars, some of which are rare. The sugars are of no further interest to us here but we shall discuss the aglycones. The aglycones of the different heart glycosides are either cardenolides with 23 C atoms or bufadienolides with 24 C atoms (Fig. 82). The chemical difference between the two lies in the structure of the lactone ring attached to C 17. In the case of the *cardenolides*, it is a 5-ring and in the case of the bufadienolides a 6-ring. The aglycones of the digitalis glycosides belong to the cardenolides. An example is digitoxigenin. However, cardenolides are found in numerous other genuses, such as *Strophanthus, Nerium* and *Convallaria*.

As the name implies, *bufadienolides* are found in the secretions of toads *(Bufo)*. They are also present in plants. One such is hellebrigenin from the rhizomes of the black hellebore *(Helleborus niger)*. The bio-

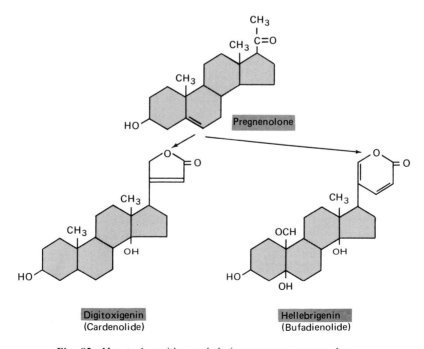

Fig. 82. Heart glycosides and their precursor pregnenolone.

synthesis of the cardenolides and bufadienolides is only partially under-stood. A steroid with 21 C atoms, pregnenolone (Fig. 82), is derived from unknown intermediates. C^{14} labeled pregnenolone is converted to digitox-igenin and other cardenolides in the leaves of *Digitalis lanata,* and to hellebrigenin in the leaves of *Helleborus atrorubens.* In the case of the car-denolides, this conversion involves the addition of 2 C atoms, which are provided by malonyl CoA. The formation of the bufadienolides requires the addition of 3 C atoms. Their origin is unknown.

c. Steroid alkaloids

The steroid alkaloids are a group of secondary plant products, which are usually classified structurally with the alkaloids (page 144) but bio-synthetically with the terpenoids.

Two groups of steroid alkaloids are known, those with 27 C atoms and those with 21. We shall deal briefly here only with the 27 C steroid alkaloids. They fall into the two subgroups of the solanum alkaloids and the veratrum alkaloids. The former subgroup is to be found especially in *Solanaceae,* in the genus *Solanum,* the latter especially in *Liliaceae,* in the genus *Veratrum.* Demissidine (Fig. 83) is an example of a solanum alkaloid. It occurs as the glycoside demissine in the wild potato *Solanum demissum* and is responsible for its resistance to the Colorado beetle. The cultivated potato can be made resistant to attack from the Colorado beetle by crossing in the genes for demissine.

The biosynthesis of the steroid alkaloids takes the usual paths of tri-terpene synthesis, i.e. via farnesyl pyrophosphate, squalene, and lanosterol or cycloartenol. The nitrogen probably comes from ammonia or ammonium compounds.

4. Diterpenes

Phytol and the gibberellins are the only diterpenes that will be men-tioned here. Phytol is a component of the mixed terpenes, namely, of the chlorophylls. A carboxyl group attached to its pyrrole system IV is esterified with the alcohol phytol (Fig. 27).

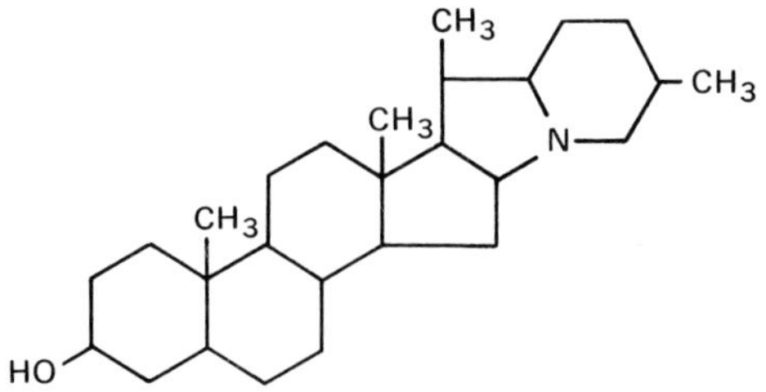

Fig. 83. Demissidine.

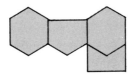

Fig. 84. Gibbane.

The *gibberellins* are a group of phytohormones, which all bear the gibbane skeleton (Fig. 84). The individual members of the group are designated as gibberellin A_1, A_2, A_3, etc. In fungi and higher plants new gibberellins are still being found. In experiments gibberellin A_3 = gibberellic acid, whose presence in higher plants has often been demonstrated, is often employed.

At first glance, it is difficult to recognize the gibberellins as diterpenes. However, any element of doubt about this has been dispelled by isotope experiments *in vivo* and in a cell-free system. In the fluid endosperm of the wild cucumber *Echinocystis macrocarpa* the following synthetic pathway has been demonstrated (Fig. 85): geranylgeranyl pyrophosphate $-(-)$ kaurene $-(-)$ kauren-19-ol $-$ gibberellin A_5. Presumably the synthesis then proceeds via A_1 to gibberellic acid.

5. Tetraterpenes: Carotenoids

a. Chemical constitution

The carotenoids are divided between the two large groups of the carotenes and the xanthophylls. There is also a third, much smaller group

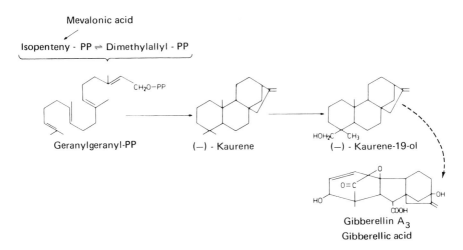

Fig. 85. The biosynthesis of the gibberellins. (from Hess 1968).

of the carotenoid acids. It is most likely that carotenoid acids are degradation products of tetraterpenes.

   Carotenes (Fig. 86) are hydrocarbons with 40 C atoms. They arise by
tail-to-tail addition of two units, both of which consist of four 5 C building
blocks. The individual carotenes exhibit varying degrees of unsaturation.
The presence of a certain minimal number of conjugated double bonds
leads to their being yellow or orange. In the biosynthetic sequence of the
carotenes, the first colored material to appear is ζ-carotene (cf. Fig. 88).

   Xanthophylls (Fig. 86) are oxidation products of the carotenes. They
are derived from the corresponding carotins by the introduction of oxygen functions. Both the carotenes and the xanthophylls may be open
chain or cyclic compounds. The cyclic molecules may contain one or two
ring systems. In the case of the carotenes two types of rings are found, α-
ionone and β-ionone rings, which differ from each other with respect to
the position of their only double bond (Fig. 86). Carotenes with at least
one β-ionone ring are provitamins A. In animals and man such carotenes
can be cleaved in the middle of the molecule. The split halves containing
the β-ionone ring can then be converted into vitamin A. Thus, one

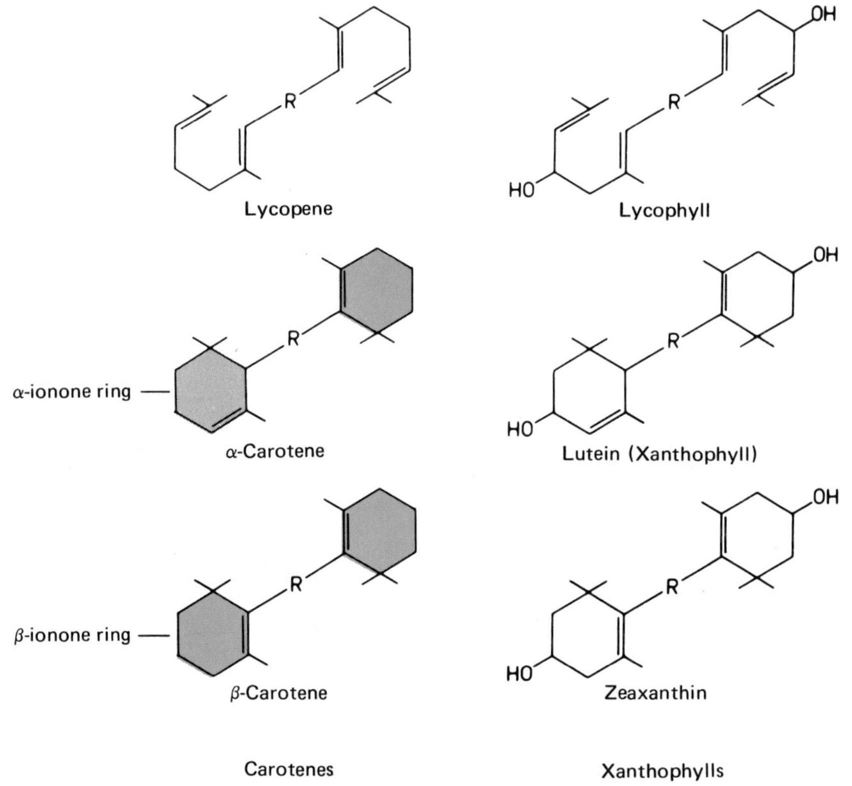

Fig. 86. Carotenes and xanthophylls.

molecule of β-carotene with its two β-ionone rings furnishes two molecules of vitamin A, whereas one molecule of α-carotene with its one β-ionone ring provides only one molecule of vitamin A. Of the *carotenoid acids* only crocetin will be mentioned. It occurs in a glycosylated form as a yellow pigment in the stigma of the saffron *Crocus sativus* and in the flowers of *Nemesia strumosa*. Crocetin is a dicarboxylic acid with 20 C atoms (Fig. 87). It is formed from carotenes or xanthophylls by symmetric oxidative degradation from both ends of the molecule.

b. Biosynthesis

The biosynthesis of the carotenoids can be considered in several stages:

(1) *The formation of the 40 C skeleton* (Fig. 88). The formation of the 40C skeleton of the carotenoids results from tail-to-tail addition of two units of geranyl geranyl pyrophosphate. In principle the process is similar to that in which squalene is formed by tail-to-tail addition of two molecules of farnesyl pyrophosphate. However, one important difference deserves mention. In the case of squalene the addition occurs together with reduction but in carotenoid biosynthesis this reduction does not take place. As a result the 40 C product bears a double bond in the middle of the molecule. It has been demonstrated many times that this initial 40 C product is phytoen.

(2) *Dehydrogenations* (Fig. 88). The key substance phytoen is subjected to a series of dehydrogenations which lead to phytofluen, ζ-carotene, neurosporene and, finally, lycopene. As already mentioned ζ-carotene is the first colored carotene in this sequence. Apart from both ends of the molecule lycopene consists of a sequence of completely conjugated double bonds. Its name comes from its occurring in certain breeds of tomatoes *(Lycopersicon esculentum)*.

(3) *Cyclization* (Fig. 89). Lycopene and all preceding intermediates in the biosynthetic pathway are open chain. It is still unsettled whether the cyclization of the ends of the chain occurs at the level of lycopene or of its precursor neurosporene. We will simply state one of the two possibilities: the cyclization occurs with lycopene.

(4) *Oxidations.* So far in the biosynthetic sequence we have encountered only carotenes. Xanthophylls are derived form them by the introduction of oxygen functions. It is almost certain that these oxidations occur after cyclization.

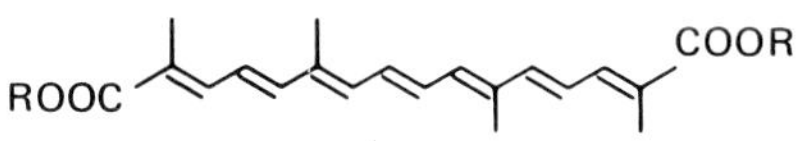

Fig. 87. Crocetin and crocin. Crocin is the yellow pigment of the stigmata of the meadow saffron. R = H: crocetin, R = gentiobiose: crocin.

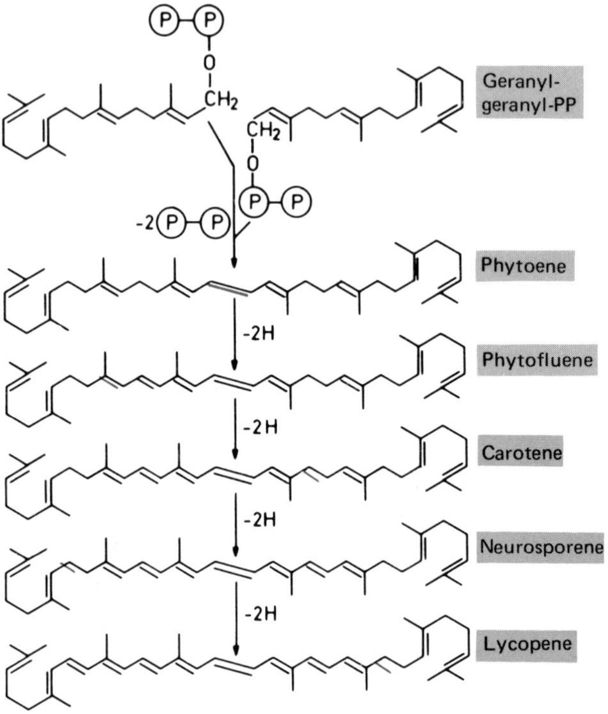

Fig. 88. Formation of the 40-C skeleton of the carotenoids and its dehydrogenation.

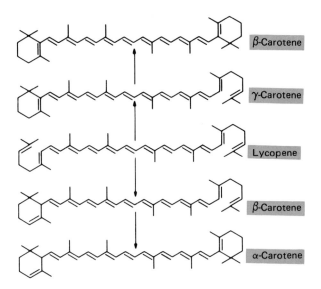

Fig. 89. One possible mode of cyclization of carotenoids.

6. Polyterpenes

Of the polyterpenes rubber, gutta-percha, and balata, our discussion will be restricted to the technically most important product, rubber. About 2,000 species of higher plants produce rubber but only a few of them, mainly from the families *Apocynaceae, Asclepiadaceae, Compositae, Euphorbiaceae,* and *Moraceae,* do so in sufficient quantity as to make its extraction technically worthwhile. A few of the more important species are listed in Table 5. The main source of rubber is the rubber tree *Hevea brasiliensis. Achras sapota,* the chicle of which provides the basic ingredient of chewing gum, is also included in the list.

Table 5. Several of the more important rubber plants (from Paech 1950)

Species	Family	Indigenous to	Type of growth
Achras sapota	*Sapotaceae*	Tropical America	Tree
Castilloa elastica	*Moraceae*	Central America	Tree
Ficus elastica	*Moraceae*	Asia, Africa	Tree
Hevea brasiliensis	*Euphorbiaceae*	South America	Tree
Manihot alaziovii	*Euphorbiaceae*	South America	Tree
Parthenium argentatum	*Compositae*	Mexico, Texas	Bush
Taraxacum koksaghyz	*Compositae*	Central Asia	Plant

With some exceptions, rubber is formed in jointed or unjointed latex tubes. With progressive differentiation the protoplasmic content of these latex tubes is converted into latex, a mixture of mitochondria, ribosomes, and proteins. The cell nuclei usually lie in a plasma membrane. In addition, the most varied materials are present, such as, for example, alkaloids and also polyterpenes such as rubber. Not every latex contains rubber. When present, it is suspended as droplets in the latex. A feature of rubber suppliers is that their latex also contains the complete set of enzymes necessary for the conversion of acetate to rubber.

Rubber is built up from 500 to over 5,000 5 C units. The double bonds of these 5 C units are present in the *cis* form. In gutta-percha, which usually exists as shorter chains, the double bonds are found to be *trans* (Fig. 90).

The latex which, as mentioned, contains all of the necessary enzymes, provides a suitable cell-free system for the investigation of the biosynthesis of rubber. Experiments of this kind have been carried out by the groups of Bonner and Lynen, among other. It could be shown that a total synthesis of rubber occurs in the latex, provided that the rubber which is already present in the latex is removed and that dimethylallyl pyrophosphate is added as primer. If rubber particles are present in the

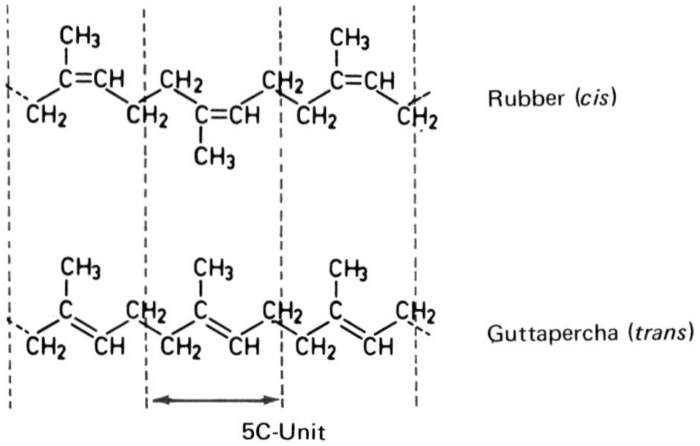

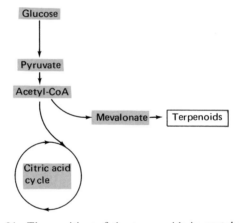

Fig. 90. Rubber and guttapercha.

Fig. 91. The position of the terpenoids in metabolism.

latex no new synthesis occurs, instead an extension of preexisting chains takes place by attachment of IPP.

In conclusion, let us summarize the position of the terpenoids in the *total scheme of metabolism* (Fig. 91). They are formed from acetyl CoA via mevalonate and IPP. Nothing known is with certainty about their degradation in higher plants. It is known that microorganisms can degrade geraniol to acetate, acetyl CoA, and a substance with 5 C atoms (dimethylacrylyl CoA), all substances that can be easily and completely oxidized to CO_2 and H_2O.

Phenols

A. Chemical Constitution

Phenols are substances that bear at least one hydroxyl group or functional derivative of it attached to an aromatic ring system. The number and variety of the phenols make them an exceedingly important group of "secondary" plant substances. A few groups of the phenols are presented in summary form in Figure 92. Detailed structural formulae are to be found in the sections on biosynthesis.

Simple phenols consist of an aromatic ring that bears one or more hydroxyl groups. In addition, the ring system may bear other substituents, especially methyl groups.

Phenol carboxylic acids are simple phenols which bear a carboxyl group as a substituent.

Phenylpropane derivatives possess the C skeleton of phenylpropane, i.e. an aromatic system to which a side chain of three C atoms is attached. Examples of these are cinnamic acids, cinnamaldehyde, cinnamyl alcohol, coumarins, and, also, the high polymer lignin.

Flavan derivatives are characterized by the flavan skeleton. It consists of an aromatic ring A, an aromatic ring B, and a central, oxygen-contain-

C - Skeleton	Group	Several examples
⬡	Simple Phenols	Hydroquinone Arbutin
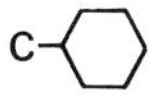 C–⬡	Phenolcarboxylic acids	p - Hydroxybenzoic acid Protocatechuic acid Gallic acid
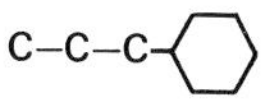 C–C–C–⬡	Phenylpropanes	Cinnamic acids Cinnamic alcohols Coumarins Lignin
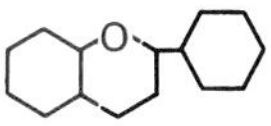 (flavan skeleton)	Flavan derivatives	Flavanones Flavones Flavonols Anthocyanidins

Fig. 92. Survey of a few groups of phenols.

117

ing heterocycle. Several groups of flavan derivatives or flavanoids, e.g. the flavanones, flavanols, anthocyanidins, and flav-3,4-diols, can be distinguished, depending upon the state of oxidation of this heterocycle.

All of these substances occur widely as glycosides or sugar esters, which are deposited in vacuoles.

B. Biosynthesis (general)

Aromatic systems are formed by three different routes in higher plants:

(1) *The shikimic acid pathway.* This is the most important biosynthetic route.

(2) *The acetate-malonate pathway.* This is used to synthesize the aromatic ring A of the flavan derivatives. Otherwise this route is more important for microorganisms.

(3) *The acetate-mevalonate pathway.* In principle, we have already discussed this route. It is concerned with the formation of cyclic terpenes which can be dehydrogenated to aromatic systems. An example of such a terpene with aromatic character is thymol. This pathway is relatively unimportant in higher plants. Of the three pathways listed we have still to discuss pathways 1 and 2.

1. The Shikimic Acid Pathway (Fig. 93)

The shikimic acid pathway was discovered by Davis in investigations with bacterial auxotrophs. However, it is found not only in microorganisms but also in higher plants. Most of the enzymes of the shikimic acid pathway have been demonstrated in a cell-free system, even in higher plants. The pathway is named after an intermediate, shikimic acid. Its importance lies not only in its furnishing phenols but especially in the provision of the aromatic amino acids, phenylalanine, tyrosine, and tryptophan.

The shikimic acid pathway begins with phosphoenolpyruvate which is obtained from glycolysis, and D-erythrose-4-phosphate, which comes from the pentose phosphate cycle. The two are linked to form an intermediate with 7 C atoms which cyclizes to 5-dehydroquinic acid. The latter exists in equilibrium with quinic acid. The pathway proceeds via 5-dehydroshikimic acid and shikimic acid to 5-phosphoshikimic acid. An additional phosphoenolpyruvate unit is now attached to the last-mentioned compound. The product of this reaction is converted, in several steps, to chorismic acid.

With chorismic acid an important junction in the shikimic acid pathway has been reached. For as the Greek name implies (chorizo = split), the synthetic route divides into two branches after this substance. One branch leads via anthranilic acid to tryptophan, and from it to the phytohormone indole-3-acetic acid. The stages along this route are well-defined in microorganisms but are still disputed in higher plants.

The second branch leads from chorismic acid first to prephenic acid. After this substance the pathway forks again: via phenylpyruvate to phenylalanine and via p-hydroxyphenylpyruvate to tyrosine. These two aromatic amino acids are closely related to each other since phenylalanine can be oxidized to tyrosine. However, this last reaction does not seem to be very important in higher plants. On deamination, phenylalanine yields cinnamic acid and tyrosine p-coumaric acid, a derivative of cinnamic acid.

Summarizing, it can be said that the shikimic acid pathway furnishes

(1) the aromatic amino acids tryptophan, phenylalanine and tyrosine;

(2) cinnamic acids, which are derived from phenylalanine and tyrosine (the cinnamic acids serve as starting materials for the biosynthesis of the remaining phenylpropanes, as we shall see);

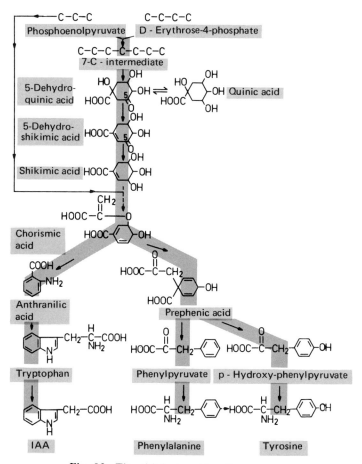

Fig. 93. The shikimic acid pathway.

(3) Phenol carboxylic acids. It should be mentioned that these can also be obtained by other routes branching off from, for example, shikimic acid, 5-dehydroshikimic acid or quinic acid. However, this means of forming phenol carboxylic acids seems to be less important in higher plants (cf. page 128).

(4) p-Benzoquinones. They originate from a more complicated biosynthetic pathway which starts from p-hydroxyphenylpyruvate. Substances of central importance such as plastoquinone (page 41) and ubiquinone (page 84) belong to this group.

2. The Acetate-Malonate Pathway (Fig. 94)

The synthesis of phenols via the acetate-malonate pathway shows similarities to the synthesis of the fatty acids. In the case of the latter acetyl CoA was the initiator, and here various other acyl CoAs serve this purpose. Three units of malonyl CoA are added to the initiator with accompanying decarboxylation. We may remember that in the case of the fatty acids malonyl CoA was also added with decarboxylation until the final chain length was obtained. In the present case, a polyketoacid is formed which can cyclize in different ways. We are interested here only in the so-called *1-6 C acylation* which gives rise to the phenols with the hydroxyl pattern of phloroglucinol. They differ in the nature of their R substituent and, in addition, can be subjected to further modifications.

3. Precursors and Intermediates

Let us interrupt our discussion of phenol biosynthesis at this point to consider some of the difficulties which arise in *in vivo* isotope experi-

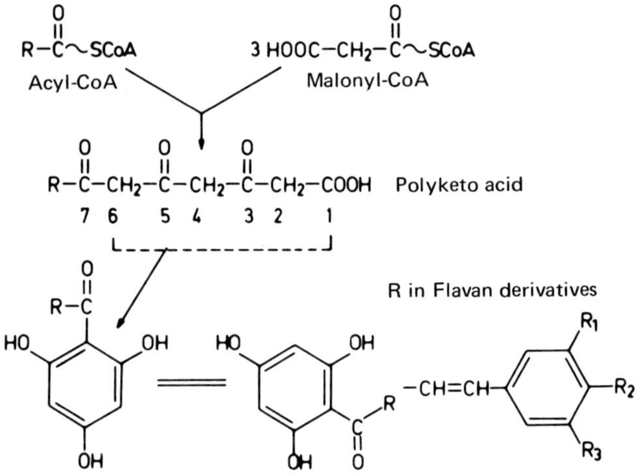

Fig. 94. The acetate-malonate pathway. 1-6-C acylation.

ments. Davis has emphasized that great caution must be exercised when interpreting tracer experiments. In such experiments one supplies a supposed intermediate in radioactive form to the organism and then checks whether it is used in the synthesis of the natural product under investigation. If this should be the case, one can still not say with any certainty that this represents a stage of the biosynthesis under natural conditions. This is because micoorganisms and plants in particular are equipped with biochemical capabilities which allow them to convert substances which do not lie directly on a biosynthetic pathway into useful biosynthetic intermediates. Following a proposal of Davis, it is important to distinguish between precursors, natural intermediates, and obligatory intermediates.

A *precursor* is any substance in the organism under investigation that can be converted into the natural product being studied. It may itself lie directly on the synthetic pathway or be capable of being converted into a compound which does.

A *natural intermediate* is a precursor which has been shown to be present in the organism being investigated.

An *obligatory intermediate,* finally, is the only intermediate possible in a biosynthetic pathway.

It is relatively easy to determine whether a substance serves as a precursor or not. By means of tracer experiments and experiments in cell-free systems, it is also usually possible to determine whether this substance is a natural intermediate. To identify a substance as an obligatory intermediate is the most difficult step. This is because it must always be borne in mind that the organism may be provided with an unknown, but more important, means of synthesizing the natural product under investigation. The biosynthesis of secondary plant materials is a field where such surprises can never be excluded.

The individual members of the shikimic acid pathway discussed above are examples of obligatory intermediates. The shikimic acid pathway is certainly the main route to the synthesis of the aromatic amino acids. Any doubts that might have existed have been removed by experiments with bacterial mutants. If, as a result of a mutation, the biosynthetic pathway is blocked beyond a certain substance and aromatic amino acids are then no longer formed, the substance affected must be an obligatory intermediate along the route to the synthesis of these amino acids.

C. Biosynthesis (particular)

1. Cinnamic Acids

The parent compound is cinnamic acid itself. Its ring system can, however, be substituted, leading to a large number of derivatives which are called "cinnamic acids". The formulae of the most important of them are given below (Fig. 95). We shall come across their pattern of substitu-

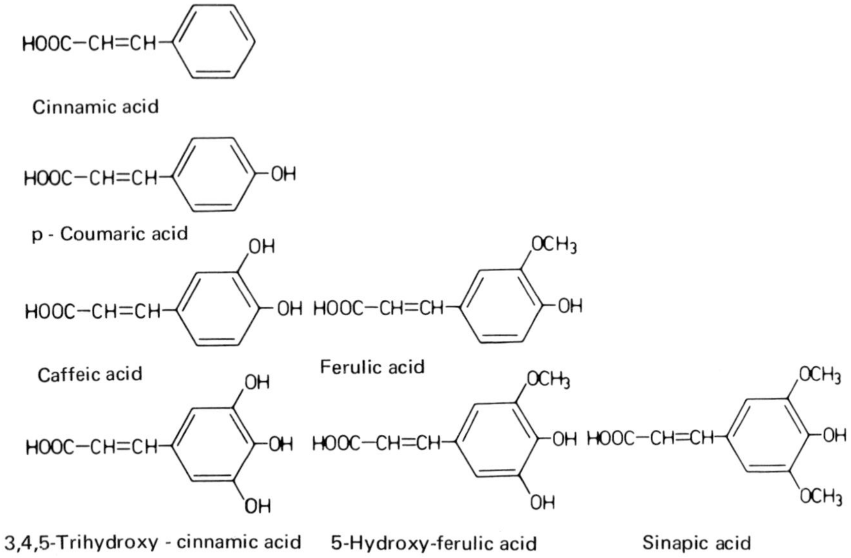

Fig. 95. Cinnamic acids.

tion again in a number of other phenols since the cinnamic acids are the starting materials for the synthesis of these phenols.

First let us mention the biosynthesis of the cinnamic acids themselves (Fig. 96). As mentioned earlier they are derived from phenylalanine and tyrosine. By oxidative deamination phenylalanine is converted to cinnamic acid and tyrosine to p-coumaric acid. Since ammonia in the form of ammonium ions is set free in this reaction the enzymes concerned are called ammonium lyases. The tyrosine-ammonium lyase seems to be particularly important in grasses but is also to be found in the rest of the plant kingdom. However, the phenylalanine-ammonium lyase (PAL) is the more important of these two enzymes. We shall come across it again as the key enzyme of phenylpropane synthesis.

Cinnamic acid can be hydroxylated to p-coumaric acid. Other memebers of the cinnamic acid family can be derived from the latter by simple substitution steps, all of which have been carried out in a cell-free system. Cinnamic acids are found free in plants to only a small extent. Usually they are bound to a sugar either as glycosides or as esters. They may also occur as depsides. These are phenols that bear a carboxyl group linked to each other or to related substances through an ester linkage. Caffeic acid is very often found in the form of chlorogenic acid, a depside formed between caffeic acid and quinic acid.

2. Coumarins

If cinnamic acid is oxidized in the o-position to the side chain and then a lactone ring is formed with the elimination of water, coumarin is

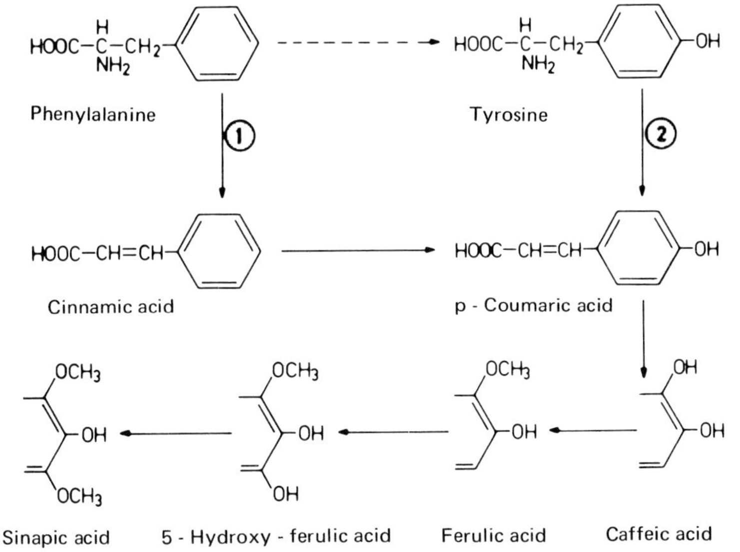

Fig. 96. The biosynthesis of cinnamic acids. 1 = Phenylalanine-ammonium-lyase, 2 = Tyrosine-ammonium-lyase.

produced (Fig. 97). This two-step sequence is useful as a memory aid. In actual fact, the biosynthesis is somewhat more complicated. Coumarin is the parent compound of the family of the coumarins in the same way as cinnamic acid is the generic name for the cinnamic acid family. These coumarins show the same pattern of substitution in their aromatic ring system as that found in the cinnamic acids.

Coumarins are physiologically highly active. For example, they inhibit the growth of microorganisms. Several coumarins, such as coumarin itself and scopoletin, are inhibitors of germination of seeds and cell elongation. At the biochemical level, a few coumarins stimulate the IAA oxidases. These are enzymes which oxidatively degrade the phytohormone indole-3-acetic acid. Perhaps some of the physiological effects of these coumarins are due to a reduction in the level of IAA thus brought about. On the other hand, other coumarins are known which inhibit the activity of the IAA oxidase.

The biosynthesis of the coumarins can be illustrated by considering the case of coumarin itself (Fig. 98). It starts from cinnamic acid.

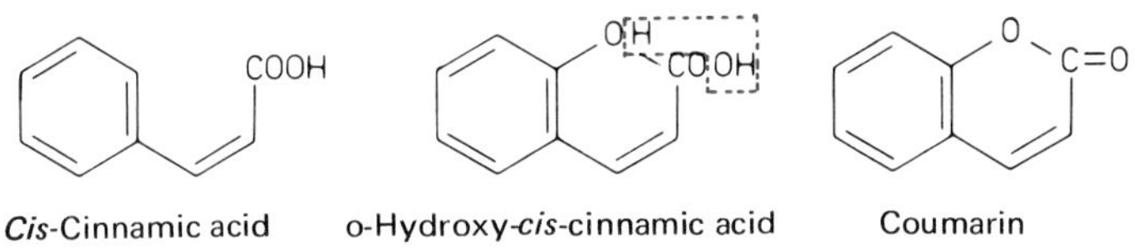

Fig. 97. A memory aid for the structure of the coumarins.

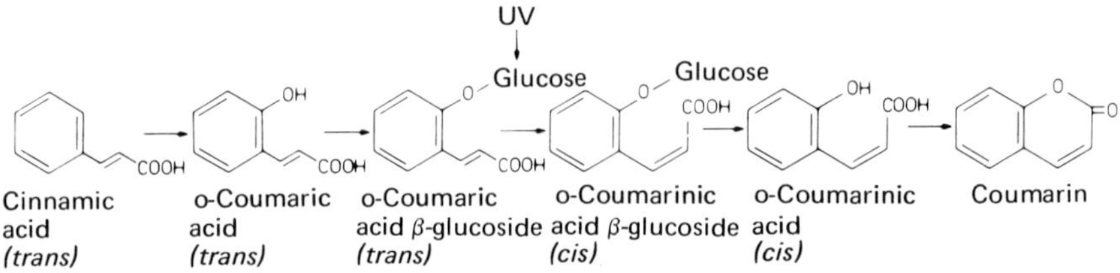

Fig. 98. The biosynthesis of coumarins.

However, we must first make one important point: the double bond in the side chain of the cinnamic acids gives rise to stereoisomerism. The naturally occurring cinnamic acids are predominantly *trans*. The *cis* configuration is, however, also found. The equilibrium between the *trans* and *cis* forms can be displaced within the plant in favor of "*cis*" by UV light.

The starting material for coumarin biosynthesis is, then, *trans*-cinnamic acid. It is converted to *o*-hydroxy *trans*-cinnamic acid, also known as *o*-coumaric acid, by the introduction of a hydroxyl group in the *o*-position to the side chain. This is glucosylated to give *o*-coumaric acid β-blucoside which still has the *trans* configuration. In the tissue it is converted to the corresponding *cis* compound, *o*-coumarinic acid β-glucoside, by UV light, i.e. nonenzymatically. This compound is designated "bound coumarin." This is because no coumarin is usually found in the well-known coumarin plants such as the melilot (*Melilotus albus*), the woodruff (*Asperula odorata*), and in feather-grass (*Hierochloe odorata*), only *o*-coumarinic acid β-glucoside.

Coumarin is formed only as a result of injury or drying out, when *o*-coumarinic acid β-glucoside and a β-glucosidase, from which it is normally spatially separated, are brought in contact. The β- glucosidase hydrolyses the "bound coumarin" to *o*-coumarinic acid which spontaneously lactonises to give coumarin itself.

All findings up to now indicate that all other coumarins are formed in the same way from the corresponding substituted cinnamic acid. *p*-Coumaric acid furnishes umbelliferone, caffeic acid aesculetin and ferulic acid scopoletin (Fig. 99). Thus, our reaction scheme (Fig. 98) is generally valid, the appropriate ring substituents have simply to be inserted. The mode of biosynthesis provides us with an explanation of the fact that the most important cinnamic acids and coumarins exhibit the same pattern of substitution.

3. Lignin

In terms of amount, lignin is the second most important organic substance after cellulose. That is not a chance happening: the woody material lignin is the most important structural substance of plants and, thus, is universally distributed in the plant kingdom from the mosses upward.

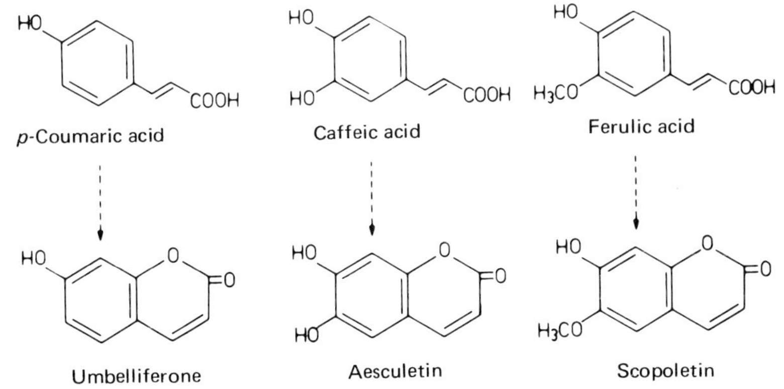

Fig. 99. Formulae and outlines of the biosynthesis of several additional coumarins.

Originally, it was lignin that made possible the transfer of plant life from water to land. In tubular plants it is usually found in the xylem, the individual elements of which show cell walls incrustated with lignin.

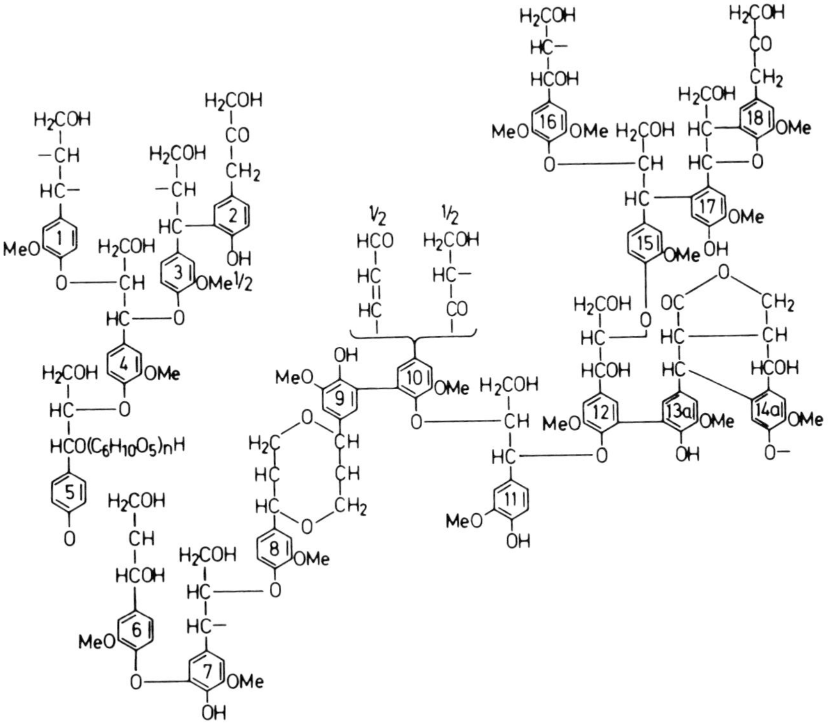

Fig. 100. Partial formula of spruce lignin. The network must be pictured in three dimensions (from Freudenberg and Neish 1968).

Freudenberg, Neish, and Brown, among others, made important contributions to the elucidation of the constitution and biosynthesis of lignin. It is a highly polymeric substance in which phenylpropane units are linked to form a three-dimensional network. (Fig. 100). The substitution patterns of *p*-coumaric acid, ferulic acid, and sinapinic acid can be recognized in these phenylpropane residues. However, it is not the acids but the corresponding alcohols, *p*-coumaryl alcohol, coniferyl alcohol and sinapyl alcohol, which are incorporated into lignin (Fig. 101). The relative amounts of these three components can vary greatly, depending on the age and nature of the plant. Thus, there is not simply one lignin but many. Conifer lignin contains predominantly coniferyl alcohol, and also some *p*-coumaryl and sinapyl alcohols. A part of the structure of spruce lignin is represented in Fig. 100. In the lignin of angiosperms such as the beech coniferyl and sinapyl alcohols are found in roughly equal amounts, together with a trace of *p*-coumaryl alcohol. The lignin of monocotyledons, especially of grasses, contains all three building blocks

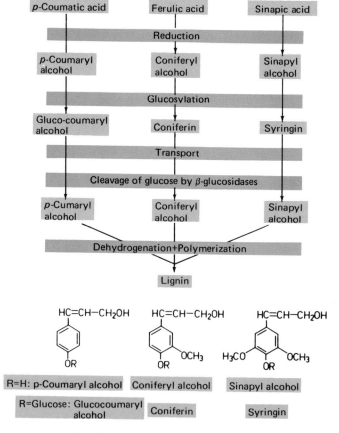

Fig. 101. Outline of lignin biosynthesis.

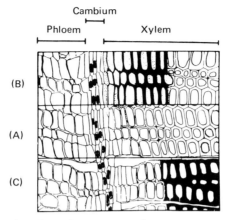

Fig. 102. Cross section through the trunk of a spruce at the boundary between phloem and xylem. (A) unstained, (B) stained for β-glucosidase, (C) stained for lignin. β-glucosidase activity is detectable in the rows of cells where lignification is beginning. About 9 rows inwards from the cambium, lignin formation is complete and β-glucosidase activity is extinguished (from Freudenberg and Neish 1968).

but the high content of p-coumaryl residues is particularly striking. Perhaps there is a connection here with the activity of tyrosine-ammonia lyase, which is particularly widespread in grasses and which deaminates tyrosine to p-coumaric acid. The latter is then converted into p-coumaryl alcohol as we shall see presently.

The biosynthesis of lignin (Fig. 101) starts from the cinnamic acids. p-Coumaric acid, ferulic acid, and sinapic acid are converted to the corresponding alcohols, p-coumaryl, coniferyl, and sinapyl alcohols. The reduction which this conversion implies takes place with the help of NADPH +

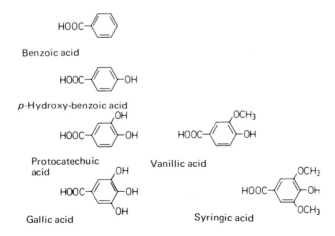

Fig. 103. Phenol carboxylic acids.

H^+ and the substrates for the reaction are the coenzyme A esters of the cinnamic acids, not the free acids themselves. The alcohols are then glucosylated to glucocoumaryl alcohol, coniferin, and syringin, respectively. The glucosides are the form in which the alcohols are transported.

Lignin synthesis in the xylems starts a few cell layers inside the cambrium. Once the glucosides arrive there the glucose is split off. The β-glucosidase responsible for this can be detected in the zones where it occurs by color reactions (Fig. 102). The alcohols that are set free are reduced enzymatically to organic radicals, which then polymerize. The reducing enzymes are, perhaps, phenol oxidases or peroxidases. The whole process is known as "dehydropolymerization." In conifers, lignin formation is completed about 10 cell layers inside the cambrium. The thick cell walls are incrustated through and through with lignin, as can be shown by appropriate color reactions (Fig. 102).

4. Phenol Carboxylic Acids and Simple Phenols

In the biochemical transformations so far discussed the C skeleton of the cinnamic acids serving as starting material has been maintained unchanged. The derivation of the two groups of phenols to be discussed in this section from the cinnamic acids involves the partial or total degradation of the side chains of the latter.

The phenol carboxylic acids show again the substitution pattern of the cinnamic acids from which we can infer a close biogenetic relationship between the two groups of substances (Fig. 103). Phenol carboxylic acids, such as protocatechuic acid and gallic acid, occur widely. A group of tannins, the gallotannins, are polymers consisting of many units of gallic acid.

Simple phenols occur in higher plants much less frequently than was earlier supposed. The most important are hydroquinone and its glucoside arbutin (Fig. 104). Both also occur as their methyl ethers. They are known

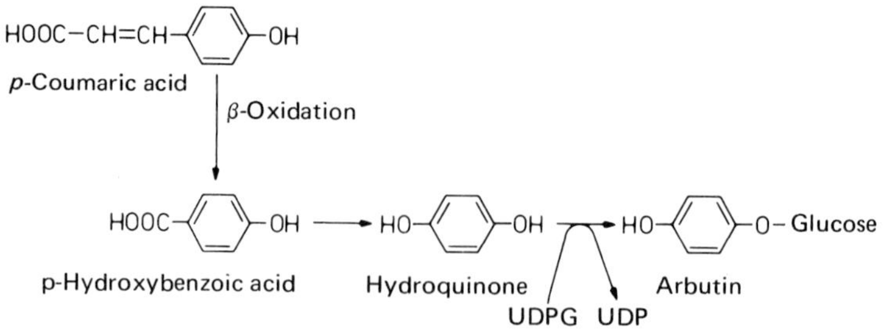

Fig. 104. The biosynthesis of phenol carboxylic acids and simple phenols.

to occur in the leaves of *Ericaceae* and pears. The darkening of pear leaves in the autumn is due to an oxidation of hydroquinone to quinone.

The biosynthesis of the phenol carboxylic acids and that of the simple phenols are closely related and should be discussed together for this reason (Fig. 104). Phenol carboxylic acids can be derived from different stages of the shikimic acid pathway (page 119).

Phenol carboxylic acids can also be supplied by the acetate-malonate pathway, at least in microorganisms. In higher plants, however, they are more likely to be derived from cinnamic acids. Let us consider as an example the conversion of *p*-coumaric acid into *p*-hydroxybenzoic acid and, subsequently, hydroquinone, a sequence of reactions discovered by ZENK.

First, *p*-coumaric acid is subjected to a β-oxidation which reminds us of the β-oxidation of the fatty acids and it is possible that the mechanism of the two reactions is the same. At any rate an aromatic acid is produced, the side chain of which is 2 C atoms shorter. It is called *p*-hydroxybenzoic acid and is an example of a phenol carboxylic acid. In a completely analogous manner protocatechuic acid is obtained by β-oxidation of caffeic acid, vanillic acid from ferulic acid, etc.

Now let us consider the simple phenols. In the case we were studying *p*-hydroxybenzoic acid is decarboxylated to hydroquinone and can be glucosylated to arbutin by means of UDPG. For the formation of the methyl ethers of hydroquinone and arbutin, *S*-adenosyl methionine, "active methionine," serves as the methyl group donor.

5. Flavan Derivatives

a. Chemical constitution

The flavan derivatives or flavanoids constitute the largest group of phenolic substances. They derive their name from the yellow coloring (latin flavus = yellow) of some of the substances of the group. The flavan derivatives can be divided into a number of subgroups according to the oxidation state of the central heterocycle. Only the more important of them can be mentioned here (Fig. 105). Each subgroup comprises a multitude of substances which differ from each other with respect to the pattern of substitution of the skeleton, particularly in the substituents of their B ring.

Most flavan derivatives occur in plants as glycosides. The sugars are attached to the hydroxyls of the A ring and of the heterocycle, attachment to the hydroxyl of the C atom 3 occurring particularly frequently.

Chalcones are not flavan derivatives since they lack the characteristic central heterocycle. They are converted spontaneously into true flavan derivatives, the flavanones, the reaction occurring particularly readily in acidic medium. They play a central role in the biosynthesis of the flavan derivatives. Chalcone glycosides occur in relatively high quantities in the flowers of several of the *Compositae* and *Leguminosae,* which, as a result,

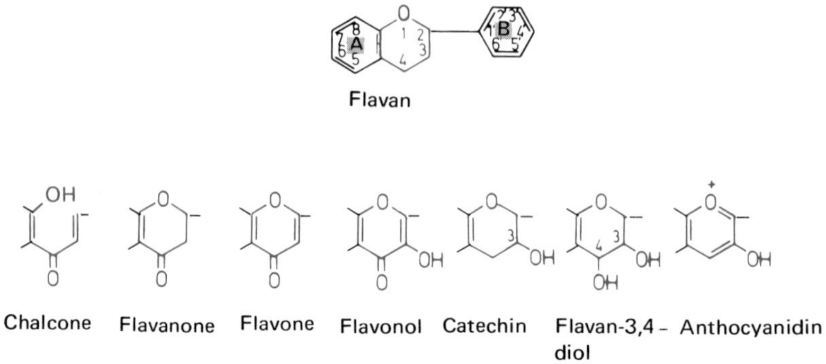

Flavan

Chalcone Flavanone Flavone Flavonol Catechin Flavan-3,4- Anthocyanidin
                                                                      diol

Fig. 105. Survey of several flavan derivatives. Above, the basic skeleton. Below, the central heterocycle of several groups of flavan derivatives.

are yellow. An example is butein (Fig. 106), which is found in the flowers of different Compositae.

Flavones possess one double bond more than the flavanones in their central heterocycle. The white "meal" on shoots and leaves of Primulaceae, such as the bird's eye primrose *Primula farinosa,* consists, in part, of the parent compound, flavone itself. However, derivatives of flavone such as apigenin and luteolin (Fig. 106) are much more widely occurring.

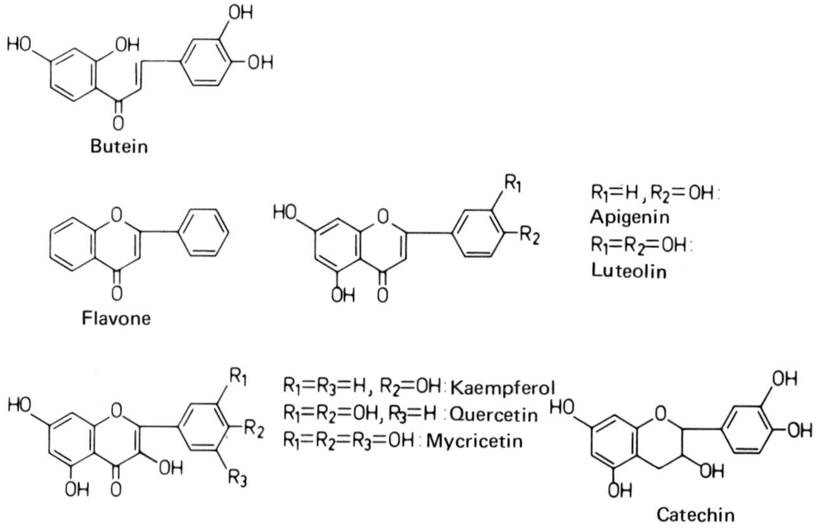

Butein

Flavone

$R_1=H, R_2=OH$:
Apigenin
$R_1=R_2=OH$:
Luteolin

$R_1=R_3=H, R_2=OH$: Kaempferol
$R_1=R_2=OH, R_3=H$: Quercetin
$R_1=R_2=R_3=OH$: Mycricetin

Catechin

Fig. 106. A few flavan derivatives (excepting anthocyanidins).

If a hydroxyl group is introduced into the 3-position of the central ring of flavones, *flavonols* are obtained. Biosynthetically, however, they are not produced in this way (Fig. 110). Flavonol glycosides lend a whitish to pale yellow tinge to flowers. Their occurrence is in no way limited to the flower, they are found in all other parts of the plant. Their ubiquitous occurrence suggests a central function for the flavonols, although up to now the hypotheses based on this supposition are poorly founded. Thus, it is assumed that some flavonols are inhibitors, others activators, of the IAA oxidase (page 198).

Examples of frequently occurring flavonol aglycones are kaempferol and guercetin. Myricetin is rarer (Fig. 106).

Flavan-3-ols and Flavan-3,4-diols are the precursors of the second large group of tannins. They polymerize to the so-called "condensed tannins." We have already mentioned the first group of tannins, the gallotannins. Flavan-3-ols are better known under the name *catechins* (Fig. 106). They are widely distributed in the plant kingdom. The same is true of the flavan-3,4-diols, which belong to the so-called "leuco anthocyanins." This is because flavan-3,4-diols can be very easily converted into anthocyanins by boiling with alcoholic hydrochloric acid. However, this conversion does not play any role in the biosynthesis of the anthocyanins.

Thus, we come to the *anthocyanins.* They are flavan derivatives, the heterocycle of which exhibits an oxonium structure in an acidic medium. Anthocyanins are well-known red and blue flower pigments, which usually are dissolved in the cell sap of the epidermis. They are also to be found, however, in the vegetative parts of the plant and in leafy plants often have an ornamental function. Up to now we have spoken of anthocyanins and they are glycosides. Sugar-free anthocyanins; that is, anthocyanin aglycones are called *anthocyanidins:* anthocyanin = anthocyanidin + sugar. Apart from a few exceptions of uncertain nature, anthocyanidins occur in plants as glycosides. As mentioned at the beginning the hydroxyl on C atom 3 is glycosylated preferentially.

Let us consider the anthocyanidins in more detail. The formulae of the more important of them are shown in Figure 107. The difference between the individual anthocyanidins lies in the pattern of substitution of ring B. Only in rare cases, which are not mentioned here, are anthocyanidins found which differ from each other with respect to the pattern of substitution in the other two ring systems. The isolated anthocyanidins are colored in different shades of red and blue. We shall come back to this point later.

First let us take another glance at the structural formulae of the individual anthocyanidins. We notice that the pattern of substitution in ring B is the same as that which we have already encountered with the cinnamic acids, the coumarins and the phenol carboxylic acids. In retrospect we may notice that this observation holds true for the whole group of the flavan derivatives, though the parallels are usually easier to see than here

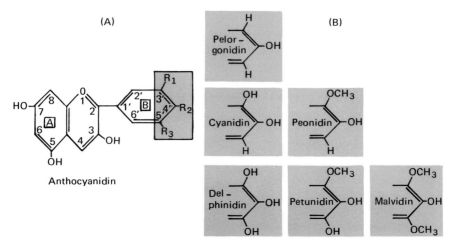

Fig. 107. The most important anthocyanidines. (A) the basic structure; (B) the B ring, with respect to which the individual anthocyanidins differ from each other.

where the number of anthocyanidins is so large. We must then pose the question: do cinnamic acids also perhaps play a role in the biosynthesis of flavan derivatives, which have a quite different structure?

b. Biosynthesis of the anthocyanidins and other flavan derivatives

Analysis of Building Blocks

Usually the elucidation of a biosynthetic pathway begins with an analysis of the building blocks. The structure of the substance whose biosynthesis is to be investigated allows inferences to be made as to possible precursors. These presumed precursors are then fed to the organism in radioactive form. After a certain incubation time the end product is isolated and tested for radioactivity. If the substance is found to be radioactive it is broken down into fragments with appropriate methods and subsequently the radioactivity of each of the fragments is investigated. In this way it can be ascertained whether a given substance can serve as precursor, i.e. as a building block, in the synthesis of the substance under investigation.

This kind of analysis of the anthocyanidins showed (Fig. 108) that ring A is built up from three acetate units. Ring B and the C atoms 2, 3, and 4 of the heterocycle are supplied by phenylpropanes, e.g. the C skeleton of phenylalanine and of cinnamic acids. In cases where hydroxyl groups of ring B occur as their methyl ethers, the methyl group is derived from active methionine, *S*-adenosyl methionine. This completes the analysis of the building blocks. Now the difficulties begin, for we have to account for the intermediates.

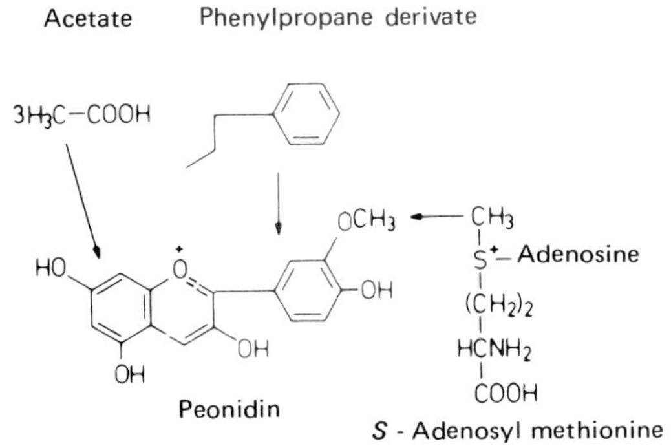

Fig. 108. Analysis of the building blocks of the anthocyanidin peonidin.

Biosynthesis of the anthocyanidins

Three groups of problems can be distinguished:

(1) *Formation of the 15 C Skeleton (Fig. 109).* An acceptable working hypothesis is that malonyl CoA and the CoA esters of cinnamic acids participate in the formation of the 15 C skeleton. The mechanism corresponds to that found for the acetate-malonate pathway: cinnamic acid CoA serves as starting material, to which three units of malonyl CoA are added successively with decarboxylation. The product is a substance with 15 C atoms which contains only one aromatic ring that is destined to become the B ring and which is, in fact, the aromatic ring of the starting cinnamic acid. The structure, which is later to be the A ring and which is formed from decarboxylated malonate = acetate is, for the moment, present as an open chain. An enzyme that is capable of converting *p*-coumaric acid into its CoA-ester has been isolated from cell cultures of parsley. A chalcone synthetase which links *p*-coumaryl-CoA to malonyl-CoA to form the corresponding chalcone has also been detected recently in a cell-free system. These experimental findings demonstrate the essential correctness of the hypothesis we have been discussing.

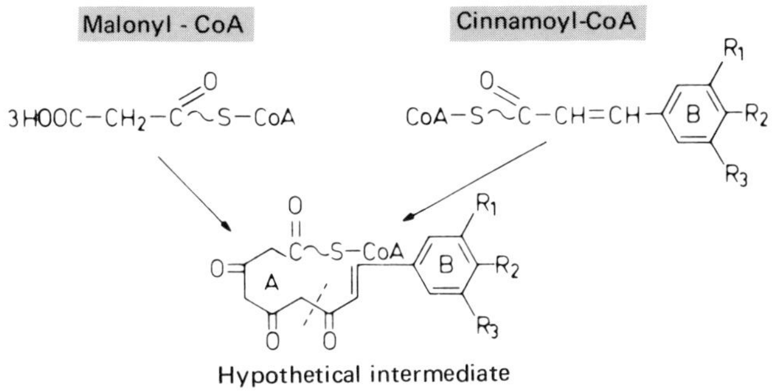

Hypothetical intermediate

Fig. 109. Hypothesis concerning the formation of the 15-C skeleton of the flavan derivatives.

(2) *Modifications of the Heterocycle (Fig. 110).* In the next step the open chain structure is closed to form ring A. We obtain the kind of substance which we recognize, namely, a chalcone. Isotope experiments, especially those of Grisebach, have removed all doubt that such chalcones are the parent substances of all flavan derivatives. The chalcones are converted to flavones and all of the subsequent reaction sequences up to the anthocyanidins consist only of appropriate modifications of the central ring system. We should realize that the elucidation of the biosynthesis of the anthocyanins implies at the same time an understanding of the biosynthesis of all flavan derivatives. This is because the individual flavan derivatives are either precursors of the anthocyanins or they can be derived from such precursors.

(3) *Timing of the Substitution in Ring B (Fig. 111).* Two possibilities are discussed. One of them is that at the start of anthocyanin synthesis the appropriately substituted cinnamic acid is utilized. The substituents of the cinnamic acids then become the substituents of ring B automatically. We have established a number of times that substituted cinnamic acids can be

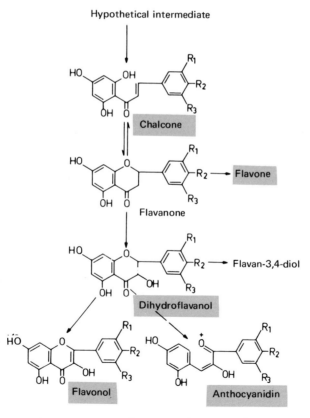

Fig. 110. The biosynthesis of the most important flavan derivatives.

implicated in synthesis: in the synthesis of the coumarins, lignins, phenol carboxylic acids and simple phenols. We shall become acquainted with other examples when we consider alkaloid biosynthesis. Recognition of this fact makes the striking similarity between the pattern of substitution of the cinnamic acids and that of the anthocyanidins easy to understand.

A second possibility is that the substituents of ring B are introduced at a relatively late stage of the biosynthesis, perhaps at the level of the dihydroflavanols.

The problem remains unsolved. Some biochemical findings support one mechanism, others support the other. Perhaps both possibilities are operative, depending on the plant. However, data on the occurrence of different flavan derivatives in one plant and genetic findings are not compatible with the second possibility.

Let us summarize: biosynthetically the flavan derivatives are hybrid substances. Their A ring is derived from acetate, their B ring and the C atoms 2, 3 and 4 of the heterocycle from phenylpropanes. The 15 C skeleton is probably synthesised from malonyl CoA and cinnamic acid CoA compounds, the reaction sequence being analogous to the acetate-malonate pathway.

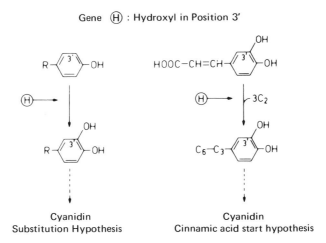

Fig. 111. Two hypotheses concerning the timing of the substitution of ring B of the anthocyanidins. According to the substitution hypothesis, substituents are introduced into the ring system of the cinnamic acids or of the 15-C precursors of the anthocyanidines under genetic control. According to the cinnamic acid start hypothesis, the genes bring about the same effect in a different way: they restrict the initiation of anthocyanidin synthesis to quite specific cinnamic acids with the appropriate pattern of substitution. According to the substitution hypothesis, gene H would be responsible for the incorporation of a hydroxyl group in the aromatic ring system. The cinnamic acid start hyspothesis states that gene H would lead to activated caffeic acid being linked to three units of active acetate (malonyl-CoA) to give a 15-C body which would then be converted into cyanidin.

6. Flower Pigmentation

We are now familiar with the carotenoids and the flavan derivatives and they represent the most important flower pigments. Carotenoids are responsible for yellow to red colors and flavan derivatives for whitish, yellow, red, and blue colors. Anthocyanins are the red and blue flavan pigments.

So far the pigments have been classified according to their chemical structure. They can also be distinguished according to cytological criteria and Seybold has suggested that they be divided into chymochromic, plasmochromic and membranochromic pigments. *Chymochromic pigments* are soluble in the cell sap of the vacuoles. They include the glycosylated flavan derivatives, especially the glycosides of the flavones, flavonols, and anthocyanidins, and also the glycosylated carotenoid acids such as the crocetin derivatives mentioned earlier. *Plasmochromic pigments* are found in the plastids. Thus, chlorophylls, carotenes and xanthophylls are plasmochromes. Finally, *membranochromic pigments* impregnate cell walls. Instances of these are the different phenolic substances which color heartwood. Red pigments of unknown structure found in the cell walls of peat mosses also belong in this category.

The realization of a particular flower coloration is determined by a number of factors, the most important of which are considered below.

(1) *Nature of the Pigments.* Frequently, members of several pigment groups occur simultaneously in flowers. Thus, in breeds of the pansy *(Viola tricolor)* chymochromic anthocyanins in the epidermis are imbedded in plasmochromic carotinoids in the subepidermis. And even when the pigmentation of a flower is due to only one pigment group, several members of this group are usually present. For example, several anthocyanidins in different states of glycosylation and sometimes also bearing different cinnamic residues are frequently found in one flower. The individual, isolated anthocyanidins are colored differently. The coloring becomes deeper with the increasing number of hydroxyl groups. Pelargonidin is salmon pink, cyanidin red and delphinidin blue. Methylation of the hydroxyl groups shifts the color a shade to the red. Peonidin is red, petunidin purple, and malvidin rose pink. As mentioned, that is true of isolated anthocyanidins. In the flower, the difference in the extinction between pelargonidin on the one hand and the other anthocyanidins on the other is particularly important.

(2) *Amount of Pigments.* This can be considerable, in some breeds of the pansy *(Viola tricolor),* for example, approaching 30% of the dry weight of the flower, and this leads to a corresponding intensity of pigmentation.

(3) *pH Value.* Anthocyanins are red in acid, blue in alkali. At least this is true of the compounds in a test tube. In nature, the pH value is not as important as it was formerly assumed to be. An illustration of this is the

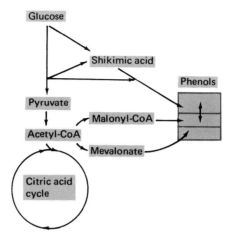

Fig. 112. The position of the phenols in metabolism.

cornflower, which, as is known, is blue although the pH value of its cell sap may often be as acid as 4.9.

(4) *Complex Formation with Polyvalent Metal Ions.* As Bayer and Hayashi showed complex formation with metal ions such as Fe^{+++} and Al^{+++} is much more important for pigmentation than the pH value. Anthocyanidins with two adjacent hydroxyls in the B ring—cyanidin, delphinidin, and petunidin—form blue colored metal complexes by means of these hydroxyls. Chelates of this kind which are often also linked to carbohydrate carriers, are responsible for the blue coloration of the cornflower at mildly acid pH.

(5) *Copigments.* Other substances, some of which are colorless themselves can affect the coloring of the anthocyanins. Such substances are known as copigments. Examples of them are flavanaols and catechins. The copigment effect usually brings about a deepening of the color. The causes are still a matter of dispute.

This completes our discussion of the phenols, the second largest group of secondary plant substances after the terpenoids. Listing the biosynthetic pathways in order of increasing importance for higher plants, phenols are formed by the acetate-mevalonate pathway, the acetate-malonate pathway, and, particularly, the shikimic acid pathway (Fig. 112). The cinnamic acids occupy a central position in phenol metabolism, participating as they do in the biosynthesis of all other important phenols.

Amino Acids

In the discussion of translation, the amino acids were taken for granted. In this chapter we must discuss several aspects of their biosynthesis. Their name is derived from the fact that in all frequently occurring amino acids an amino group is found in the α-position to a carboxyl group. We shall concern ourselves first with the origin of this reduced N and then see how it is transferred to the C skeleton, i.e. how the amino acids are formed.

A. The Reduction of Nitrogen

Air consists of nitrogen to about 78% by volume in the form of N_2. This nitrogen cannot be utilized directly by higher plants. Only certain microorganisms are capable of fixing it and reducing it to the level of NH_3 = ammonia. Some of them are free-living microorganisms, some are symbionts of higher plants. Blue algae, fungi, and bacteria come into this category. Of the symbiotic bacteria the best-known is the genus *Rhizobium* which is found in the tubers of the Leguminosae. Higher plants can obtain nitrogen compounds from the symbionts mentioned above. However, their most important source of nitrogen is the nitrate of the soil. Even this nitrate is supplied largely by bacteria. NH_3, set free by decomposing organic material, is oxidized by nitrite bacteria to the level of NO^-_2 and further by nitrate bacteria to the level of NO^-_3. Although higher plants can also take up NH_3 in the form of ammonium compounds, it is predominantly nitrate that serves as starting material for the synthesis of amino acids and, then, additional N-containing compounds.

For this to happen nitrate must be reduced. This reduction is carried out in several steps, each of which represents a 2 electron transition. The first step is the conversion of nitrate to nitrite. Two futher reductions follow, leading to intermediates which have still not been identified. In the last reduction step ammonia is finally formed (Fig. 113).

Of this total of four reduction steps the first, the conversion of nitrate to nitrite, is the most studied. The enzyme involved is nitrate reductase. This is a flavoprotein with FAD as coenzyme but which also contains molybdenum. It draws its electrons from $NADPH + H^+$, which can itself be derived from the primary processes of photosynthesis. It can, however,

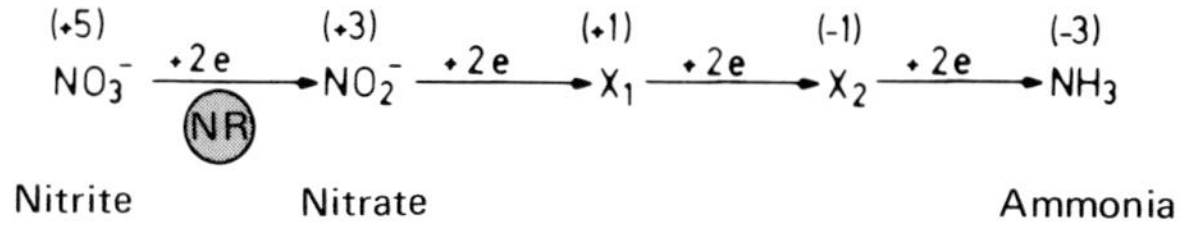

$$(+5) \quad\quad (+3) \quad\quad (+1) \quad\quad (-1) \quad\quad (-3)$$

$$NO_3^- \xrightarrow[\text{NR}]{+2e} NO_2^- \xrightarrow{+2e} X_1 \xrightarrow{+2e} X_2 \xrightarrow{+2e} NH_3$$

Nitrite   Nitrate   Ammonia

Fig. 113. The reduction of nitrate. NR = nitrate reductase.

be replaced by $NADH + H^+$. It is assumed that NADPH or NADH first passes its electrons to FAD, which then transfers them to molybdenum, and, finally, to nitrate. The subsequent reductions leading ultimately to NH_3 have been relatively poorly studied.

The assimilation of S

C,O,H,N,S, and P are, from a quantitative standpoint, the predominant elements of organic matter. As we have just sketched the mode of assimilation of N by higher plants, it may not be out of place to mention a few facts about the assimilation of S.

Just as in the case of N, the source of the sulphur of higher plants is an oxidized form, namely sulphate. And just like nitrate, sulphate must first be reduced. Ultimately, sulphur is present in the doubly negative form as S^{--}. The first step in the assimilation of S is the fixation of sulphate. This is brought about by sulphate reacting with ATP to liberate pyrophosphate. An adenosine-phosphate-sulphate compound is formed and to the ribose of this compound another phosphate residue from another ATP molecule is attached. The product thus obtained is 3'-phosphoryl-5'-adenosine-phosphoryl-sulphate or simply "active sulphate" (Fig. 114). In this way sulphur is fixed and activated. It is this bound form of "active sulphate" which is subjected to reduction to the level of S^{--}. It is likely that 2 electron transitions are also implicated here. The mechanism is still unknown.

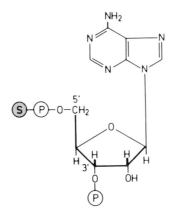

Fig. 114. Active sulphate.

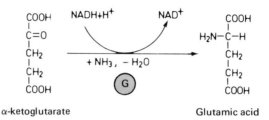

Fig. 115. Reductive amination. G = glutamic acid dehydrogenase.

HS-Groups or sulphydryl groups are biologically important. The amino acid cysteine bears an HS-group and it is in this form that HS-groups are incorporated into proteins, including enzyme proteins. HS-Groups can be functionally important for enzymes and one then speaks of HS-enzymes, of which the fatty acid synthetase is an example.

B. Reductive Amination

Let us return to nitrogen. It had been reduced to the level of NH_3. It is now necessary to convert this ammonia N into amino N. This happens in a reductive amination step (Fig. 115): α-ketoglutarate, which we recognize as an intermediate of the citric acid cycle, accepts the N in a reaction in which reduction and splitting off of water occur simultaneously. The product is glutamic acid. The enzyme that catalyzes the reaction and on that has been shown to be present in many plants is glutamic acid de-hydrogenase. The hydrogen donor is $NADH + H^+$. One could imagine that other keto acids might be reductively aminated just like α-ketoglutarate, e.g. oxalacetate. However, only the reductive amination of α-ketoglutarate seems to be of significance in higher plants. Thus, glutamic acid dehydrogenase is the key enzyme of amino acid metabolism.

C. The Formation of Glutamine

Glutamic acid is a dicarboxylic acid, one of the carboxyl groups of which can be converted to the corresponding amide (Fig. 116). Ammonia is also used in this reaction, which requires ATP. It is catalyzed by glutamine synthetase, an enzyme widely distributed in plants. Glutamine is an important substance in N metabolism. It furnishes one of the N atoms, which are found in the amino acid arginine and in the pyrimidine and purine bases. It seems likely that asparagine, the corresponding amide of aspartic acid, is of less importance. It is probably formed in a similar way.

COOH   ATP   ADP+P   COOH
  | |
H₂N—C—H   H₂N—C—H
  | |
 CH₂   CH₂
  | +NH₃ |
 CH₂   CH₂
  | GS   C₂O
COOH   C—NH₂

Glutamic acid   Glutamine

Fig. 116. Formation of glutamine. GS = glutamine synthetase.

D. Transamination

We have seen how N is reduced and fixed in the form of NH_2-groups in certain compounds. Let us consider first glutamic acid. It can pass on its amino group to a number of α-keto acids, e.g. pyruvate and oxalacetate (Fig. 117). In this reaction glutamic acid is converted to the corresponding α-keto acid, α-ketoglutarate, and the α-keto acids to the corresponding amino acids, i.e. pyruvate to alanine and oxalacetate to aspartic acid. The reversible transfer of an amino group to an α-keto acid is known as transamination and the enzymes which catalyze the reaction as transaminases. The coenzyme of the transaminases is pyridoxal phosphate, which mediates the transfer in the form of pyridoxamine phosphate (Fig. 117).

Glutamic acid is the central product of reductive amination. Transamination from glutamic acid to α-keto acids leads to the formation of

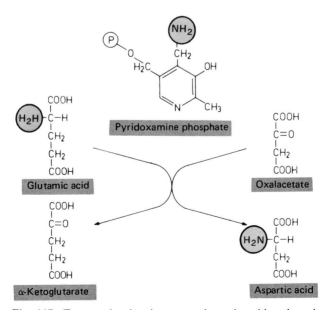

Fig. 117. Transamination between glutamic acid and oxalacetate.

other amino acids as outlined above. These amino acids can then, in turn, donate their amino group to α-keto acids in a transamination reaction, giving rise, finally, to a whole spectrum of amino acids.

E. The Origin of the C Skeleton of the Amino Acids

The origin of the C skeleton of a few of the amino acids has already been indicated. Thus, the C skeleton of glutamic acid is derived from α-ketoglutarate, that of aparagine from oxalacetate and that of alanine from pyruvate. All three of these α-keto acids are derived from the degradation of carbohydrates and are familiar to us. Glutamic acid, aspartic acid, and alanine are, however, the precursors of three groups of amino acids. Thus, one speaks of a glutamic acid family, an aspartic acid family, and a pyruvate family. In addition, there are other families of amino acids, which can be shown to be linked to carbohydrate metabolism. Of these we have already become acquainted with the shikimic acid family (phenylalanine and tyrosine on the one hand and tryptophan on the other). A survey of the different families and their connection to carbohydrate metabolism is presented in Fig. 118.

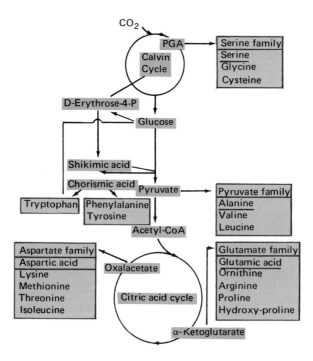

Fig. 118(a). The relationship of the individual amino acid families to metabolism. The more complicated route to histidine is not shown.

In summary, the C skeleton of the amino acids is derived from carbohydrate metabolism. This means that whole families of amino acids can be traced back to one intermediate. The introduction of the amino group into the C skeleton can be considered to occur in a sequence of three main steps: reduction of nitrate, reductive amination of α-ketoglutarate and transamination.

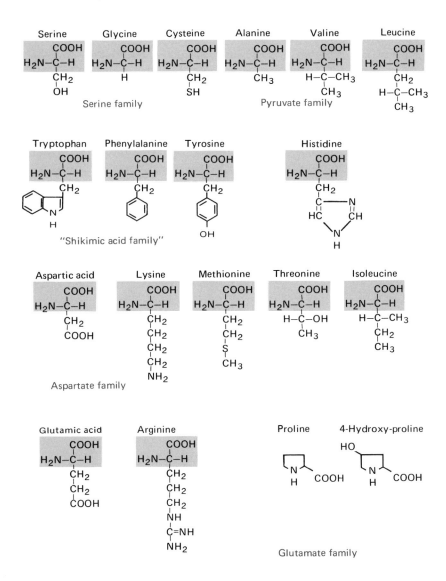

Fig. 118(b). The most important amino acids arranged according to the families of Figure 118 (A).

Alkaloids

The alkaloids are a group of basic, secondary plant substances, which usually possess a N-containing heterocycle. Their basic character is due to the ring N (alkaloid = alkali-like). Biosynthetically the alkaloids are derived almost exclusively from the amino acids ornithine, lysine, phenylalanine, tyrosine, and tryptophan.

At the present time about 3,000 alkaloids are known which are distributed among almost 4,000 plant species. In view of their profusion, it is not surprising that not all of the compounds that are usually considered as alkaloids meet fully the definition given above. Alkaloids are also not limited to plants; a few have been demonstrated in animal organisms. Furthermore, not all compounds so classified are basic. Exceptions to this rule are, for example, nicotinic acid, colchicine and the betacyanins. Finally, the N is not always contained in a heterocycle, as can be seen from the structural formula of colchicine. However, these exceptions are usually included in the alkaloids because their biosynthesis leaves no doubt as to their close relationship to "true" alkaloids.

As far as the biosynthesis is concerned Winterstein and Trier pointed out in 1931 that, according to their structure, alkaloids must be derivatives of amino acids. Thus, an amino group supplies the N of the alkaloids. Today no one doubts the correctness of this statement. Indeed, their biogenesis from amino acids is that which allows the heterogeneous collection of alkaloids to be considered as a group. That does not exclude the possibility that sometimes building blocks other than amino acids, in particular isopentenyl pyrophosphate, can be drawn upon for the biosynthesis of alkaloids.

In this connection brief mention must be made of the terpenoid alkaloids and the steroid alkaloids. They are substances whose C skeleton is furnished exclusively by 5 C units. They contain N, the origin of which is disputed. On account of their N they are usually regarded as alkaloids, although they are not derivatives of amino acids. It would be equally correct to regard them as terpenoids (page 110).

Finally, the name protoalkaloids is given to a group of N-containing substances of simple structure. Examples of these are "biogenic" amines which arise from the decarboxylation of amino acids and their oxidized, alkylated, or acylated derivatives.

In this chapter we shall not be further concerned with terpenoid alkaloids, steroid alkaloids, and protoalkaloids. We can choose only a few of the large number of groups of the true alkaloids for consideration and for each group only a few of its many members. A survey is given in Fig. 119. It is common knowledge that alkaloids can affect the animal and human organism in different ways, frequently via the nervous system. Much less is known about the function the alkaloids carry out in the

Structure	Group	Precursor
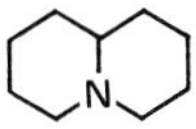	Quinolizidine - A.	Lysine
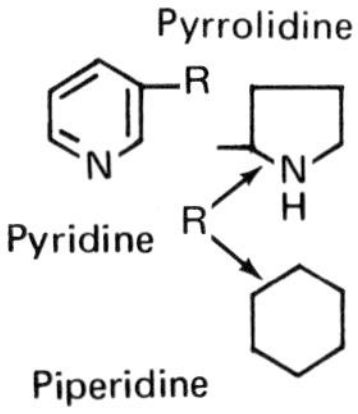 Pyrrolidine / Pyridine / Piperidine	Nicotiana - A.	Nicotinic acid Ornithine Lysine
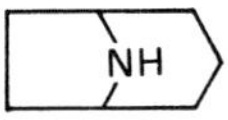	Tropane -A.	Ornithine
See text	Amaryllidaceae — A. Colchicine	Phenylalanine Tyrosine
See text	Betacyanins Betaxanthines	Tyrosine
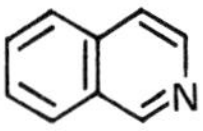	Isoquinoline	Tyrosine
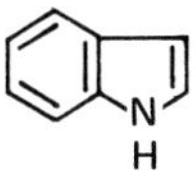	Indole -A.	Tryptophan
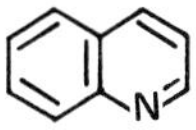	Quinoline	Tryptophan
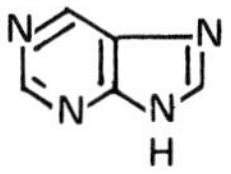	Purine	See text

Fig. 119. Survey of several groups of the alkaloids.

plants which produce them. It is assumed that some alkaloids serve to protect the plant against predators and that others serve as N-reserves which
can be mobilized when needed. By and large, however, the alkaloids seem
to be excretion products. These excretions can then be used secondarily
in individual cases for particular functions.

A. Derivatives of the Aliphatic Amino Acids, Ornithine and Lysine

Both amino acids can be precursors in the synthesis of N-containing
ring systems. Ornithine furnishes a five-membered ring and lysine a six-
membered ring. The five-membered ring occurs most frequently in the
alkaloids as a pyrrolidine system, the six-membered ring as a piperidine or
pyridine system. In the case of both amino acids two pathways lead to the
ring systems. In the first type (a) a biogenic amine arises by decarboxylation: putrescine from ornithine and cadaverine from lysine. On the second pathway (b) an amino group is first removed oxidatively. Thus, a
biogenic amine is not formed. Later the two routes coincide (Fig. 120).
Although it was formerly assumed that only the first pathway is found in
nature, it has since been learned that both routes are used and in some
cases route (b) is even preferred.

1. Quinolizidine Alkaloids

The quinolizidine alkaloids contain either one or two quinolizidine
systems. They occur most abundantly in the *Papilionaceae* in the genus
Lupinus. For this reason they have also been given the name lupin

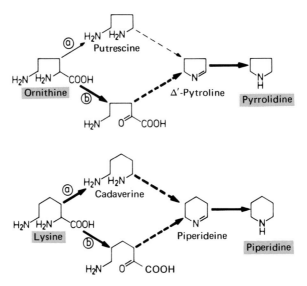

Fig. 120. Possible cyclization reactions of ornithine and lysine.

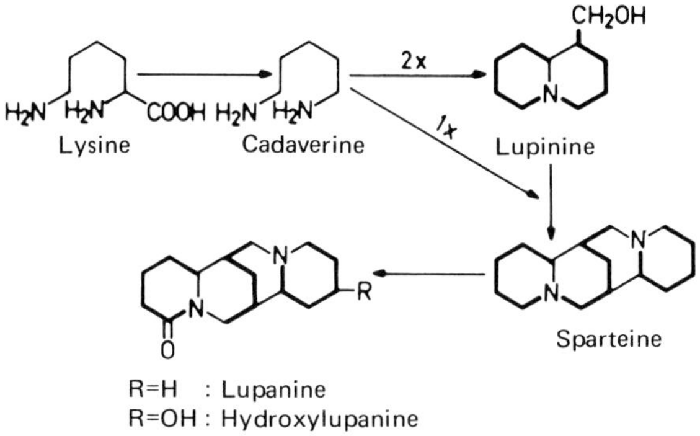

Fig. 121. The most important lupin alkaloids and their biosynthesis.

alkaloids from their occurrence in lupins. At the end of the 1920s, Sengbusch selected mutants that were poor in alkaloids from millions of lupin plants and that could be used as cattle fodder. They have become well-known as "sweet lupins." The most important lupin alkaloids are lupinin with one, and sparteine, lupanin, and hydroxylupanin with two, quinolizidine systems (Fig. 121).

The starting material for the synthesis of the lupin alkaloids is the amino acid lysine, which is first decarboxylated to give its biogenic amine cadaverine. Two units of cadaverine are then joined via a still hypothetical intermediate to give lupinin. Addition of another cadaverine unit to lupinin gives sparteine, which then can be oxidized to lupanin and, further, to hydroxylupanin. The C skeleton of the quinolizidine alkaloids is derived entirely from lysine. We shall now consider two further groups of alkaloids, the nicotiana alkaloids and the tropane alkaloids, which derive only a part of their C skeleton from the aliphatic amino acids or-nithine or lysine.

2. Nicotiana Alkaloids and Nicotinic Acid

The most important nicotiana alkaloids (tobacco alkaloids) are nicotine, nornicotine, and anabasine (Fig. 122). Nicotine and nornicotine are the chief alkaloids of *Nicotiana tabacum,* and anabasine of *Nicotiana glauca.* Both nicotine and nornicotine consist of a pyridine ring to which a pyrrolidine ring is attached. In the case of nicotine this pyrrolidine ring bears a methyl group on its ring-N. Anabasine also consists of a pyridine

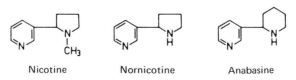

Fig. 122. The most important tobacco alkaloids.

ring but the substituent in this case is a piperidine ring. The pyridine ring of the nicotiana alkaloids is supplied by nicotinic acid. Thus, we must first deal with the biosynthesis of nicotinic acid (Fig. 123). Depending on the organism it can arise in two different ways, which meet in the precursor quinolinic acid. In animals and fungi quinolinic acid is supplied by the degradation of tryptophan. In bacteria and higher plants a glycerol derivative combines with a derivative of aspartic acid to form quinolinic acid. The later is then converted into nicotinic acid mononucleotide with accompanying decarboxylation and this can be converted either directly or indirectly via NAD^+ into free nicotinic acid. The cycle from and to nicotinic acid mononucleotide is called the pyridine nucleotide cycle.

The formation of nicotinic acid—or, more precisely, of quinolinic acid—via two different pathways is an example of biochemical convergence. Now to the biosynthesis of the pyrrolidine ring of nicotine and nornicotine. Ornithine comes into play here. It furnishes the pyrrolidine component by the route already outlined, and this is then combined with nicotinic acid to give nicotine (Fig. 124). It is not known with certainty at what stage the methyl group is introduced. Very probably ornithine or putrescine is methylated and the methylated precursor is then icorporated into nicotine. Nicotine can be converted into nornicotine by demethylation.

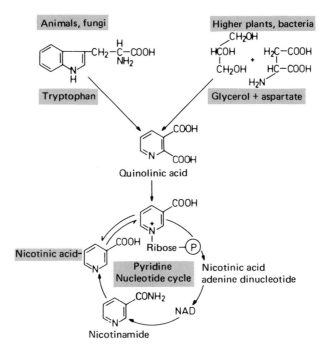

Fig. 123. The biosynthesis of nicotinic acid.

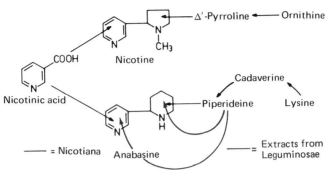

Fig. 124. Scheme of tobacco alkaloid biosynthesis.

Anabasine has two N-containing six-membered rings. One might suppose that they are both derived from the same precursor nicotinic acid. At least in the genus *Nicotiana* this is not so. The piperidine component comes from lysine and is then linked to nicotinic acid to form anabasine (Fig. 124). It was surprising to find that anabasine is formed in cell-free systems from peas and lupins since neither of them contains anabasine. Even more surprising was the fact that this anabasine is formed from two units of cadaverine, the biogenic amine derived from lysine (Fig. 124). Probably the anabasine synthesis in cell-free systems of peas and lupins is due to an enzyme which acts quite unspecifically. At least the experiment shows us how careful one should be in interpreting *in vitro* experiments on higher plants.

One is tempted, particularly in the case of higher green plants, to attribute any other synthetic performance one chooses to the leaves, the organs of photosynthesis. That is particularly true of the synthesis of substances that are exclusively found there. However, it could be shown that nicotine—and also the tropane alkaloids—are synthesized in the roots and nicotine is then transported from them to the leaves. Only a little nicotine is formed in the shoots of *Nicotiana tabacum*. The demethylation of nicotine to nornicotine occurs, however, in the leaves. Anabasine, which occurs in *Nicotiana tabacum* as a minor alkaloid, is also formed predominantly in the roots. On the other hand, in *Nicotiana glauca,* in which anabasine is the major alkaloid, synthesis occurs preferentially in the shoots. We should note from these findings not only that the root has considerable synthetic capabilities, but we should also heed the warning that one should be well informed in which plant organ, at which stage of development, and under what conditions a particular substance is formed before continuing with further experiments.

3. Tropane Alkaloids

The name comes from the tropane skeleton which occurs in two variations (Fig. 125). Variant 1 is found in the *Solanaceae,* i.e. in the

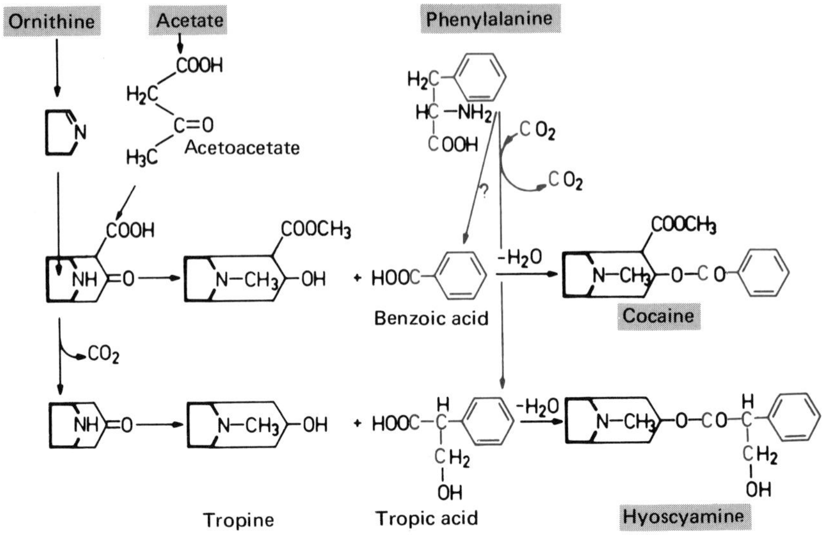

Fig. 125. Several tropane alkaloids and their biosynthesis.

genuses *Atropa, Datura, Hyoscyamus,* and *Mandragora,* among others. In this case a pyrrolidine ring is linked at two points with a 3 C chain. The resulting tropane skeleton is esterified with tropic acid. Well-known examples are hyoscyamine, the ester of the base tropin with tropic acid, and scopolamine, the ester of the base scopolin with tropic acid.

Variant 2 is characteristic of the genus *Erythroxylon* (family *Erythroxylaceae*). Here, a pyrrolidine ring is united with a 4 C chain. Subsequent to a few further modifications compounds such as cocaine are obtained.

The biosynthesis starts from the amino acid ornithine which is converted into the pyrrolidine system by the two paths sketched above. The 3 C or 4 C chain must now be built on. The starting material for this is, in both cases, acetate. It is probable that two units of acetate first form acetoacetate which can then be linked with the pyrrolidine ring to give variant 2 of the tropane skeleton. Variant 2 can then be converted to cocaine. However, decarboxylation of variant 2 gives variant 1 of the tropane skeleton which can then be transformed into hyoscyamine or scopolamine. For this tropic acid, which is derived from phenylalanine, is necessary (Fig. 125).

B. Derivatives of the Aromatic Amino Acids, Phenylalanine and Tyrosine

1. Amaryllidaceae Alkaloids and Colchicine

The starting materials for the biosynthesis of the approximately one hundred known alkaloids of the family *Amaryllidaceae* are the amino acids phenylalanine and tyrosine (Fig. 126). Tyrosine is decarboxylated to

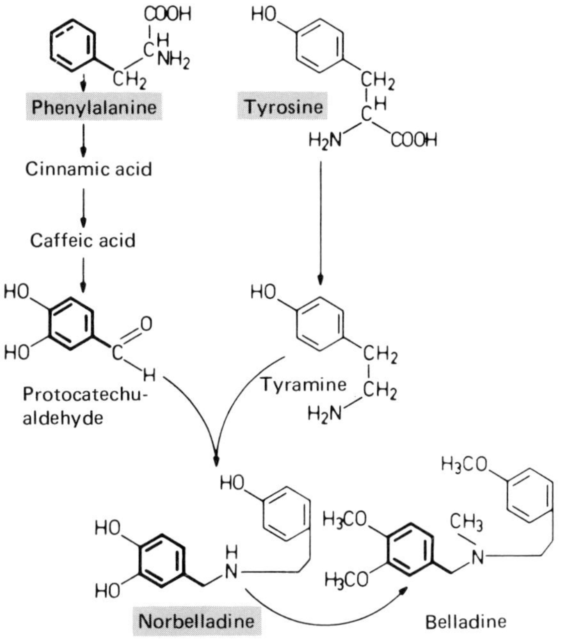

Fig. 126. Several Amaryllidaceae alkaloids and their biosynthesis.

its biogenic amine, tyramine. This supplies the N of the Amaryllidaceae alkaloids. The other half of the molecule is furnished by a C_6-C_1 body which can be derived from phenylalanine. Phenylalanine is converted into caffeic acid by the route already familiar to us. Probably a β-oxidation then occurs to give protocatechuic acid, followed by a reduction to protocatechualdehyde which is coupled with tyramine to give the key substance of the Amaryllidaceae alkaloids, norbelladine.

The most important of the other Amaryllidaceae alkaloids can be derived from norbelladine. Methylation gives belladine. The methylation of the hydroxyls can be carried out in a cell-free system from *Nerine bowdenii*. The enzymes involved are not very specific, they methylate a large number of other phenolic hydroxyls. As an example of further derivatives of norbelladine only galanthamine from the Caucasian snowdrop *(Galanthus woronowii)* will be mentioned here. Colchicine, the alkaloid of the meadow saffron *(Colchicum autumnale)*, is known to the plant breeder as an agent for producing polyploid plants. Scrutiny of the formula does not reveal any similarity with that of the Amaryllidaceae alkaloids. However, colchicine must be mentioned at this point since its biosynthesis resembles that of the Amaryllidaceae alkaloids in important points (Fig. 127). The starting materials are again the amino acids phenylalanine and tyrosine. Again, tyramine provides the one half of the molecule and phenylalanine the other. In this case, it is converted into

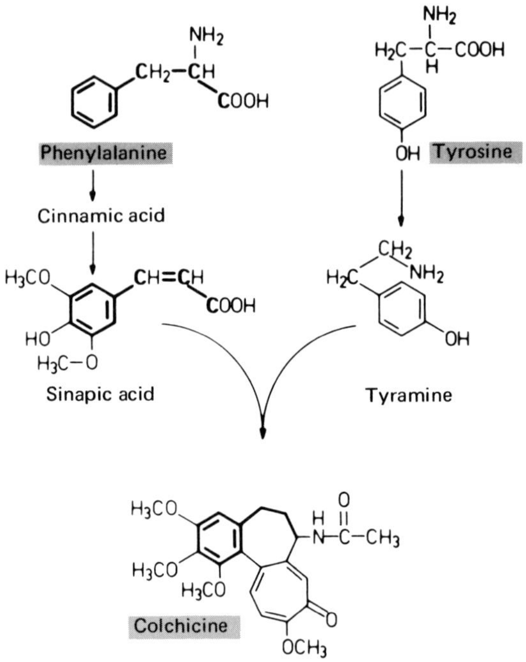

Fig. 127. Colchicine and its biosynthesis.

sinapic acid. The latter is, however, not subjected to β-oxidation. Rather, its entire phenylpropane skeleton is coupled with tyramine. Colchicine is ultimately formed via intermediates, which are still hypothetical. In addition to the coumarins, lignin, phenol carboxylic acids, and simple phenols and, perhaps, flavan derivatives, the Amaryllidaceae alkaloids and colchicine provide us with a further example of how the cinnamic acids, with their characteristic ring substituents, can be channelled into certain synthetic or degradative pathways. The substituents reappear in the end product.

2. Betacyanins and Betaxanthins

In all families of the order *Centrospermae* with the exceptions of the *Caryophyllaceae* and the *Molluginaceae,* red and yellow chymochromic pigments are found, which form a pigment group of their own. They are called betacyanins and betaxanthins after their occurrence in the genus *Beta* (beet).

Betaxanthins are yellow pigments (xanthos = yellow). An example is indicaxanthin from the Indian fig *Opuntia ficus-indica.* It consists of a pyridine derivative to which a proline ring is attached (Fig. 128). Biosynthetically, the pyridine ring is furnished by dihydroxyphenylalanine (DOPA) after certain modifications and the proline ring by the amino acid

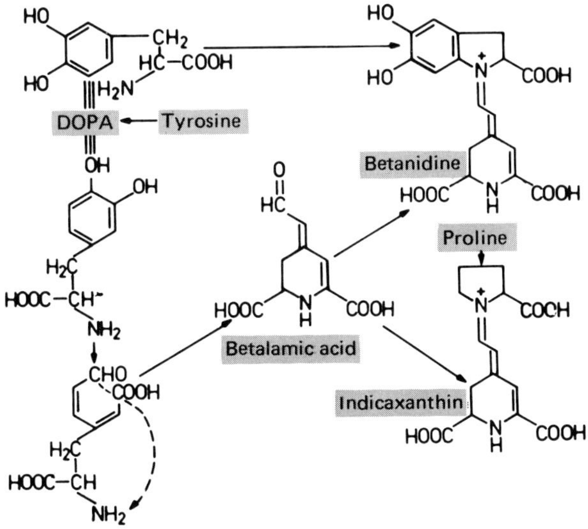

Fig. 128. Betacyanins and betaxanthins and their biosynthesis.

proline. DOPA is an oxidation product of tyrosine. In other betaxanthins, proline is replaced by other amino acids.

Betacyanins are red pigments. They occur in the cell sap as glycosides. Their aglycones are known as betacyanidines. Betacyanidines consist of a pyridine derivative to which an indole derivative is attached. Betanidine from the red beet (Beta vulgaris) is an example (Fig. 128). In biosynthesis both parts of the molecule, the pyridine as well as the indole component, are furnished by DOPA.

3. Isoquinoline Alkaloids (Benzylisoquinoline Alkaloids)

The group is characterized by the isoquinoline skeleton. More important than the simple representatives of the group which bear only the iso-quinoline skeleton are derivatives in which a benzyl residue is attached to the isoquinoline skeleton. Accordingly, these substances are called benzylisoquinoline alkaloids. Among them are the major alkaloids of the genus *Papaver*.

The biosynthesis starts from the amino acid tyrosine which is first hy-droxylated to DOPA (Fig. 129). Two units of DOPA provide the skeleton of the benzyl-isoquinoline. Norlaudanosoline is a key substance which has also been shown to be present, among others, in the genus *Papaver*. The rest of the benzylisoquinolines can be derived from it. Only two kinds of derivatives will be mentioned here, papaverine and morphine alkaloids.

If all the hydroxyls of norlaudanosoline are methylated and, in addi-tion, the N-heterocycle is dehydrogenated, then the compound produced is papaverine. If, on the other hand, only two of the four hydroxyls are

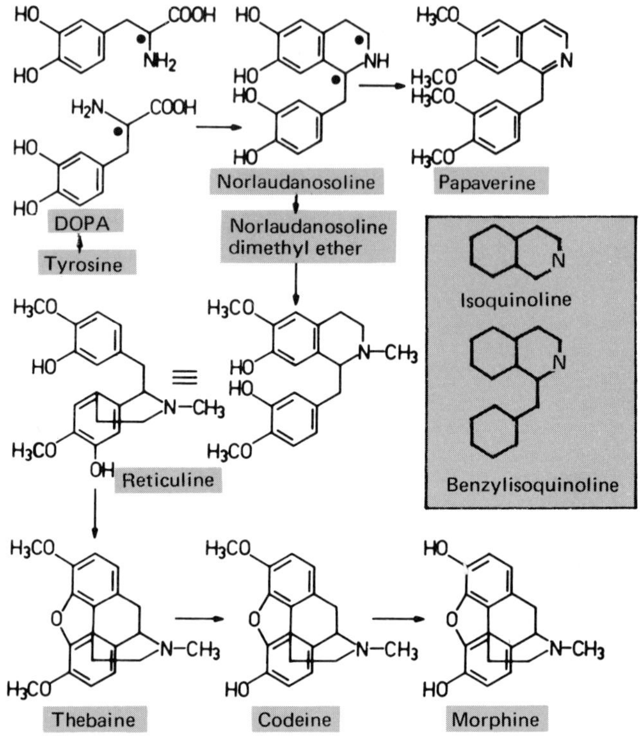

Fig. 129. A few benzylisoquinoline alkaloids (including papervine and morphine alkaloids) and their biosynthesis.

methylated, then norlaudanosoline dimethyl ether is obtained which on N-methylation gives (−) reticulin. The latter is further modified in a biosynthetic sequence leading to thebaine, codeine, and, finally, morphine.

Which pathway is taken depends quite decisively on the kind of O-methylation to which norlaudanosoline is subject. If all the hydroxyls are methylated — and this can be brought about by quite unspecific methylating enzymes — then papaverine and its derivatives will be formed. If only certain hydroxyls are methylated in a quite specific manner than the *morphine alkaloids* will be produced, the most important representatives of which are thebaine, codeine, and morphine. Incidentally, there is good biochemical and genetic evidence that thebaine is first demethylated to codeine and this is then demethylated to morphine, as shown. A similar case with which we are familiar is the demethylation of nicotine to nornicotine.

C. Derivative of the Amino Acid Tryptophan: Indole Alkaloids and Derivatives

The presence of an indole nucleus, which is provided by the amino acid tryptophan, is characteristic of this group. Frequently one or two 5 C building blocks are attached to the indole nucleus, leading to alkaloids of complicated structure. In addition, the indole ring may also be opened and, subsequently, a new kind of ring closure may occur, leading to quinoline systems.

Physostigmine from the Calabar bean *Physostigma venenosum* (Fig. 130) is an indole alkaloid of simple structure. It is derived from tryptophan via its biogenic amine, tryptamine.

Let us now proceed to indole alkaloids of more complicated structure. In fungi of the genus *Claviceps,* particularly in "ergot," the sclerotia of *Claviceps purpurea,* alkaloids are found that are characterized by the "ergoline skeleton" (Fig. 130). Surprisingly, these *ergot alkaloids* or, better, *ergoline alkaloids* are also found in higher plants, namely, in the family *Convolvulaceae.* Lysergic acid (Fig. 130) is a well-known example which occurs as amide or peptide derivative in both *Claviceps* and *Convolvulaceae* (e.g. in the genus *Ipomaea*).

As far as the biosynthesis is concerned it was soon established that the indole nucleus is derived from the amino acid tryptophan. It was much more difficult to determine the origin of the residual C skeleton but it has been found in recent years that it is furnished by mevalonate or isopentenyl pyrophosphate. Thus, in the ergoline alkaloids the indole nucleus is linked with a 5 C unit derived from terpenoid metabolism.

In the families of the *Apocyanaceae, Loganiaceae, Rubiaceae,* and *Euphorbiaceae,* indole alkaloids of complicated structure are found which can be arranged in different groups. We may take reserpine from *Rauwolfia serpentina* as an example. A sinapic acid residue is present as part of the molecule, in addition to the basic skeleton of tryptophan and two 5 C units (Fig. 130).

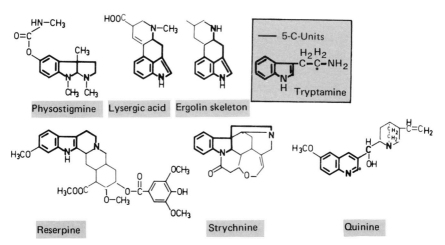

Fig. 130. Indole alkaloids. Their biosynthesis starts from tryptophan which is decarboxylated to tryptamine.

Finally, substances of still more complicated structure, such as strychnine from *Strychnos nux-vomica,* belong to the indole alkaloids. Nonetheless strychnine still possesses the indole nucleus (Fig. 130). Now this indole ring can itself be reconstructed. That happens in the genus *Chinchona* of the family *Rubiaceae.* Some *Chinchona alkaloids* still bear the indole nucleus; in others such as the pharmacologically important quinine the indole nucleus has been converted to a *quinoline system* (Fig. 130). The conversion of the indole system to a quinoline system may be envisaged to occur in two stages: first, the *N*-containing five-membered ring of the indole is opened and then the product is rearranged to form a *N*-containing six-membered ring.

D. Purine Alkaloids

Purine rings are familiar to us from the purine bases of the nucleic acids. Their synthesis proceeds according to the same principles in microorganisms, animals and plants. A characteristic is that the free purine skeleton is not formed first and then converted into its nucleotide but that the purine ring is built up, piece by piece, on a ribose phosphate unit. Thus, a purine nucleotide is first formed.

The biosynthesis begins with 5-phosphoribosyl pyrophosphate which is first converted into 5-phosphoribosylamine (Fig. 131). In higher plants the amino group of this compound is acquired from asparagine by transamination. This N later becomes N-9 of the purine ring. In the next step glycine is bound in peptide linkage to this amino group. All additional C and N atoms are then built on, atom by atom. The C atoms come from the "C_1 pool" of the cell. By this is meant formyl groups which are transferred from tetrahydrofolic acid and CO_2 which is transferred from biotin. The N atoms come from the amino groups of glutamine and aspartic acid. The first purine nucleotide to be formed on this biosynthetic pathway is inosine-5'-phosphate.

Inosine-5'-phosphate can be converted to the nucleotides of adenine and guanine, well-known components of nucleic acids. Here, however, we are concerned with other possible means of utilizing this substance. For when one talks of purine alkaloids, one thinks primarily of caffeine, theophylline, and theobromine (Fig. 131). These methylated purine alkaloids are widely distributed, especially caffeine. Its occurrence in coffee (among others, *Coffea arabica*), tea (*Camellia sinensis*), and cocoa (*Theobroma cacao*) is well-known.

Only the main features of the biosynthesis of the three purine alkaloids mentioned are known. It is likely that the starting material is, again, inosine-5'-phosphate which is first converted into xanthosine-5'-phosphate. This nucleotide contains the purine base xanthine. It is possible that the base is first liberated and then methylated to give the purine alkaloids.

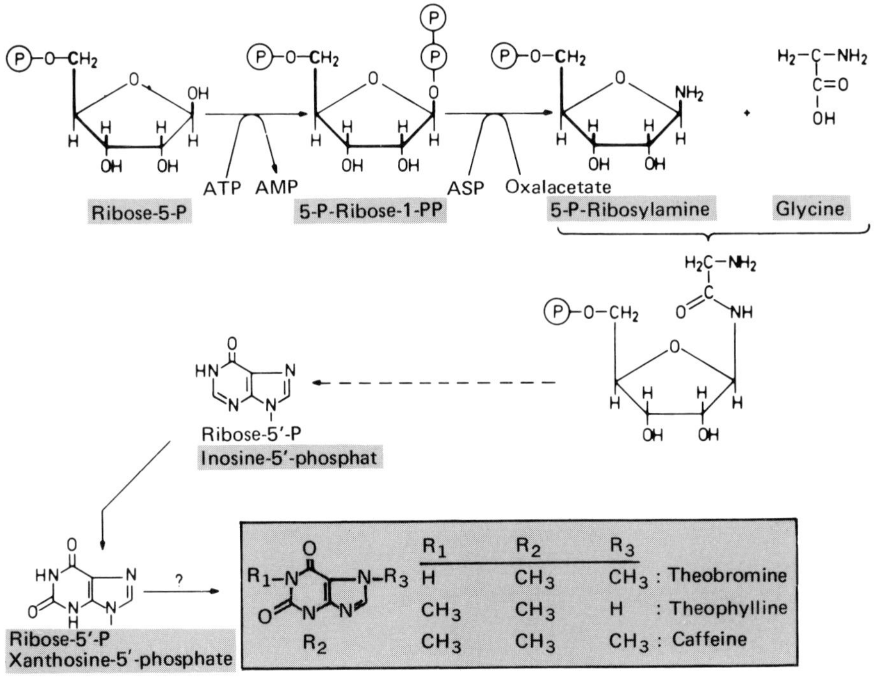

Fig. 131. The biosynthesis of the purine nucleus and purine alkaloids.

Another possible mode of synthesis of the methylated purines is also discussed. We have mentioned earlier that tRNA is characterized by the occurrence, in its structure, of rare bases. Among them are certain methylated purines. It has been demonstrated that the methylation of the purines occurs after their incorporation into the tRNA molecule. Only after degradation of tRNA can the methylated purines as such be set free.

Whether this second mode of synthesis of caffeine, theobromine, and theophyllin actually occurs in nature is rather doubtful. There are, however, other purine derivatives, the cytokinins, which show close relationships to tRNA and are of great interest to physiologists. They are a group of phytohormones, of which isopentenyl adenine (IPA) is an example. Several investigations have shown that the isopentenyl unit is built on to an adenine residue, which is already incorporated in a tRNA molecule (also cf. page 209).

E. Biochemical Systematics

The systematics of plants is the science of comparing them and, by extension, the science of plant relationships. It can be practiced at very different levels, cytological, anatomical, morphological. At the same time it is necessary to strive for an integration of all the data obtained by the different disciplines.

Now the formation of each character, whether cytological, anatomical, or morphological in kind, is ultimately due to certain biochemical processes. Thus, it seems reasonable to try to develop systematics on a chemical and biochemical basis. An examination of the stock of materials contained in plants can serve this purpose. An inventory of this kind is still in the very early stages owing to the profusion of plants to be examined. In consequence, unexpected findings can always turn up. For example, the finding of ergoline alkaloids in the family *Convolvulaceae* was quite unexpected. Up to then they had been regarded as typical of the genus *Claviceps,* i.e. fungi. However, it has become evident that many plant materials occur widely distributed throughout the whole system. This is true also of secondary plant materials, such as caffeine or anthocyanins. In a very few cases, however, it seems that the occurrence of secondary plant materials is limited to certain systematic units and, thus, is useful as a systematic criterion. An example: the betacyanins occur in all families of the order *Centrospermae* with the exception of the *Caryophyllaceae* and the *Molluginaceae*. This confirms the view of several systematizers that the *Caryophyllaceae* and the *Molluginaceae* ought to be considered separately from the other families of the *Centrospermae*. This does not mean that the remaining families that all carry betacyanins are a closed unit. The *Cactaceae,* for example, are different from the rest in that they have a differently formed perianth and a different kind of placentation. This brings us to a point which is often overlooked: a chemical criterion is only *one* of many possible criteria. It must be seen in relation to other criteria. Now, of course, merely to note the distribution of a material is an inadequate criterion. For just as there is convergence at the morphological-anatomical level (one thinks, for example, of the appearances of succulents in *Cactaceae, Euphorbiaceae, Stapeliaceae,* among others), so there is also convergence in the formation of chemical characters. An example of this is the biosynthesis of nicotinic acid, which is synthesized by different routes in different groups of organisms. A further example is the biosynthesis of anabasine, the pyridine ring of which is formed in the genus *Nicotiana* in a way different from that in a cell-free system from *Leguminosae.* Such appearances of convergence make it very doubtful whether the sheer existence of a material in a plant can be used as a systematic criterion. If at all possible the biosynthetic pathway should be used as a criterion. Then, of course, the chemosystematics would become a truly *biochemical systematic.*

It is possible to do no more than simply point out here that chemosystematics can also be pursued with the aid of DNA hybridization. The better DNA preparations from two organisms can be hybridized, the more closely related they are—there are, of course, exceptions to this rule. Thus, chemosystematic investigations can be carried out on the genetic material itself. Furthermore, the products of transcription and translation, the polypeptides or proteins, can also be evaluated chemosystematically. In this connection serological methods have shown themselves to be particularly successful, once technical difficulties had been overcome and the serological data had been evaluated in relation to all other available data, and not just on their own.

The alkaloids represent the third, large group of secondary plant materials after the terpenoids and phenols. From the biosynthetic standpoint they are derivatives of amino acids, the amino groups of which furnish the N of the alkaloids. The C skeleton can, however, be considerably enlarged by the incorporation of other kinds of components, such as isoprenoid units.

Porphyrins

Porphyrins are a small but important group of secondary plant substances. We dealt with their structure when we discussed the chlorophylls. The biosynthesis of the porphyrins is interlinked with the citric acid cycle and amino acid metabolism. (Fig. 132). The precursor succinyl CoA is derived from the citric acid cycle and the precursor glycine from amino acid metabolism. The two are combined to give an unstable intermediate which loses CO_2 to form δ-aminolevulinic acid. Two molecules of δ-aminolevulinic acid are combined to yield porphobilinogen, a pyrrole system, which is the building block of the porphyrins. Four such porphobilinogen molecules become linked in a series of steps to form the porphyrin system. The first porphyrin to appear in the course of biosynthesis is uroporphyrinogen III. Of the subsequent intermediates protoporphyrin IX is worthy of mention. This is because introduction of Mg^{++} on the one hand leads to Mg-protoporphyrin IX and then further to the chlorophylls, whereas introduction of Fe^{++} on the other yields Fe-

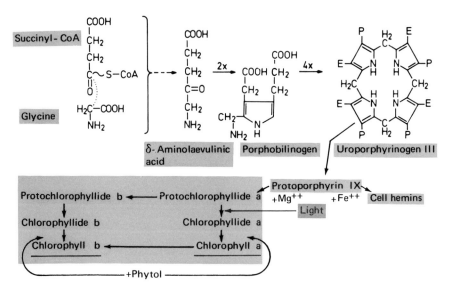

Fig. 132. Scheme of porphyrin synthesis, E = acetate, P = propionate.

protoporphyrin IX and, subsequently, to cell hemins, haem, the cytochromes, cytochrome oxidases, peroxidases, and catalases.

The pathway from Mg-protoporphyrin IX to the chlorophylls is still not known in complete detail. For that reason we must be content with mentioning the names of a few probable intermediates (Fig. 132). The biogenetic relationship between chlorophylls a and b is also still a matter of dispute. Chlorophyll a may be converted into b, but it is also possible that the biosynthetic pathways separate from each other earlier, at the state of protochlorophyllide a. For physiologists, an important point about these last stages of the route to the chlorophylls is that the biosynthesis of the chlorophylls is light-dependent in higher plants. The light-controlled reaction is the conversion of protochlorophyllide a into chlorophyllide a.

Phycobiliproteins. The bile pigments of animal organisms are open chain tetrapyrroles. They arise by degradation of cell hemins. This degradation entails opening of the porphyrin ring which thus leads to open chain tetrapyrroles.

Open chain tetrapyrroles of this kind are also found in plants. Those from blue algae and algae in which they are combined with proteins are

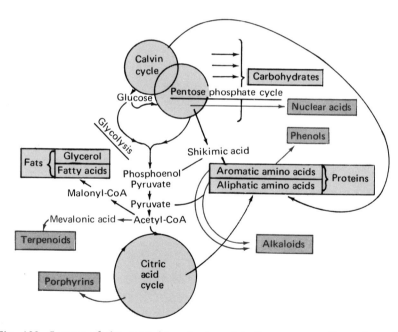

Fig. 133. Survey of the most important metabolic processes. Secondary plant materials are indicated in red. In the interests of clarity not all metabolic interconnections are shown. Thus, the participation of malonate and mevalonate in the synthesis of the phenols is omitted. The nucleic acids, which here are included under the designation "secondary" on the basis of their biosynthesis, emphasize how artificial is the classification into "primary" and "secondary" substances (page 100).

particularly well-known. With reference to their occurrence in algae and as protein complexes they are known as phycobiliproteins. The phyco-cyanins and the phycoerythrins belong to this group. The biosynthesis of the phycobiliproteins is still not understood. Perhaps they are derived from cell hemins in plants too. Phycobiliproteins are of interest not only as pigments of blue algae and algae. This is because the *phytochrome system* (Fig. 180), which is of crucial importance for morphogenesis in higher plants, is a phycobiliprotein. The biosynthesis of the porphyrins and, in all probability, of the phycobiliproteins too, two functionally important groups of secondary plant substances, is fed from two sources: from the citric acid cycle in the form of succinyl CoA and from amino acid metabolism in the form of glycine.

Cell Division

A. Development—Growth and Differentation

In the preceding chapters we have been concerned with metabolic reactions in higher plants. This has provided us with a basis from which to consider the processes of development which will now be presented.

The multicellular system of a higher plant grows by sexual propagation from a single cell, the zygote. This growth denotes only one aspect of development, however. We will remember that the individual cells of a plant can be very different from each other. This development of dissimilarity, this differentiation, is the second aspect of development.

Thus development is based on growth and differentiation. Now we need to define these terms adequately. The definition one chooses for growth depends on how the question is formulated—and on the approach of the questioner. Sometimes it is practical to choose increase in dry weight, increase in fresh weight or increase in volume as the parameter. On other occasions biochemical parameters, such as DNA or protein synthesis, may be preferred.

In higher plants *growth* is often designated as an *irreversible increase in volume.* If this definition is accepted here, then one reservation must be borne in mind: in higher plants it is usual to distinguish growth by cell division from subsequent growth by cell elongation. This is important because the definition mentioned seems to be most appropriate for the increase in volume during cell enlargement, which can often be considerable. But this kind of growth is not only "growth," it is simultaneously an early process of differentiation. For this reason growth by cell division and growth by cell enlargement will be discussed separately.

Following a proposal of Becker *differentiation* can be defined as *variedly progressive change of cells in the course of the development of an individual.* It is obvious that what is or is not really progressive is debatable, particularly today.

B. Cell Division

1. The Mitotic Cycle

Growth as a result of division is based on mitotic cell divisions. The events of mitosis will be taken for granted here. In the stages of prophase, metaphase, anaphase, and telophase each chromosome is split lengthwise into two homologous chromatids. These chromatids are divided between two daughter nuclei, which are being formed in such a way that one of the two homologous chromatids is allotted to each of the daughter nuclei. Appropriate subsequent changes in the cytoplasm give rise to daughter cells, each of which carries one of the two daughter nuclei.

In every reputable textbook it is stated that, in contrast to meiosis, mitotic division entails exact chromosome duplication. Thus, each daughter cell carries the same genetic information as the parent cell. Since we already know that DNA is the genetic material of higher organisms, we must direct our attention to the DNA when considering evidence for this statement.

If each daughter cell is to contain a complement of DNA identical with that of the parent cell, then the DNA of the parent cell must be reduplicated identically before mitosis. For the moment let us ignore the word "identical" and first ask only whether a DNA reduplication of this kind can be apprehended quantitatively. This is indeed so. In the meristem zones of plants, cells pass through a mitotic cycle or cycle of DNA synthesis from one mitosis to the next (Fig. 134). Directly after a division there is a postmitotic phase without DNA synthesis. This is followed by a phase of DNA in which the DNA has been shown by quantitative measurements to double. After that there is another phase without DNA synthesis which precedes the next mitosis and which is thus called premitotic. It leads to mitosis and subsequent division. The daughter cells which arise can pass through the same cycle again, provided they have retained their ability to divide.

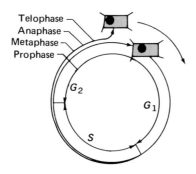

Fig. 134. Outline of the mitotic cycle (DNA synthesis cycle). G_1 = postmitoic phase with DNA synthesis. (G = gap), S = phase of DNA synthesis, G_2 = premitotic phase without DNA synthesis (modified from Bielka 1969).

2. The Autocatalytic Function of DNA: Replication

a. Semiconservative DNA replication

We have just established that the DNA content of a cell does actually double before a mitotic division. The next thing we have to do is to investigate whether this doubling represents an *identical* replication. For only this would ensure an identical distribution of genetic properties among the daughter cells. At the molecular level this means that two identical DNA double helices must come into existence from one DNA double helix in the process of DNA replication. Several models of such a replication process have been developed, of which that of a semiconservative DNA replication has turned out to be correct. The kind of semiconservative DNA replication proposed begins with a local unwinding of the double helix into its two single strands (Fig. 135). On the parts of the single strands a second, complementary strand is then synthesized. Unwinding and formation of the complementary strands proceed until, finally, there are two identical DNA double helices instead of one.

b. DNA synthesis in a cell-free system

Enzymes have been isolated from the most diverse sources that can support DNA synthesis in a cell-free system. They are called *DNA polymerases* or DNA replicases. They need, in addition to other cofactors, DNA as "primer" and deoxyribonucleoside-5'-triphosphates as DNA building blocks. The triphosphates align themselves on the DNA single strand introduced as primer according to the rules of base pairing: an

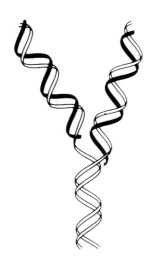

Fig. 135. Semiconservative DNA replication. New DNA strands indicated in black (after Stahl 1969).

adenine as nucleoside triphosphate matches a thymine in the strand and a cytosine as nucleoside triphosphate pairs with a guanine in the strand, etc. The DNA building blocks, having thus been brought into position, are joined together to form a complementary DNA strand by the DNA polymerase with the liberation of pyrophosphate (Fig. 136).

Kornberg's success in 1967 in carrying out the synthesis of biologically active DNA in a cell-free system was hailed as a triumph of molecular genetics. In his experiments, the DNA polymerase from *E. coli,* which functions particularly well, was presented with the DNA of the phage ϕX 174 as primer. The genome of this phage consists of a DNA single strand closed to form a ring. This DNA single strand initiated DNA synthesis in the cell-free system — and the newly formed DNA behaved exactly like the original ϕX 174 DNA in an infectivity test. The biological activity, as well as other details the experiments were designed to provide, presented clear-cut evidence that semiconservative DNA reduplication had occurred.

According to the findings in a cell-free system it seems that Kornberg's DNA polymerase forms only quite short pieces of DNA. A

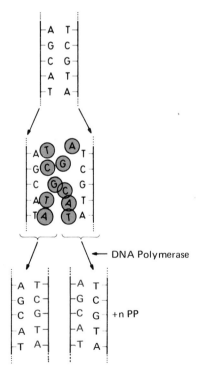

Fig. 136. Semiconservative DNA replication in a cell-free system containing DNA polymerase. Encircled letters = nucleoside-5-triphosphates (modified from Hess 1968).

second enzyme, the *DNA ligase,* links these pieces into longer strands or together to form a ring as in the case of the phage ϕX 174.

It seems likely that the DNA polymerase first investigated by Kornberg is important *in vivo* primarily for the repair of damaged DNA strands. Normal replication is carried out by other DNA polymerases which, in principle, work quite similarly.

c. The Meselson—Stahl experiment

We have just mentioned a cell-free system derived from viruses and microorganisms. However, evidence for the semiconservative nature of DNA reduplication has also been obtained in higher plants. One piece of evidence consisted in repeating an experiment which had previously been carried out by Meselson and Stahl on bacteria and which is named after them the Meselson-Stahl experiment.

Cells from tobacco callus tissue can be grown in liquid nutrient media as well as on solid. In a liquid medium the number of the cells doubles in about two days. Thus, in about two days the cells pass through one round of DNA synthesis. Now the cells were first maintained for several rounds of DNA synthesis in a medium with compounds containing "heavy" nitrogen, N^{15}. This leads to N^{15} being incorporated in the purine and pyrimidine bases of the DNA. Finally, all of the DNA is present as "heavy" DNA with N^{15} bases. This heavy DNA can be separated from normal DNA with N^{14} bases in a cesium chloride gradient in the ultra-centrifuge.

After all of the DNA had become "heavy" the cells were transferred to a nutrient medium that once again contained nitrogen in the form of the normal N^{14} isotope. At certain time intervals samples were removed from which the DNA was isolated and its density investigated in a cesium chloride gradient (Fig. 137). After two days in the N^{14} medium all of the DNA is present in a "medium heavy" form. Meselson and Stahl had already demonstrated previously in their experiment with bacteria that the double helices of such medium-heavy DNA consist of one N^{15} and one N^{14} single strand, i.e. they represent N^{14}/N^{15} hybrids. On the fourth day, i.e. after two rounds of DNA reduplication, normal and medium-heavy DNA are found in the ratio of 1:1. With each succeeding round of DNA reduplication the proportion of hybrid DNA is found to decrease. The results of the experiment can only be understood on the basis of a semiconservative DNA reduplication.

d. Taylor's experiments on Vicia faba

The Meselson-Stahl experiment brings evidence for the semiconservative nature of DNA reduplication at the molecular level. However, in Taylor's experiments a similar demonstration was possible at the level of the chromosomes by using the technique of autoradiography.

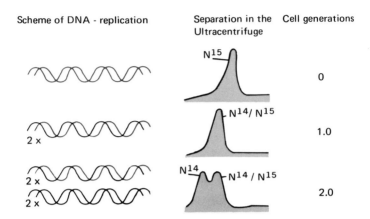

Fig. 137. The Meselson-Stahl experiment. The tracings of the curves in the middle column represent the amounts of N^{15}, N^{14}, and hybrid DNA found in each generation (modified from Hess 1970).

Root tips of the broad bean (*Vicia faba*) were cultivated in a medium containing thymidine-H^3 and colchicine. Thymidine-H^3 is incorporated into the DNA of the chromosomes of the actively dividing cells of the root meristem. Thus, the chromosomes become radioactively labeled. Colchicine allows the reduplication of the DNA and of the chromosomes but blocks spindle formation and, thus, nuclear division. This means that the daughter chromosomes remain united in the "old" nucleus, with the result that the chromosomal descendants can be easily followed. After an initial chromosome reduplication in the medium with thymidine-H^3 both daughter chromosomes were labeled (Fig. 138). If, after this first reduplication, the root tips were transferred to a medium without thymidine-H^3 and there were allowed to undergo a second reduplication, then of the total of four chromosomes two were radioactively labeled and two were not.

Now to the interpretation of these findings. The chromosome under investigation originally consisted of two strands. How these two strands are constructed is still a matter of great controversy. For our purposes we shall simply assume that these two strands formed a DNA double helix or a succession of DNA double helices which are connected to each other by protein links. During the first reduplication in the presence of thymidine-H^3 a new complementary strand which is radioactively labeled by uptake of thymidine-H^3 is formed on each single strand of the double helix. These new, radioactive strands can be detected by autoradiography and lead to the identification of two labeled chromosomes after the first reduplication.

The second reduplication took place in a medium without thymidine-H^3. If, now new single strands are synthesized that are complementary to each of the single strands present after the first reduplication, then four

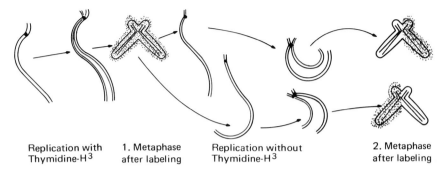

| Replication with Thymidine-H^3 | 1. Metaphase after labeling | Replication without Thymidine-H^3 | 2. Metaphase after labeling |

Fig. 138. Taylor's autoradiographic investigations on *Vicia faba*. The semiconservative replication of only one chromosome is depicted. Radioactively labeled strands are red. Blackening of the photographic plate by tritium radiation = red dots (modified from Taylor et al. 1957).

double strands are obtained. Two of these are not labeled and two consist of one labeled and one unlabeled strand. The last two pairs can be detected autoradiographically. Thus, after the second round of reduplication, two labeled and two unlabeled chromosomes are found. Taylor's experiments support a semiconservative DNA reduplication on the basis of a "single strand model" of chromosome structure. According to this model, a chromosome before reduplication consists, in principle, of a DNA double helix plus protein, an assumption on which our interpretation has been based.

3. Plant Tumors: Crown Galls

Within the intact plant, considered as a unit, active cell division is limited to certain zones around the meristems of the tips of shoots and roots. If one goes downward from the shoot tips and upward from the root tips, cell division becomes progressively less active and finally ceases altogether. After they have ceased to divide, the cells become visibly different from each other; they begin to differentiate.

The manner in which this transition from cell division to differentiation is regulated is largely unknown. We shall discuss a few relevant findings later (page 231). If this regulatory mechanism is disturbed, then cell division can occur without subsequent differentiation. The disturbances can be of quite different kinds. For example, one can take tissue from the whole plant and grow it in a defined culture medium. Under certain conditions, such tissue cultures can maintain their ability to divide almost indefinitely (page 246).

Other examples of faulty regulation are tumors. Complexes of uninhibited cell division such as these can be due to different causes. For example, genetic tumors are well-known, particularly in hybrid species of the genus *Nicotiana*. They occur particularly frequently when one of the partners to the cross was *N. langsdorffii.* In these and other hybrid species the necessary interplay between the two genomes united by the cross is so greatly disturbed that active cell division can no longer be adequately controlled. Other tumors are caused by viruses, e.g. by the virus *Aurigenus magnivena,* which is transmitted by grasshoppers. However, the best known are the crown galls.

Crown galls are plant tumors that are induced by *Agrobacterium tumefaciens.* Their occurrence is in no way limited to roots; they can occur on any part of the plant (Fig. 139). A precondition is a wound. In the first place it is necessary for the bacteria to penetrate into the plant, but, in addition, the wound leads to a conditioning of the plant cells. For if the bacteria are introduced without injury into the plant no tumor is formed. What this conditioning is, is unknown. At the site of the wound, a wound callus is first formed, as is usual after an injury. Later the tumor grows out of this callus. Wound callus and tumor can be readily distinguished from each other. This is because call divisions in cellus tissue are oriented periclinally, whereas in tumor tissue no orientation at all is to be found.

The crown galls are characterized as true tumors by unlimited cell division. This is true also of isolated tumor tissue grown in culture. Furthermore, tumor tissue can be freed from bacteria by exposing it to temperatures of about 40. At this temperature *Agrobacterium* is killed, whereas the plant cells are not significantly damaged. The resulting bacteria-free tumor tissue also maintains its ability to divide in culture and does so without it being necessary to add factors that stimulate division and delay differentiation. In contrast, normal plant tissue grown in culture needs the addition of such factors, for example IAA, if it is to maintain its

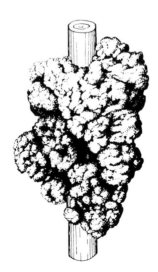

Fig. 139. "Crown gall" on a tomato shoot (from Nultsch 1968).

ability actively to divide. Thus, in comparison with normal tissue, tumor tissue is permanently altered. DNA is certainly involved in this alteration, as experiments by Bopp on *Bryophyllum daigremontianum* have shown. If wounds on one half of a leaf of *Bryophyllum* are infected with *Agrobacteria* then wound calluses are first formed, out of which tumors develop. If the other half of the leaf is similarly infected but, in addition, the inhibitor of DNA synthesis, 5-fluorodeoxyuridine, is also injected, only a wound callus and no tumor is formed (Fig. 140).

DNA is thus implicated, but at first it was not known whether it was the DNA of the plant or of the bacterium. Many experiments strongly suggested the existence of a tumor-inducing principle (TIP) which was supposed to be released by the bacterial cells and initiate the change (transformation) in the plant cells. It now seems likely that the principle is indeed bacterial DNA. As has been shown, DNA migrates from the bacteria into the plant cells where it replicates itself and induces the formation of bacterial-specific proteins. It seems likely that this DNA is ultimately responsible for causing the tumors.

Let us summarize. The cell material of the plant is supplied by growth during mitotic divisions. During the mitotic cycle the DNA of the chromosomes is reduplicated semiconservatively. Mitoses are divisions that parcel the genetic material in identical complements, at least as far as chromosomal heredity is concerned. The ability to divide is under the control of the organism. This becomes most clear when this control breaks down as it does in tumor formation.

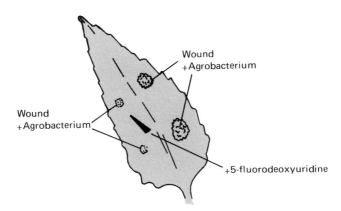

Fig. 140. Evidence for the participation of DNA in the formation of crown galls. Both halves of the blade of a leaf of *Bryophyllum daigremontianum* are wounded and infected with *Agrobacterium tumefaciens*. The left half of the leaf is treated, in addition, with 5-fluorodeoxyuridine, a competitive inhibitor of thymidilate synthesis. On this side no tumors form, only a wound callus (modified from Bopp 1963).

Differential Gene Activity As Principle Of Differentiation

A. Totipotency

Almost the entire association of cells of higher plants arise from the zygote by mitotic divisions. The only exception is the haploid material of the micro- and macrogametophytes which arises from haploid cells furnished by meiosis. We have just established that mitosis divides the genetic material into identical complements. Thus, all diploid cells of higher plants should be equipped with the same stock of genes. How then can one explain the fact that the individual cells of plants can differentiate in quite different ways? How is it that different types of cells, tissues and organs can come into being when all cells have the same genetic equipment?

The easiest assumption to make is that in mitosis the genetic material is not divided identically. Or even if this is so, changes in the genetic stock of the cells might follow the cessation of cell division. In both cases the multicellular plant would be a mosaic of genetically different cells and associations of cells. This genetic diversity would then make the different types of differentiation intelligible. As a matter of fact there are examples of genetically nonidentical mitoses. However, these do seem to be special cases. Usually all cells of a higher plant are equipped with all of the genetic information. The cells are genetically totipotent or omnipotent. The evidence for this was obtained in experiments concerning regeneration. It is known for instance that a complete, new begonia plant can develop from a single epidermal cell of a begonia leaf. Thus, this epidermal cell, which is highly differentiated, still contains all of the genetic information (Fig. 141).

Even more convincing are experiments with isolated, single cells. Carrots and tobacco have been grown in tissue culture, tissue being taken from the phloem in the former case. Single cells from such tissue cultures grew into reproductive carrot and tobacco plants (Fig. 142). Thus, inspite of their being differentiated, the cells studied had remained genetically totipotent.

Indeed one can go a step further. Cell walls may be removed from plant cells by treatment with appropriate enzymes (pectinases, cellulases, etc.) and protoplasts are then obtained. Isolated protoplasts such as these can also be grown into whole plants. This was achieved not only with standard test objects, such as tobacco plants and carrots, in regeneration experiments but also with petunias.

The protoplasts used in the experiments just mentioned above were isolated from the mesophyll of leaves, i.e. from highly specialized cells. The result of the experiments constitute a further piece of evidence for totipotency. However, protoplasts are of the greatest interest for entirely different reasons which ought at least to be mentioned here: they promise to be useful objects for genetic manipulation, for somatic hybridization as well as transformation (p. 7).

An experiment of the kind, which has been possible with animal cells since the beginning of the 1960s, was first successfully carried out with plants ten years later by Cocking and, subsequently, by other groups of workers: a fusion of protoplasts of different species. At first only the cytoplasma could be made to fuse, but not the nuclei. However, this was a necessary precondition for true hybridization involving nuclear fusion between diploid, vegetative cells (*"somatic" hybridization*). Indeed, in 1972 protoplasts from *Nicotiana langsdorfii* were successfully fused with protoplasts from *Nicotiana glauca* and entire hybrid plants were grown from they hybrid protoplasts obtained. The hybrid plants could not be distinguished from hybrids, which may be obtained from the two species by conventional crosses (cf. p. 169). With that the first somatic hybrid had been constructed. Since then Gamborg and his colleagues have also succeeded in growing somatic hybrids from grain and leguminosae, at least to

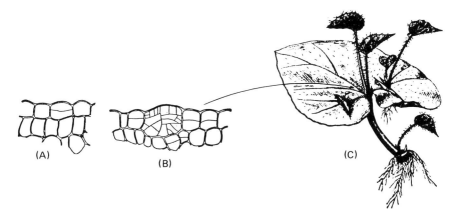

Fig. 141. Regeneration of begonias. A begonia leaf is spread out and it develops roots (c). The veins of its leaf are partially served. In the region of the sites of the cuts new begonias regenerate. Moreover, a complete, new plant can arise from a single epidermis cell (a,b) (modified from Strasburger 1967).

the stage of the first divisions. Now one speculates about somatic hybrids between, say, a potato and a tomatoe. Combinations of this kind no longer seem impossible.

Protoplasts may well also offer advantages in transformation experiments. This is because they can take up DNA particularly readily after removal of the cell wall. Once the protoplasts have integrated the added DNA it ought to be possible to grow them into transformed plants. In this way one hopes to increase considerably the proportion of plants that are transformed since this is usually very low.

B. Differential Gene Activity: the Phenomenon

All diploid cells are still equipped with all of the genetic information and can use it for the formation of characters. That is not to say that all of the genes need to be active simultaneously. It is possible to envisage a situation in which, in one particular tissue and at one particular stage of development, certain genes are active, and in another tissue at another stage of development other genes are brought into play. A quite specific event or set of events in differentiation would then correspond to each such pattern of active genes. This possibility provides an alternative explanation of differentiation: instead of a difference in the stock of genes a difference in gene activity appears that results in corresponding differences in the formation of characters.

We must still define some of the terms we will be using. Gene activity can be classified into primary and secondary processes. Under *primary*

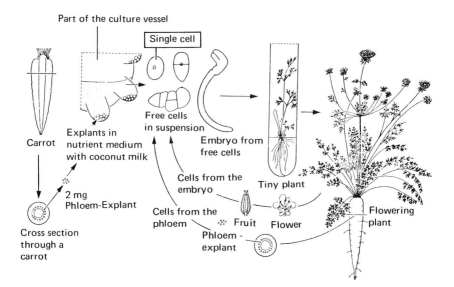

Fig. 142. Regeneration of reproductive carrots from related single cells (from Steward et al. 1964).

gene activity the production of mRNA is understood, i.e. transcription. All subsequent stages leading to character formation, including translation, are secondary. If the primary activity at a gene locus shows differences that are specific for the tissue and the stage of development, one speaks of a *differential gene activity*. Differential gene activity is not the only driving force behind differentiation but it is a very considerable one.

For the moment we shall concern ourselves only with evidence for differential gene activity in higher plants and set aside the further question concerning the causes of differential gene activity.

1. RNA Synthesis on Giant Chromosomes

Differential gene activity can be demonstrated on the chromosomes themselves provided suitable objects, such as giant chromosomes and lampbrush chromosomes, are chosen and adequate methods are used.

Giant chromosomes are found, *inter alia,* in the salivary glands of *Diptera* such as *Drosophila, Chironomus, Acricotopus,* and *Sciara.* They are bundles of a few thousand homologous chromatids laid down in parallel and, at the same time, slightly skewed like the strands of a rope (Fig. 143). Each chromatid appears to be divided lengthwise: chromomeres appear to be strung on the chromonema like a set of beads.

In the giant chromosomes the homologous structures of the chromatids lie side by side. Thus, there is a lengthwise arrangement of the giant chromosome into transverse bands and connecting pieces corresponding to the lengthwise arrangement of the chromatids. This pattern can be seen in the microscope. Much detailed work has shown that genes are located on the transverse bands. If these gene loci switch over to the state of primary activity, this leads to a structural change in the transverse bands concerned. The DNA of the transverse bands unfolds into a so-called puff (Fig. 144). A particularly extensively developed puff is known as a Balbiani ring. mRNA is formed on the unfolded DNA of such puffs. Thus, puffs are the sites of higher, primary gene activity.

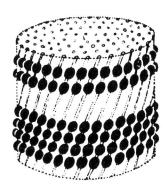

Fig. 143. Sketch of a giant chromosome (from Kuhn 1965).

Puffs are not fixed structures. In the course of development puffs that are already in existence on a given giant chromosome contract and new ones are formed. Differences in puff formation on any given giant chromosome are also found between one tissue and another. In puff patterns of this kind, which are specific for the tissue and stage of development, differential gene activity becomes microscopically visible.

Although they occur pretty infrequently giant chromosomes are also found in higher plants as, for example, in the cells of the suspensor of the bean *Phaseolus vulgaris*. These plant giant chromosomes are, unfortunately, much more difficult to work with than those of the *Diptera*. Nonetheless, Nagl was able to show extensive parallels between them. Plant giant chromosomes also show a pattern of local swellings of chromosome structure specific for a given phase of development, which corresponds to a puff pattern (Fig. 145). RNA is formed in these puffs just as in the case of the *Diptera*. These puff-like swellings contract if the temperature is too high or too low and after treatment with actinomycin C_1. At the same time RNA synthesis at these sites also ceases. Thus, in higher plants too specific puff patterns are found as a microscopically visible expression of differential gene activity, although they are poorly defined compared with those of the *Diptera*.

2. Phase-specific mRNA

The phase and tissue specific puff patterns are microscopically visible patterns of active genes. However, gene activity means production of mRNA. mRNA of one particular composition should correspond to one particular pattern of active genes and mRNA of another composition to another pattern of active genes. Thus, it should be possible to demon-

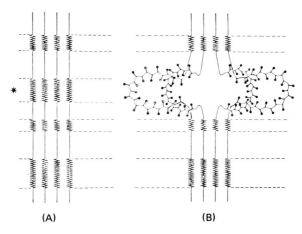

(A)   (B)

Fig. 144. Sketch of puff formation. Only four of very many chromatids are shown (from Beermann (1966).

strate differential gene activity by comparative analysis of the mRNA at different phases of development or from different tissues.

Seedlings are suitable objects for investigations of this kind. At each stage of development of the seedling many new genes must be activated if development is to proceed. Using a very sensitive method it has been possible to show that the structure of the mRNA of seedlings changes from one state of development to another.

Experimentally, the mRNA of peanut cotyledons (*Arachis hypogaea*) of different ages was isolated. The mRNAs from the different stages of development were then compared with each other. This was done by using the so-called double labeling technique. We will consider only one series of experiments. First, a control experiment (Fig. 146a). Uridine-H^3 was fed to one preparation of peanut seedlings and uridine-C^{14} to another. On the second day after germination the nucleic acids from both preparations were isolated, mixed, and applied to a MAK column (cf. p. 24). This separation of the *mixture* is the crucial point of the method. The nucleic acid preparations that are to be compared are, in this way, subject to the same aribtrary influences that can affect the results of any "standardized" procedure. The nucleic acids are eluted from the column in the following order: tRNA, DNA and complexes of DNA and RNA, rRNA, and mRNA. Each RNA group consists of many different kinds of molecules. This is indicated by the "peak" profile for each group. Using special measuring instruments the RNA labeled with uridine-H^3 can be recorded separately from that labeled with uridine-C^{14}. The elution profiles for H^3-mRNA and C^{14}-mRNA are seen to coincide to a large extent. This coinci-

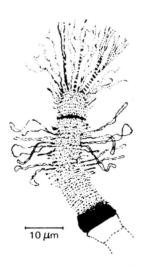

10 μm

145. A giant chromosome from the suspensor of *Phaseolus coccineus*. The local loosenings of the structure correspond to the formations of puffs (from Nagl 1970).

dence established the reliability of the method. Where no difference was expected none was found.

Now to the main experiment (Fig. 146b): using the technique outlined above, H^3-mRNA from two-day old cotyledons was compared with C^{14}-mRNA from 14-day old cotyledons. Great differences in the elution profiles are found, particularly in the region of rRNA and mRNA. What is of interest for us is that the mRNA from different states of seedling development are different. Thus, differential gene activities can also be demonstrated at the level of mRNA.

3. Phase-and Tissue-specific Protein Patterns

One can still go to a step further and carry out comparative analyses of the first secondary gene products, the polypeptides or proteins. Consequent on the introduction of a suitable method, namely, zone electrophoresis on different supports, phase- and tissue-specific protein patterns have been detected in a wealth of experiments in recent years. As an example tissue-specific patterns of soluble proteins from different parts of the tulip are shown (Fig. 147). Enzymes, of course, also number among the proteins. We have already dealt with isoenzymes, multiple

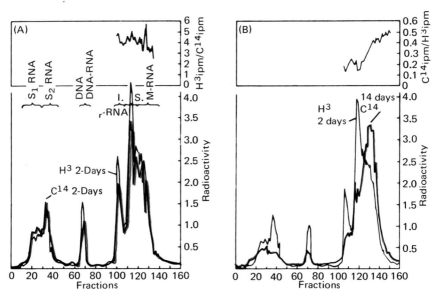

Fig. 146. Evidence for different mRNAs in cotyledons of the ground nut (*Arachis hypogaea*), which are phase-specific. (A) = control: H^3- and C^{14}-nucleic acids from cotyledons on the second day after germination. (B) = experiment: H^3-nucleic acids from cotyledons on the second, C^{14}-nucleic acids from cotyledons on the fourteenth, day after germination. Considerable differences appear between the H^3- and C^{14}-profiles, particularly in the region of mRNA (right side of diagram) (modified from Key and Ingle 1966).

forms of one particular enzymatic activity. Phase- and tissue-specific patterns for the most diverse isoenzymes have been discovered in a great number of plants, those for amylases, catalases, peroxidases and various dehydrogenases and aminopeptidases being reported particularly frequently since they are easy to detect. In Figure 148 tissue-specific peroxidase isoenzyme patterns are shown.

Thus, differential gene activity is also demonstrable at the level of the proteins. One can depart still further from primary gene activity and consider the formation of visible characters. Character formation proceeds in an orderly sequence, first the development of character A, then that of character B, etc. According to our general concept differential gene activity should lie behind this visible sequence of events. Experimental evidence for this was obtained in a number of cases using antimetabolities of

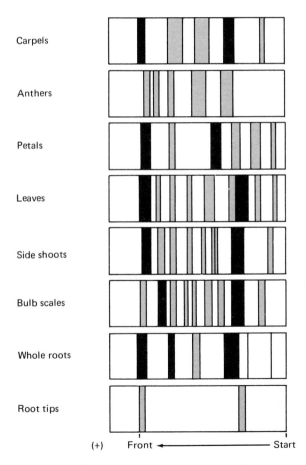

Fig. 147. Tissue-specific protein pattern of the tulip. Separation by zone electrophoresis (modified from Steward 1965).

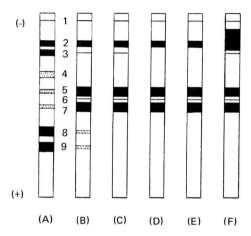

Fig. 148. Tissue-specific isoenzyme pattern of the peroxidase of the petunia *(Petunia hybrida).* (A) flower buds, (B) young leaves, (C) old leaves, (D) young shoots, (E) old shoots, (F) root, (from Hess 1967).

transcription and translation. Using them, the development of particular characters in such a sequence, such as that of character B, could be prevented. We shall consider an example in detail later (page 294).

Regulation

A. States of Activity of the Gene

Our analysis has progressed to the point where we can say that in sexual propagation the entire cellular material of a multicellular organism is derived from the zygote by mitotic divisions. During the mitotic cycle the genetic material localized in the chromosomes is reduplicated semiconservatively. Thus, all daughter cells carry in their chromosomes the same genetic information as the parent cell. That this is so and, furthermore, that the total genetic capacity can be maintained during subsequent differentiation is supported by regeneration experiments.

This implies that the differentiation of cells can, at least in most cases, not be due to changes in the genetic stock. Instead we have presented evidence to show that differential gene activity, which is phase- and tissue-specific, is one of the driving forces of differentiation. The next question must be: how is this differential gene activity brought about? This question is of more profound importance than we may at first surmise. Up to now we have spoken only of active genes. Particular stages of development or particular tissues were distinguished, in each case, by specific patterns of active genes. The pattern of active genes varies from one stage of development to another and from one tissue to another. A transition of this kind implies that active genes must be inactivated and hitherto inactive genes must be activated. Applying this consideration to a given stage of development or tissue, the effect of a given stimulus should lead one to distinguish the following four states of activity of the genes:

(1) *Active genes* are active before the stimulus and remain so afterwards.

(2) *Inactive genes* are inactive before the stimulus and remain so afterwards.

(3) *Activatable genes* are potentially active genes; they are inactive before the stimulus and active afterwards.

(4) *Inactivatable genes* are repressible genes, potentially inactive genes; they are active before the stimulus and inactive afterwards.

Our question concerning the causes of differential gene activity is thus a question of which factors activate and inactivate the genes. We shall deal with this regulation of gene activity in this chapter. It can only be appreciated in the context of the entire system of regulation of metabolic processes.

B. Regulation: Point of Departure

By 1903 the plant physiologist Klebs had already established that the development of an organism is accomplished by an interplay between a "specific structure" and "external" and "internal" conditions. The "specific structure" has been identified as the genetic material although this would not have met with the complete approval of Klebs. Today developmental physiologists speak of a reaction norm laid down by the genetic material and within which external and internal factors can exert a regulatory influence on growth and differentiation.

We must also bear in mind that we have to deal with multicellular organisms and not with a single cell. That compels us to distinguish between intercellular and intracellular regulation or intercellular and intracellular factors of regulation.

It has been emphasized many times that one or more metabolic reactions lie behind the formation of each character. Thus, we can equate regulation with "regulation of metabolic processes." And if we further recall that one or more genes lies behind each metabolic process then we have once again established a connection with gene activity. Table 6 provides a guide to the most important kinds of regulation. Regulation of gene activity occupies a central position. Even in the case when the activity of an enzyme, and not that of the gene at all, is affected, the altered metabolic situation can lead secondarily to fluctuations in gene activity. In view of the elaborate interlinking of intercellular processes such secondary effects on the genetic material are almost inevitable.

Table 6. Survey of the types of regulation in plants

Regulation by internal factors
1. Intracellular regulation
   a. Regulation of gene activity
      aa. Regulation of transcription
         Substrate induction
         End product repression
         Repression by histones
      bb. Regulation of translation
   b. Regulation of enzyme activity
      Enzymatic regulation
      Isosteric effects: competitive inhibition
      Allosteric effects: end product inhibition and end product activation
2. Intercellular regulation
   Phytohormone

Regulation by external factors

C. Regulation by Internal Factors

First let us consider regulation by internal factors, i.e. factors that are already present in the organism. In doing so we must distinguish between intracellular and intercellular regulation, as mentioned above.

1. Intracellular Regulation

a. Regulation of gene activity

Regulation of gene activity is usually understood to include not only the regulation of transcription but also that of translation. This is because the two processes are usually coupled in series. They lead to the formation of polypeptides or proteins, among which the enzyme proteins merit special interest. In many, but not all, cases the regulation of gene activity is the regulation of enzyme synthesis.

aa. Regulation of transcription

In 1961, Jacob and Monod published a hypothesis for the regulation of transcription which has since become known as the *Jacob-Monod model*. All subsequent investigations of the regulation of gene activity amount, in essence, to attempts to prove the Jacob-Monod model — or a matter of equal importance for the progression of our understanding — to refute it. The Jacob-Monod model has been shown in many cases to be applicable to bacteria. According to the Jacob-Monod model the genes can be classified into three groups, regulator genes, operator genes, and structural genes. The relationships between the three groups are characterized by a rigorous hierarchy. At the summit of this hierarchy are the *regulator genes*. They control the activity of the operator genes. Since they are usually spatially separated from the operator genes, they exercise their control function by the production of inhibitors of gene activity, which are known as *repressors*. Several of these repressors are proteins as was first established by Gilbert in 1966/67 for the repressor of the lac operon of *E. coli* (see below).

The *operator genes* occupy the second place in the gene hierarchy. As mentioned above their activity is controlled by the regulator genes. The operator genes, in turn, control the activity of structural genes. They are located in direct proximity to the structural genes.

The *structural genes* are already familiar to us. Via transcription and translation, they supply polypeptides, which are needed for well-defined metabolic reactions. In microorganisms, all of the structural genes that contribute to the synthesis of one end product lie one after the other on the genome in many cases. Moreover, the activity of the whole group is controlled by an operator gene which is located right next to the first structural gene of this group. Such a group of functionally related structural genes, together with the operator gene that controls them, is referred to as an *operon*.

Interplay between regulator genes, operator genes, and structural genes can lead to an activation of genes via substrate induction and to an inactivation of genes via end product repression. The decisive role falls to the repressors, the state of activity of which is controlled by low molecular weight metabolites, the so-called *effectors*.

Substrate Induction (Fig. 149). First, the general scheme of substrate induction: the repressor which is produced by the regulator gene is active and blocks the subordinate operator gene. This brings the activity of the entire operon to a halt. Not one of the structural genes of the operon can form mRNA. The repressor can, however, be inactivated by a low molecular weight effector. The effector brings about a change in the steric configuration of the repressor. The repressor, thus altered, no longer fits on the operator gene. Thereby the blockade of the entire operon is lifted. The structural genes proceed to form mRNA. The effector, which was mentioned, can be a substrate of one of the enzymes that are coded for by the operon in question, hence the expression substrate induction. Frequently the term *adaptive enzyme formation* is still used, because the enzyme formation results from adaptation to a new metabolic situation, namely, the supply of the enzyme substrate.

Substrate induction thus permits a quite rapid adaptation — the duration of the enzyme synthesis itself causes a certain lag. It is often found in microorganisms that must adapt to radically changed environmental conditions at short notice. The best-known example is the lac operon of *E. coli* in which three structural genes are linked together, one of which codes for β-galactosidase. This is an enzyme which is capable of cleaving the β-galactosidic linkages of substrates such as lactose. In this case the repressor has been shown to be a protein. It is inactivated by lactose.

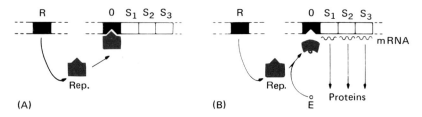

Fig. 149. Outline of substrate induction. A repressor Rep which is coded for by the regulator gene R blocks the operator gene O and, thus, the activity of *all* of the structural genes ($S_1 - S_3$) of the operon (A). The repressor is inactivated by a low molecular weight effector E. E is the substrate of an enzyme which is encoded in the structural genes of the operon. After the inactivation of the repressor, the blocked operon becomes active and its structural genes code for proteins. Among the latter is the enzyme that can utilize the effector as substrate (B) (modified from Hess 1968).

Therefore, the addition of lactose to *E. coli* leads to the adaptive synthesis of β-galactosidase by the mechanism outlined above.

In higher plants there are many findings that can be interpreted on the basis of substrate induction. An example is the induction of thymidine kinase in various *Liliaceae* by its substrate thymidine. Thymidine kinase converts the nucleoside d-thymidine into its nucleotide d-thymidine-5'phosphate with the help of ATP. Thymidine-5'-phosphate is then utilized in the synthesis of DNA. Usually the DNA building block thymidine-5'-phosphate is supplied by another route (Fig. 14). The activity of thymidine kinase can assume importance however if thymine is predominantly available in the form of its nucleoside thymidine and DNA must be synthesized. That is the case in a particular phase of pollen development in several *Liliaceae*. First thymidine accumulates in the anthers and a little later the activity of thymidine kinase increases sharply. This increase in activity is due to new synthesis of the enzyme. It can be triggered by supplying thymidine experimentally. When this was done a further interesting finding was made: exogenously added thymidine induces the synthesis of thymidine kinase in isolated flower buds of *Lilium longiflorum* only in the phase of development in which the enzyme is induced under natural conditions (Fig. 150). Thus, the potential for substrate induction is limited here by an unknown regulatory mechanism of a higher order.

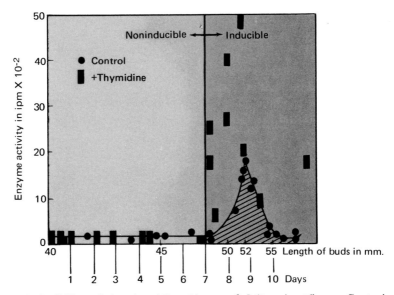

Fig. 150. Inducibility of the thymidine kinase of *Lilium longiflorum*. Control: enzyme activity during the development of buds under natural conditions. + Thymidine: inducibility of enzyme activity by supply of thymidine (modified from Hotta and Stern 1963).

The example of thymidine kinase showed us that:

(1) At least the principles of substrate induction are realized in higher plants too.

(2) Substrate induction is subject to a temporal control of an unknown kind. Substrate induction *and* a temporal control of substrate induction of a higher order have been discovered in a few further instances, such as different kinds of nitrate reductase and an enzyme involved in anthocyanin synthesis in *Petunia hybrida*. In addition, there is quite a number of enzymes for which substrate induction *or* a temporal control of synthesis of a higher order has been established. We shall become acquainted with examples of the latter type later.

End Product Repression. Again we will first outline the general scheme of end product repression (Fig. 151). The repressor that the regulator gene codes for is initially inactive in this case. The operator gene is not blocked, all of the structural genes of the operon form mRNA. Then the repressor is activated by a low molecular weight effector. The effector alters the steric configuration of the repressor in such a way that the repressor now fits on the operator gene like "the key in a lock." This results in *all* of the structural genes of the operon being blocked. They cease to produce mRNA for certain enzymes.

The effector here is an end product of a reaction or chain of reactions, the enzymes of which are furnished by the operon under consideration. In *E. coli,* for example, all of the structural genes for arginine are linked together in an arginine operon. Thus, this arginine operon supplies all of the enzymes that are necessary for arginine synthesis. The end product of this enzymatic activity, arginine, serves as effector to activate the repressor of the arginine operon.

End product repression also permits a relatively rapid adjustment to changed conditions. If an excess of a substance is present in the bacteria, this substance can block its own, further synthesis by end product repression, provided that the following two conditions are met:

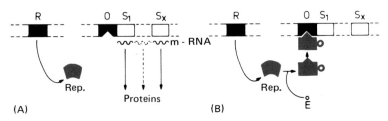

Fig. 151. Outline of end product repression. The repressor Rep which is coded for by the regular gene R is initially inactive. The structural genes ($S_1 - S_x$) code for proteins, including enzymes, which furnish specific end products (A). One of these end-products acts as effector (E) to activate the repressor. The repressor now blocks the operator gene O which controls the operon. The activity of the structural genes ceases (B) (modified from Hess 1968).

(1) The mRNA that has already been transcribed from the operon under investigation

(2) and the enzymes that have already been coded from this mRNA must be eliminated.

For if mRNA and the enzymes were sufficiently stable, they could lead to a further supply of an end product which is no longer needed. As far as the mRNA of bacteria is concerned, it is, with a few exceptions, short-lived. Its half-life time (the time taken for half of a particular type of mRNA to be broken down) amounts to only a few minutes. Thus, usually the mRNA of the blocked operon is eliminated from the cell a short time after the start of end product repression.

Enzymes are usually longer-lived. However, in this case there is another easy way out of the dilemma for bacteria: they divide. From one division to the next the stock of an unwanted enzyme is diluted. In the same way any stable mRNA that might possibly be present can be diluted out by division.

Consequently, end product repression in bacteria is only fully effective when divisions still occur. In the cells of higher plants, however, a division of this kind, far less a succession of divisions, would have been impossible owing to defined processes of differentiation. Thus, to a large extent, end product repression would be made worthless. A few data support the possibility that the degradation of enzymes might take the place of cell division.

Our present state of knowledge does not yet allow us to pass final sentence on the significance of end product repression in higher plants. Nonetheless it is striking that compared with many known cases where substrate inhibition is evident in higher plants, there are only a relatively few cases of possible end product repression. Among them is the repression of phytase. This is a hydrolytic enzyme that breaks down the phosphate storage substance, phytic acid, into inositol and inorganic phosphate (Fig. 152). The mRNA for phytase is formed in the scutellum of germinating wheat embryos in the first six hours after the onset of germination. Apparently experimentally added inorganic phosphate is capable of bringing the synthesis of this mRNA to a halt. Since inorganic phosphate is one of

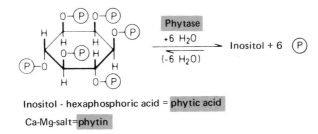

Inositol - hexaphosphoric acid = phytic acid

Ca-Mg-salt=phytin

Fig. 152. Function of phytase.

the end products of phytase activity, this might be a case of end product repression.

Repression by Histones. In the chromosomes the DNA is associated with proteins. Included among these proteins are the basic histones and, in smaller amounts, acidic and neutral nuclear proteins, including accompanying enzymes systems such as the DNA polymerase and RNA polymerase. The basic character of the histones is due to a high proportion of the basic amino acids arginine and lysine. Depending on which of these two amino acids is predominant, one speaks of arginine- or lysine-rich histones. Histones and acidic nuclear proteins are structural proteins but they also seem to participate in the regulation of gene activity. As far as higher plants are concerned, there are data in support of this supposition, especially for histones.

Repression of Transcription in a Cell-Free System. In animals as well as plants there are a few pieces of evidence that histones can inhibit transcription in a cell-free system. As far as plants are concerned the investigations of J. Bonner should be mentioned. Certain reserve globulins, which in the same or similar form are typical of Leguminosae, are formed in the cotyledons of pea seedlings. The remaining parts of the seedling carry no or only a very few of these reserve globulins. Now it is possible to isolate "chromatin" from the various parts of pea seedlings. This material consists of complexes of DNA and histones. This chromatin is added to a cell-free system that contains all of the enzymes and cofactors, including ribosomes, necessary for transcription and translation. Chromatin from cotyledons induced the synthesis of reserve globulins in this system, whereas chromatin from shoot tips, for example, did not. If, however, the histones were removed and DNA alone was added, then the DNA preparations from both cotyledons and shoot tips supported the synthesis of reserve globulins. In the chromatin from shoot tips the genetic material for the synthesis of reserve globulins was evidently repressed by histones whereas that in the chromatin from cotyledons was not (Table 3, page 29). The experimental data are of central importance in that they argue in favor of a *specific* inhibitory effect of histones (cf. p. 190). At the same time the fact should not be hidden that they have been disputed on methodological grounds.

Repression by Histones in vivo. The experiments in a cell-free system still leave several questions unanswered. Therefore attempts were made to approach the question of repression by the histones from a different angle. Thus, Fellenberg and Bopp succeeded in impairing a few morphogenetic processes which require certain genetic activities by treatment with histones. The formation of new roots in isolated pea epicotyls which is induced by IAA, the formation of wound phelloderm in *Kalanchoe daigremontianum,* and the formation of crown galls, were all inhibited by administration of histones from various sources. Certain chemical modifications of the histones (by acetylation, oxidation, phosphorylation, and thermal denaturation) also led to changes in their inhibitory effect.

Basic plant proteins and cancer in animal organisms. It was just mentioned that histones, i.e. basic proteins, can inhibit the formation of crown galls. Crown galls are plant tumors. In this connection it is interesting to note that basic plant proteins can also inhibit cancer growth in animal organisms. In the mistletoe (*Viscum album*) there are basic proteins which have been investigated in detail in the last ten years, particularly by Vester. In experiments on mice and in cell cultures these proteins proved to be powerful inhibitors of cancer. It was found with a few protein fractions that 38 molecules of protein per cell in experiments on living mice and 15 molecules of protein per cell in experiments in cell cultures still produce an inhibitory effect. According to the experiments carried out thus far these mistletoe proteins primarily inhibit transcription, but prolonged treatment also leads to the inhibition of the reduplication of DNA. Their high sensitivity to denaturing effects still prevents their application in medicine.

Thus, the basic proteins of mistletoe can inhibit transcription just like the histones. Perhaps further investigation of this system will provide hints as to the mechanism of action of the histones.

Changes in the Stock of Histones during Development. If, as is supposed, the histones do participate in the regulation of gene activity, it is conceivable that quantitative or qualitative changes in the pattern of histones appear during development and are also analytically demonstrable. Changes of this kind have actually been detected several times, during the differentiation of blossoms or the development of pollen, for example. Usually the evidence was obtained by means of certain color reactions which cannot answer the question whether quantitative or qualitative changes occurred or both together. A qualitative change could be demonstrated during meiosis in several plants: in these instances a new "meiotic" histone appeared. Such changes in the stock of histones fit very nicely with the hypothesis that histones participate in the regulation of gene activity. However, we should not forget that, taken by themselves, they do not in any way represent evidence for this.

Hypotheses Concerning the Mechanism of Action of Histone. How can one picture the mechanism of action of the histones in the repression of gene activity, or, more exactly, of transcription? Let us put forward two extreme possibilities, not forgetting that perhaps both of them might be realized.

Possibility 1. As structural proteins the histones inhibit the genetic activity of whole chromosomes or regions of chromosomes unspecifically. During division the chromosomes contract to a compact, transportable form. After division, the chromosomes then unwind. In contrast, certain chromosomes or regions of them can remain contracted until shortly before the next division. If no further division occurs the contracted condition can be maintained unchanged. Such contracted chromosomes or regions of chromosomes are called heterochromatic. In heterochromatic chromosome material gene activity is greatly diminished or has ceased. It is assumed that the histones, as structural proteins, might play a central role in the contraction of the chromosomes. In this way, they would be quite unspecific repressors of gene activity. This is because the activity of all genes in the contracted regions of the chromosomes is inhibited to a greater or lesser degree.

Possibility 2. The histones specifically inhibit quite well-defined genetic activities only. A histone would thus be tailor-made for a particular gene locus. This leads to the question as to how the histone could attain such a high specificity. Histones from sources as different as calf thymus and peas show only minor, detectable differences and so are relatively unspecific. Now it could be argued that, with improved analytical methods, specificities, which have hitherto been masked, might be discovered. Furthermore another possibility has been suggested, namely that RNA forms complexes with histones that do not need to exhibit specific differences themselves at all. The RNA component of the complex is, however, specifically related to a particular segment of DNA, i.e. a gene locus: according to the base pairing rules it is bound to the homologous base sequence in the DNA. In this way the totally unspecific histone is brought into position and can now stop transcription at the gene locus in question (Fig. 153). Thus, the RNA component would be responsible for the specificity.

Neither possibility 1 nor possibility 2 is compatible with the Jacob-Monod model. That is hardly surprising for why shouldn't such very much more complicated higher organisms be equipped with additional

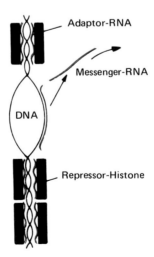

Fig. 153. Adaptor hypothesis of the repression by histones.

regulatory mechanisms? In summary, *in vitro* and *in vivo* findings suggest that nuclear proteins, particularly the histones, can inhibit transcription. How they do this is unknown. None of the current hypotheses conforms to the demands of the Jacob-Monod model.

bb. *Regulation of translation*

Findings in higher plants that can be interpreted in terms of a regulation of translation are exceedingly rare. One example is invertase (page 66) of cane sugar. As is well-known the end products of invertase activity are glucose and fructose. A few data suggest that the end product glucose induces degradation of the mRNA for invertase synthesis. As a consequence the invertase activity in glucose-treated sugar cane decreases. Transcription, i.e. the formation of the mRNA for invertase, does not appear to be impaired by glucose.

b. Regulation of enzyme activity

In discussing the regulation of enzyme activity a distinction is made between the regulation of the enzyme protein and additional regulatory mechanisms, which do not affect the enzyme protein but which exert their influence through the provision of coenzymes and other cofactors or substrates. These regulatory mechanisms which do not influence the activity of the enzyme protein itself are sometimes grouped together under the term "enzymatic regulation".

Here we intend to consider only the *regulation of the enzyme protein* in detail, and, in so doing, we must distinguish between isosteric and allosteric effects. *Isosteric effects* are already familiar to us. Particular substances which resemble structurally the normal substrates are bound to the same site as the normal substrates (Fig. 154). Since they are bound to the same site they are called "isosteric." However, inspite of their similarity

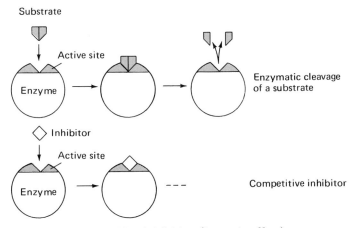

Fig. 154. Competitive inhibition (isosteric effect).

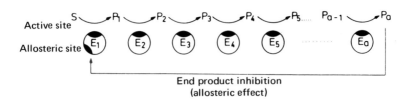

Fig. 155. End product inhibition (allosteric effect) (modified from Lehninger as presented in Bielka 1969).

these substances are so different from the normal substrate that they can-not be metabolized. This means that the enzyme protein in question is blocked. This type of inhibition is also known as *competitive inhibition.* Ex-amples of competitive inhibition of this kind are the inhibition of thymidilate synthetase by 5-fluorodeoxyuridine (page 21) and of succi-nate dehydrogenase by malonate (page 81).

In the case of *allosteric effects* substances that inhibit or stimulate ac-tivity are not bound to the substrate binding site but at another site—hence the name. If the allosteric binding site is occupied by a stimulating or inhibitory substance, changes in the configuration of the enzyme pro-tein are induced, which, in turn, lead to an increase or an inhibition of enzymatic activity. Allosteric inhibition has been investigated most thoroughly. The inhibitory substances are, in this case, end products of synthetic pathways. For this reason the term *end product inhibition* is used, which must not be confused with end product repression, which we have already discussed. An end product of this kind often inhibits only the first enzyme of the pathway specific for its synthesis (Fig. 155).

An example is the inhibition of threonine deaminase by L-isoleucine. L-Isoleucine is the end product of a synthetic pathway, the first enzyme of which is threonine deaminase (Fig. 156). In end product inhibition the steric configuration of a protein, i.e. an enzyme protein, is altered. The steric configuration of a protein is also changed in end product repression,

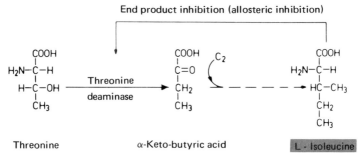

Fig. 156. End product inhibition: inhibition of threonine deaminase by isoleucine.

in this case that of the repressor—which is a protein at least in the case of the lac operon of *E. coli*. Inspite of this similarity we must take care not to confuse the two: in the case of end product inhibition the *activity* of the enzyme protein is inhibited, in end product repression, its *synthesis*.

Fine Regulation by Isoenzymes. We have just mentioned the end product inhibition of threonine deaminase by isoleucine. Isoleucine belongs, together with threonine, methionine, and lysine, to the aspartate family of amino acids (Fig. 118). The synthetic pathway to these amino acids starts with the conversion of aspartate into aspartyl phosphate which is catalyzed by aspartokinase. Later the initially undivided synthetic pathway branches out to lead to the amino acids mentioned (Fig. 157). This branching presents intracellular regulation with a serious problem. For example, more than sufficient threonine might be present in the cells.

The further supply of threonine could then be easily stopped by end product repression or end product inhibition of aspartokinase. However, the synthesis of methionine and lysine would also be stopped, although both of these amino acids might be urgently needed. The solution to the problem lies in the fact that three isoenzymes of aspartokinase have been found in *E. coli,* one of which can be blocked by threonine and a second by lysine, by end product repression or inhibition. The activity of the third isoenzyme of aspartokinase is also subject to appropriate regulation somewhat later in the synthetic pathway. Although these findings were made in bacteria there are various indications that the situation in higher plants is likely to be quite similar.

The fine regulation in the aspartate family may appear to be quite complicated enough. It is, however, still relatively simple. We have only to

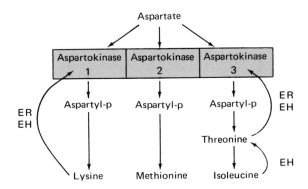

Fig. 157. Fine regulation by isoenzymes: biosynthesis of the aspartate family of amino acids. ER = endproduct repression, EH = endproduct inhibition. The synthesis of methionine is also subject to feedback mechanisms, which are, however, now shown here (cf. also Fig. 156).

think of the fact that, in addition to end product repression, there is also substrate induction and in addition to end product inhibition there is also end product activation (allosteric stimulation). All four of these regulatory mechanisms mentioned participate in the regulation of the biosynthesis of aromatic amino acids in the yeast *Saccharomyces cerevisiae*. As with the aspartate family, so here, too, incidentally, the first reaction, the linking of phosphoenolpyruvate with D-erythrose-4-phosphate to form a 7-C compound, is catalyzed by a group of three isoenzymes working in parallel (Fig. 158). The data were obtained in a yeast. However, more recent investigations suggest that similar regulatory mechanisms are also operative in higher plants.

2. Intercellular Regulation: Phytohormones

In animal organisms regulation is mediated from cell to cell by nervous or humoral means. In plants regulation via the nervous system is precluded. There remains the possibility of regulation via the vascular system in which certain regulatory factors can be transported from cell to

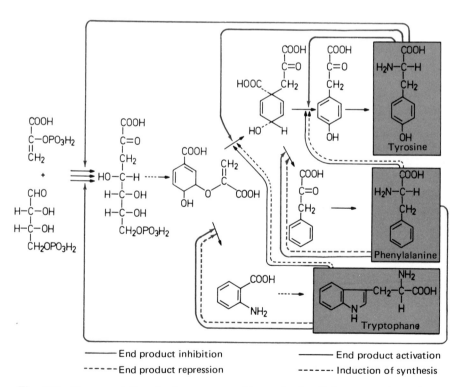

Fig. 158. Fine regulation by isoenzymes: biosynthesis of the aromatic amino acids (modified from Lingens 1969).

cell and from organ to organ. The plant hormones, the phytohormones, play a dominant role in this process. Like all hormones the phytohormones are also messenger substances, which are usually effective in small amounts and whose sites of synthesis and action are different. The transport from the site of synthesis to the site of action takes place in the vascular system of the plant. Of course, in a number of cases the phytohormones are also effective at the site of synthesis itself. Animal hormones sometimes, but certainly not always, show a quite broad action spectrum. This statement is even more true of phytohormones. One and the same phytohormone is capable of influencing a number of totally different processes. (Fig. 159). This is the one fact that all hypotheses concerning the mechanism of action of phytohormones have to take into consideration.

The four most important and chemically best known groups of phytohormones are certain indole derivatives, the gibberellins, cytokinins, and abscisins. In the following sections we shall first discuss each of these groups in turn and then consider them again in more detail with respect to their mechanism of action.

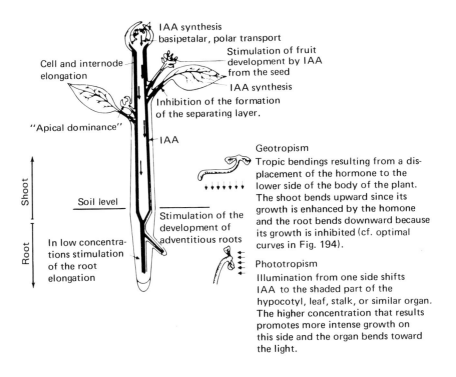

Fig. 159. Several developmental processes regulated, in part, by IAA (from Steward 1969).

$$CH_2-COOH$$

IAA

Fig. 160. IAA = Indole - 3 - acetic acid = β- Indolyl-acetic acid.

a. Indole derivatives: IAA

Chemical constitution

A few indole derivatives are phytohormones. The most important of them is β-indolylacetic acid, indole-3-acetic acid or, abbreviated, IAA (Fig. 160). It was detected in 1934 by Kogl in human urine, then in microorganisms, and, finally, also in higher plants. IAA occurs in plants either free or bound, e.g. esterifield to glucose or in peptide linkage with aspartic acid and glutamic acid.

History, method of assay

The existence of phytohormones was known even before the isolation of IAA from plant material. Final proof was provided by Went in 1928 in experiments based on preliminary studies dating back to the previous century. His experimental material was the *Avena* coleoptile, a cylindrical sheath which surrounds the primary leaf of oat seedlings. Coleoptile tips were cut and mounted on small blocks of agar. After some time the small blocks of agar were attached to one side of the coleoptile stumps. The coleoptiles then bent toward the other side. The cause of this bending was vigorous longitudinal growth in the tissue beneath the small agar block (Fig. 161).

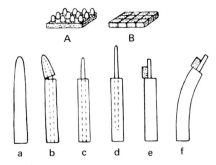

Fig. 161. The evidence of IAA in the tips of the *Avena* coleoptile. (The Avena curvature test of Went.) Coleoptile tips are placed on agar (A). After IAA has diffused into the agar, it is cut into small pieces (B). Coleoptiles are prepared by first decapitating them (b and c), then drawing out the primary leaf somewhat (d), and sticking a small piece of agar containing IAA on one side of it with a little gelatin (e). IAA migrates out of the agar in a downward, polar direction and induces vigorous longitudinal growth on the side of the coleoptile to which it is attached (f) (from Walter 1962).

Thus, a phytohormone had diffused into the agar block from the coleoptile tips and this then migrated downward in the coleoptiles and triggered the intensified longitudinal growth. Today we know that the most important phytohormone of *Avena* coleoptiles is IAA. As has been shown, particularly by the work of Pohl, IAA diffuses from the oat seed to the coleoptile tips in an inactive form, is activated there, and is then conducted downward again in the coleoptile. IAA and substances which act similarly are known as *growth substances* or *auxins* on account of their influence on growth.

IAA can be quantitatively determined in many test systems. A much used experimental object is still the *Avena* coleoptile. The Went technique is applied in the *Avena* curvature test: within certain concentration limits, the more IAA contained in the small agar blocks, the more strongly is the coleoptile bent. Another much used test is the *Avena* section test and is somewhat easier to carry out. Cylinders of defined length are cut out from the coleoptile directly beneath the tip and the primary leaves are removed from the cylinders. Hollow cylinders result, which can be immersed in IAA solutions. Within certain ranges of concentration, the more IAA present in the solutions, the more the sections elongate (Fig. 162).

Biosynthesis and degradation

Biosynthesis and degradation of IAA contribute to the regulation of the endogenous IAA level and, by means of it, to the regulation of growth. We have already outlined the biosynthesis of IAA (page 119). In microorganisms and, with high probability, also in higher plants it is derived from tryptophan. The sites of IAA synthesis are, especially, meristem tissue or young parts of the plant. In the dicotyledons the apex, i.e. the tip of the main shoot, is the most important source of IAA. Embryos also produce IAA. The development of strawberries, for example, is inhibited if the seeds, together with the embryos, are removed. If IAA is then supplied, the strawberries develop normally (Fig. 163). Deciduous leaves in the process of unfurling are also good producers of IAA.

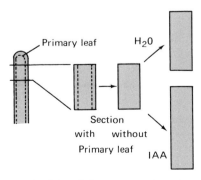

Fig. 162. Principle of the *Avena* coleoptile section test.

The degradation of IAA can be brought about by light, particularly UV, in the presence of catalysts such as riboflavin, and by enzymes. Peroxidases which are known as *IAA oxidases* rank among the IAA degrading enzymes. They require Mn^{++} and monophenols such as *p*-hydroxybenzoic acid or the flavonole kaempferol as cofactors. *o*-Diphenols such as catechol or the flavonole quercetin are inhibitors of many IAA oxidases. Perhaps one of the physiological functions of the flavonoles lies in this regulation of the activity of IAA oxidases.

Several functions of IAA

(1) *Longitudinal Growth*. We shall consider longitudinal growth and its regulation by IAA later (page 236).

(2) *Cell Division in the Cambium*. IAA is one of the factors which stimulate the activity of the cambium. This division-promoting effect of IAA is particularly important for our deciduous trees in the spring. Substances which stimulate the cambium to divide are conducted downward from the developing leaf buds by the branches and then the trunk. IAA is one of these substances.

(3) *Cell Division and Root Formation*. The formation of adventitious and side roots emanates from certain pockets of cell division which, in the case of side root formation, are located in the pericycle. The cell division leading to root formation is stimulated by IAA. For this reason preparations of IAA or synthetic growth substances producing a similar effect are used by plant growers to induce root formation in cuttings.

(4) *Cell Division in Tissue Cultures*. The functions 2 and 3 have already shown us that IAA promotes not only cell elongation but also cell division. This latter effect becomes particularly obvious in tissue cultures. This is because in many tissue cultures division only occurs if certain division-promoting substances are added. IAA is one of these substances.

(5) *Apical Dominance*. Influences emanate from the tip of the main shoot, the "apex," which inhibit the development of the side shoots. This can be readily demonstrated by removing the tip of the main shoot. If this is done the side shoots develop. This phenomenon is known as apical

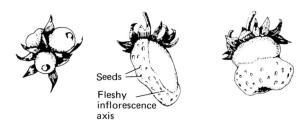

Seeds

Fleshy inflorescence axis

Fig. 163. Deformed strawberries produced after partial removal of seeds. Only those parts of the inflorescence axis that had been supplied with IAA by the seeds develop normally (from Nitsch as presented in Torrey 1968).

dominance. If now the main shoot is cut off and IAA is applied to the surface of the cut, then the side shoots still remain inhibited. Thus, IAA is one of the factors of apical dominance. According to more recent investigations IAA is reported to have only an indirect effect: it induces the formation of ethylene in the side shoots and this functions as the real inhibitor of development.

(6) *Shedding of Leaves and Fruit.* The shedding of a leaf is frequently initiated by the formation of a separation zone at the base of the leaf stalk (Fig. 164). This zone is formed by cell division occurring across the base of the stalk. The cells become more weakly attached to each other in this separation zone. This weakened attachment is usually the result of enzymatic degradation of cell wall components such as pectins, cellulose, and hemicelluloses. In this way the adherence of the cells to each other is loosened and mechanical stress leads to their being separated from each other. Often the leaf at first remains hanging on withered xylem fibers. However, this last connection is finally severed and the leaf falls. On the trunk side the wound is closed by a protective layer. As Jacobs among others has shown IAA inhibits the shedding of leaves. If the blade is removed from a *Coleus* leaf, the stalk drops off after a few days. If, however, IAA is applied to the surface of the stalk, the stalk remains attached to the shoot. This experiment has been confirmed by investigations of the output of IAA through the leaf blade. As long as the *Coleus* leaf blade dispenses sufficient IAA to the leaf stalk, the leaf remains attached to the plant. Only when the output of IAA sinks below a certain minimal value, as it does with increasing age of the leaf, does leaf shedding occur.

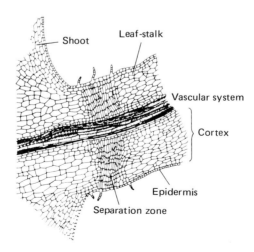

Fig. 164. A longitudinal section through a leaf-stalk of *Coleus* including the separation zone (from Torrey 1968).

The mechanism of action of IAA in the inhibition of leaf shedding is a matter of dispute. The problem is further complicated by the fact that IAA can also stimulate leaf shedding under certain conditions. Apart from this special effect of IAA there are also substances that universally promote the shedding of leaves: abscisic acid and ethylene. Their mechanism of action is somewhat better understood (page 211). Perhaps they will provide a means to elucidate the mechanism of action of their antagonist IAA.

(7) *Parthenocarpy.* IAA can induce parthenocarpy, i.e. fruit formation without fertilization, in a number of plants. Parthenocarpy in tomatoes, apples, and cucumbers among others, after treatment with IAA, is well-known.

(8) *Enzymatic Activity.* IAA can stimulate or lower the activity of enzymes. In root- and shoot-segments of peas the plants' own growth substance (IAA) and synthetic growth substances such as α-naphthylacetic acid and 2, 4-D (page 212) can "induce" the capacity to convert certain organic acids into conjugates with aspartic acid. IAA and α-naphthylacetic acid themselves are two of these organic acids, and benzoic acid is another.

Galston was able to demonstrate a lowering of enzymatic activity in tissue cultures of the tobacco plant. In medullary tissue of the tobacco plant two isoenzymes of peroxidases are detectable. If such tissue is now grown in culture two additional isoenzymes of peroxidase are found. After addition of IAA these new activities can no longer be detected.

Thus, IAA can stimulate the activity of enzymes (conjugates with aspartic acid) or lower enzymatic activity (peroxidase isoenzymes). Such a change in activity may be due to a change of gene activity but it need not be. In both examples mentioned it is not known whether a corresponding change in gene activity lies behind the change in enzyme activity or not. Several pieces of evidence suggest that the genetic material has been affected and thus implicate a new mechanism of action for IAA. Later in our discussion of longitudinal growth we shall meet with an instance of how IAA actually can activate gene material. The first case of gene activation by a phytohormone was discovered in experiments with gibberellic acid, not with IAA.

b. Gibberellins

Chemical constitution

Gibberellins are phytohormones, the chemical structure of which is characterized by the gibban skeleton and whose biological action is characterized by a powerful stimulation of the growth of certain dwarf mutants. It has already been mentioned that a gibberellin which has often been detected in higher plants and often used in experiments is gibberellic acid (cf. page 111, Fig. 165). In addition to it more than twenty

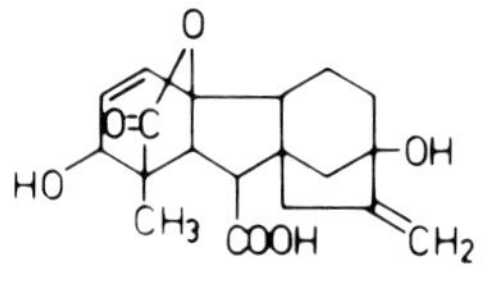

Gibberellic acid

Fig. 165. Gibberellic acid.

other gibberellins are presently known. Several different gibberellins are usually found in one plant. In contrast, often only one member of the other phytohormone groups can be detected in a given plant.

History, method of assay

A disease of the rice plant has long been known in East Asia which manifests itself in exceptionally vigorous longitudinal growth of the diseased plants. It was called Bacanae, the disease of the mad seedlings. It was demonstrated that this disease was caused by the fungus *Gibberella fujikuroi* (= *Fusarium heterosporum*). In 1926, Kurosawa succeeded in evoking the symptom of intensified longitudinal growth with culture filtrates of the fungus too. With this biological test the way to the isolation and characterization of the active principle was opened. After the second world war it was shown that the active principles, which were named after *Gibberella,* occur not only in this fungus but are also widely distributed in higher plants.

The basis of the biological assay for gibberellins is stimulation of longitudinal growth as it was in the first isolation of the material. Favorite objects for experiment are dwarf mutants of the bean and maize, for example, in which the synthesis of gibberellins is genetically blocked compared with the wild type. After treatment with gibberellins they assume normal growth. In the case of the dwarf mutants of maize the test solution can be introduced into the primary leaf. After about a week the increase in length of the primary leaf is determined (Fig. 166).

Several functions of gibberellins

(1) *Cell Division and Cell Elongation Resulting from the Stimulation of Growth.* One of the most striking effects of the gibberellins, and also the one on which the current methods of assay are based, is the stimulation of the longitudinal growth of dwarf mutants or so-called physiological dwarfs. Plants with normal longitudinal growth are influenced very much less or not at all by the gibberellins. As already mentioned, in some of these dwarf mutants synthesis of certain gibberellins is genetically blocked. In physiological dwarfs the genetic potential for normal, longitudinal growth is present but is activated only under defined external conditions. Among the latter are cold and particular light conditions, for

example. Also included among the physiological dwarfs are rosette plants which grow only a rosette of leaves, which is attached to the soil in the first year and then "shoot" the next year following the effect of the winterly cold. One of the effects of the cold consists in a change in the endogenous gibberellin level. Thus, it is not surprising that exogenously added gibberellin can compensate for the cold and can cause the rosette plants to shoot without their being exposed to the influence of the cold.

The question whether gibberellins promote longitudinal growth by stimulating cell division or cell elongation has been investigated in only relatively few plants. It was first thought that only cell elongation was affected. However, the surprising finding was made that cell division in the subapical meristem is indeed stimulated in rosette plants such as *Hyoscyamus* or *Samolus*. In other species such as the pea, cell division and cell elongation are stimulated by gibberellins.

(2) *Cell Division in the Cambium.* It has been demonstrated in a number of deciduous trees that gibberellins can stimulate cell division in the cambium in spring. Thus, IAA and gibberellins cooperate in triggering division in the cambrium.

(3) *Parthenocarpy.* This can also be induced by gibberellin treatment, e.g. in tomatoes, apples and cucumbers just like IAA.

(4) *Induction of Flower Formation.* In the case of long day plants and a number of cold-requiring plants gibberellins can induce the formation of flowers under conditions under which this does not normally occur. We shall discuss this phenomenon later (page 304).

(5) *Dormancy of Buds.* As the autumn days become shorter the shoots of our woody plants switch over to the state of dormancy. During the shortened day-time inhibitors such as abscisic acid (Fig. 167) concentr-

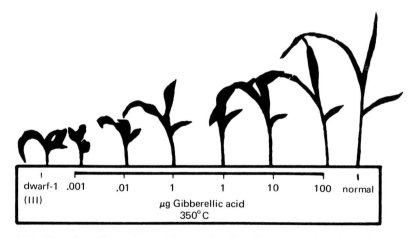

| dwarf-1 | .001 | .01 | 1 | 1 | 10 | 100 | normal |

µg Gibberellic acid
350° C

Fig. 166. Test for gibberellic acid on the dwarf mutant dwarf-1 of maize (modified from Ruhland 1961).

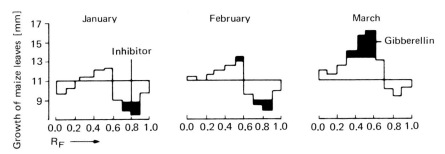

Fig. 167. Effect of cold on the content of gibberellins and inhibitors (probably abscisins) in black currants. *(Ribes nigrum)*. Shown here are so-called autobiograms: after paper chromatographic separation, the chromatograms are cut lengthwise into strips. The latter were eluted and the elutes were examined in a growth test on maize leaves for the presence of gibberellins or inhibitors. The R_f values of the individual components are shown on the abscissa: 0.0 = starting line on the chromatogram: 1.0 = solvent front on the chromatogram (from Wilkins 1969).

ate in the buds and block their activity. The inhibition is gradually overcome during the cold of winter: factors now accumulate in the shoots which promote their activity (Fig. 167). Among these are the gibberellins. An elevation of the endogenous level of the gibberellins when buds are subjected to cold has been demonstrated many times. As is to be expected, the dormancy of buds could be broken by exogenous supply of gibberellins.

(6) *Apical Dominance.* It has frequently been demonstrated that added gibberellic acid reinforces an apical dominance already present. In a few cases it seems that gibberellic acid raises the endogenous IAA level and this, in turn, leads to an intensified inhibition by IAA of the sprouting of side shoots. However, this interpretation is not universally valid.

(7) *Breaking of the Dormancy of Seeds.* A number of seeds germinate only under well-defined light conditions, many after exposure to red light, for example (page 256). In many cases such seeds can also be induced to germinate in the dark by gibberellins. The effect of the gibberellins to stimulate germination is turned to advantage in breweries, for example. Treatment of barley with gibberellic acid can lead to a uniformly high level of germination throughout the year.

(8) *Comparison with IAA.* A few developmental processes are influenced in the same way by both IAA and the gibberellins, as can be seen by comparing the list of the functions of the two. Hence, it was assumed that gibberellins exercise their function by influencing the endogenous IAA level. In a few cases this assumption may be true. However, this cannot be the rule, as is shown by a comparison of the efficacy of gibberellins and IAA on quite a large number of developmental processes (Table 7).

Table 7. Comparison of the activity of the different phytohormones in the formation of several characters. − = inhibition, + = stimulation, ● = no effect. A different behaviour from that indicated may be found depending on the object and special experimental conditions. Furthermore, it must be remembered that it is the interplay between several phytohormones that is decisive rather than the action of any given phytohormone

Character formation	Abscisic acid	IAA	Gibberellins	Cytokinins
Fall of leaves and fruits	+	−	−	−
Dormancy of buds	+	●	−	−
Germination	−	●	+	+
Cell elongation	−	+	+	(+)
Cell division	−	+	+	++
Flower formation LDP*	−	+	+	●
Flower formation SDP**	(+)?	−	●	●
Senescence	+	−	−	−
Synthesis of α-amylase in barley	−	●	+	●
Transcription and/or translation	−	+	+	+

*Long day plants
**Short day plants

(9) *Gene Activation by Gibberellic Acid.* When seeds germinate, reserve materials, which are stored in the endosperm, cotyledons, or other tissues, must be mobilized. The signals for this mobilization emanate from the embryo. In 1960, Yomo in Japan and Paleg in Australia independently discovered that gibberellic acid can be such a signal. Further experiments, particularly those of Varner in the U.S.A. finally, in 1964, provided evidence that a gibberellic acid can activate genetic material. The first report that a hormone can control genetic activity had already been provided in 1960 by Clever and Karlson in experiments on the mechanism of action of the insect molting hormone, ecdysone. Barley was the plant studied. At germination, starch, in particular, must be mobilized in barley. That happens through the hydrolytic activity of amylases (page 70). A barley grain can be halved so that one half contains the embryo and the other is embryo-free. Now the starch is hydrolyzed only in that half of the grain bearing the embryo. If, however, gibberellic acid is then added to the embryo-free half, the starch there is also degraded (Fig. 168). This finding suggests that the embryo normally releases gibberellic acid into the endosperm where it initiates the hydrolysis of starch.

This supposition has been confirmed. In the mobilization of the reserve materials of barley the following events take place: the embryo discharges gibberellic acid into the outermost layer of the endosperm, the aleuron. As a result genes for the synthesis of various hydrolytic enzymes are activated in the aleuron. The process culminates in the synthesis of

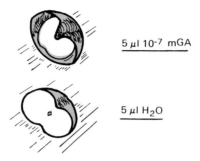

5 μl 10⁻⁷ mGA

5 μl H₂O

Fig. 168. The effect of gibberellic acid on the hydrolysis of starch in the embryo-free half of barley grains (modified from Plant Research 66).

these enzymes, one of which is α-amylase. The enzymes are then discharged from the aleuron into the internal, starch-containing layers of the endosperm. There the α-amylase participates in the hydrolysis of starch and other newly-formed enzymes hydrolyse proteins and nucleic acids.

The experimental evidence for the fact that α-amylase and a few other hydrolytic enzymes are actually formed *de novo* was crucially important. Detailed investigations have removed all doubt about that. Verifications of this kind are always necessary because some enzymes are already present in the seed and are not formed *de novo* at germination but merely activated.

c. Cytokinins (Phytokinins)

Chemical constitution

All cytokinins known up to now are, according to their chemical constitution, purine derivatives, especially derivatives of adenine, in which the amino group in position 6 bears particular substituents. Kinetin (6-furfuryl-aminopurine) and 6-benzylaminopurine are synthetic cytokinins which are often used experimentally and zeatin and N^6-($\triangle^2$-isopen-

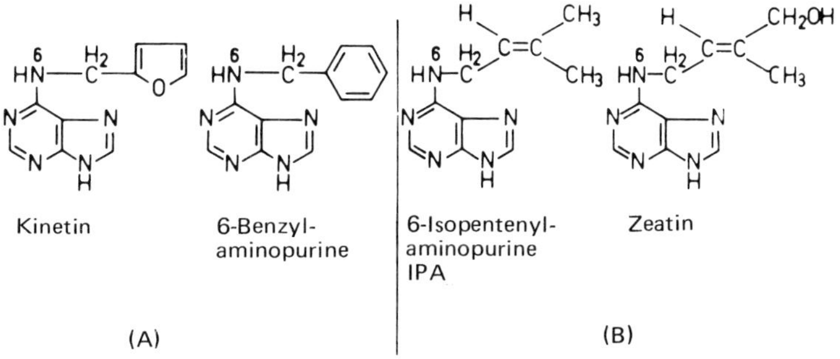

Kinetin   6-Benzyl-aminopurine   6-Isopentenyl-aminopurine IPA   Zeatin

(A)   (B)

Fig. 169. Cytokinins: synthetic (A), naturally occurring (B).

tenylamino)-purine (IPA) are examples of naturally occurring cytokinins (Fig. 169).

History, method of assay

As early as 1913 Haberlandt provided circumstantial evidence that substances from the phloem can stimulate cell division in potato parenchyma. However, the search for substances that could induce cell division only showed signs of promise after an appropriate assay had been developed. The most reliable tests can be carried out in tissue culture. In such tissue cultures, cell division is very often adjusted to an optimum with medium containing minerals, sugars, amino acids and vitamins. Growth substances such as IAA can sometimes support cell division, but not in every case.

In 1941, Van Overbeck discovered that the fluid endosperm of the coconut, the coconut milk, can induce cell division in embryos grown in artificial culture media. Subsequently, Steward and his colleagues in particular investigated the influence of coconut milk on the growth of tissue cultures. They established a division-promoting effect of coconut milk in tissue cultures too. However, in most cases it was necessary to combine the coconut milk with a growth substance such as IAA, α-naphthylacetic acid or 2, 4-D if a strong and persistent capacity to divide was to be attained.

The real advance, namely, the discovery of the cytokinins, was due to the work of another group led by Skoog. As so often happens, chance played a part. Old preparations of DNA or preparations extensively degraded by autoclaving from yeast or herring sperm were tested on the callus tissue of the tobacco plant. They induced cell division. In 1955 the isolation of the active principle, which was called kinetin, was achieved.

Kinetin is an artefact. However, chemically related substances were later discovered which stimulated cell division just like kinetin. They were given the name cytokinins (cytokinesis = cell division). The first cytokinin, zeatin, was discovered in maize in 1964. Since then zeatin, IPA and other cytokinins have been detected in various other species, in part in hydrolysates of RNA. We shall return later to their occurrence in RNA.

Biosynthesis and degradation

The purine nucleus is very probably furnished by the usual route of purine biosynthesis (page 156). The origin of the substituents on the amino group was investigated for IPA. The isopentenyl residue is derived from mevalonic acid. The site of biosynthesis is the roots.

Up to the present very little is known about the degradation in higher plants. Apparently the purine nuclei can be converted into a number of other purine derivatives.

Several functions of the cytokinins

(1) *Cell Elongation and Cell Division*. According to several findings cytokinins can stimulate cell elongation. This is only a confirmation of what we have been able to deduce from the preceding sections: phytohormones act quite unspecifically on the most varied developmental processes. However as far as the cytokinins are concerned the stimulation of cell division is by far their most important effect.

(2) *Dormancy of Seeds*. Kinetin can break the dormancy of seeds. Thus, for example, the germination of lettuce seeds (*Lactuca sativa*) can be stimulated by radiation with red light. This effect of light will be of more interest to us later (page 256). The important point in this contest is that cytokinins can replace the red light: cytokinins stimulate the germination of lettuce seeds even in the dark. If the red light and cytokinins are administered together a synergistic enhancement of the germination-promoting effect is attained.

(3) *Apical Dominance*. We are already aware of the fact that IAA and the gibberellins are factors of apical dominance. They inhibit the outgrowth of side shoots. We also know that cytokinins stimulate the formation of shoots in tissue culture. We shall also be concerned with this effect later (page 246). Only one aspect is important here: the shoots that are induced in tissue cultures by kinetin do not exhibit mutually inhibiting effects during their development. Thus, no dominance of any kind is shown by the first-formed, older shoots over younger ones, for example. In intact plants it has been established that cytokinins universally promote the outgrowth of shoots, including that of side shoots. This promoting effect can be so strong that the side shoots develop fully and, in so doing, abolish apical dominance.

In the discussion of the gibberellins we learned that a fungus can yield gibberellins, i.e. factors which function as regulators in higher plants. A parallel situation is found with respect to the cytokinins. *Corynebacterium fascians* can induce truncated (fasciated) shoots in plants. Moreover, apical dominance is abolished, leading to the formation of "broomsticks" from shoots which have grown out to the same extent. In extracts of the bacterium Klaembt was able to detect cytokinins, the identification of which was achieved by other workers somewhat later. One of the cytokinins of the bacterium is IPA with which we are familiar. Thus, the phytopathogenic bacterium produces the same substances which are formed by higher plants and used by them, among other things, for the regulation of the growth of shoots.

(4) *Delay of Senescence*. Cytokinins can delay the aging of leaves. The mark of senescence is a visible degradation of chlorophyll which is accompanied by a degradation of proteins in the leaves. In 1957, Richmond and Lang noticed that the senescence of cut leaves of the Composite *Xanthium strumarium* could be delayed by a few days by means of kinetin.

Subsequently, the phenomenon of the delay of senescence was studied in great detail by Osborne and by Mothes and his colleagues among others.

Let us consider a few examples. If tobacco leaves are cut and placed in a moist chamber they gradually turn yellow. This yellowing can be delayed by kinetin. This basic experiment can be varied in different ways, for example one half of a cut tobacco leaf can be treated with kinetin and the other half be left untreated. The latter half turns yellow quickly whereas the treated one remains green very much longer (Fig. 170). Using isotopically labeled materials, it could be demonstrated that amino acids and other substances, e.g. auxins, flow from the aging areas of the leaf to the kinetin-treated areas. If a kinetin-treated leaf is still attached to the shoot, then these substances will be transported from the shoot system into the treated leaf via the leaf stalk. This is known as an *attraction* by kinetin. However, the substances, having once flowed in, are retained in the kinetin-treated areas. Attraction is accompanied by *retention*. That is to say kinetin — and naturally occurring cytokinins too — stimulate RNA and protein synthesis. C^{14}-Labeled amino acids that had been transported to the site of kinetin treatment are incorporated there into protein. The stimulation of RNA and protein synthesis by cytokinins is probably one of the fundamental causes of the delay of senescence.

A delay of senescence can also be produced in some cases with IAA, synthetic growth substances and gibberellic acid. However, the effect of these other phytohormones is much less pronounced. With the cytokinins the effect is, on the contrary, so striking that a few assay procedures for cytokinins are based on it.

(5) *Induction of Enzymes.* Stimulation of RNA and protein synthesis may mean an activation of the genetic material. Other findings also suggest the possibility of gene activation by cytokinins. Thus there are a few investigations that suggest that cytokinins can induce the synthesis of enzymes. One such case was discovered by Borris in the corncockle (*Agrostemma githago*). In dormant seeds of the corncockle, no nitrate reductase can be detected. The synthesis of the enzyme can be induced

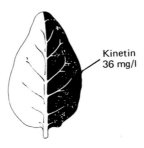

Kinetin
36 mg/l

Fig. 170. Delaying of senescence by kinetin. The half of an isolated tobacco leaf that has been treated with kinetin remains green longer than the untreated half (from Mothes as presented in Hess 1968).

by steeping embryos and then treating them with nitrate. This is a case of substrate induction, which has been demonstrated several times for nitrate reductase (page 186). However, cytokinins (kinetins and 6-benzylaminopurine) also induce enzyme synthesis. Inhibitors of transcription and translation counteract the effect of nitrate ions and cytokinins. Evidently, cytokinins can induce the *de novo* synthesis of the nitrate reductase of *Agrostemma* just like nitrate ions.

Cytokinins as components of tRNA

Kinetin, the first cytokinin, was first found in degraded DNA. It was shown to be an artefact. For example, kinetin is also obtained when adenine, a purine base of the nucleic acids, and furfuryl alcohol, which can arise in the acid hydrolysis of sugars, are autoclaved together. Nonetheless, it was subsequently shown that there is a close relationship between cytokinins and nucleic acids. The only point is that the nucleic acids concerned are certain types of tRNA, not DNA. Substances with cytokinin activity can be detected in hydrolysates of tRNA from quite different sources (bacteria, yeast, calf liver, and higher plants). The first cytokinin was identified in yeast tRNA in 1966. It was shown to be IPA. It was then shown to be present in the tRNA of a number of other organisms.

Now in any given cell there are many types of tRNA, usually more than one for each of the 20 protein amino acids. However, not all of them contain IPA. The serine tRNA of yeast, for example, contains IPA whereas alanine and phenylalanine tRNA from the same organism do not. In 1966, Zachau succeeded in localizing precisely the IPA of the serine tRNA of yeast. The IPA is located right next to the anticodon (Fig. 10). According to further investigations the cytokinin seems to be necessary for the attachment of the anticodon to the codon of the mRNA.

These findings might lead one to suspect that the mechanism of action of the cytokinins lies in their being incorporated into particular types of tRNA molecules and thus in an effect on translation. If this supposition were correct then exogenously added cytokinin ought to be incorporated into particular types of tRNA. Unfortunately, the few pieces of evidence relevant to this question, which are presently available, are contradictory. In some cases an incorporation of cytokinins into tRNA was demonstrated, in others not, although the cytokinins, which were supplied, were definitely active. What is more, there are findings which indicate that the complete cytokinins are not incorporated into tRNA. In the case of IPA residues present in tRNA it has been demonstrated several times that the isopentenyl residue is attached subsequently to an adenine moiety already present in the tRNA.

In summary, in the case of the cytokinins, the third group of phytohormones, close relationships to the nucleic acids are suspected. Evidently cytokinins can stimulate genetic activity, presumably at the level of translation. However, whether the frequently detected occurrence of cytokinins in tRNA has anything to do with the mechanism of action of the cytokinins is a matter which cannot be settled at the moment.

d. Abscisic Acid

Chemical constitution

Abscisic acid is a terpenoid (Fig. 171). Abscisic acid is the name recently proposed. In addition, the older names dormin and abscisin II are still in use. "Dormin" refers to the fact that abscisic acid brings about the dormancy of buds and "abscisin" to the fact that abscisic acid can promote the shedding of leaves and fruit. Abscisic acid seems to be ubiquitous in the plant kingdom. Chemically closely related substances exhibiting, to some extent, a similar effect have also been detected repeatedly.

History, method of assay

Different findings had led to the suspicion that there must be certain substances peculiar to plants that can promote the shedding of leaves and cause the dormancy of buds on deciduous trees. A team of workers led by Carns and Addicott worked on the physiology of the shedding of bolls of the cotton plant. In 1963, a substance was isolated which stimulated capsule shedding and, in 1965, its chemical structure was elucidated. The substance was given the name abscisin II. Simultaneously another team of workers led by Wareing and Cornforth were engaged in the isolation and identification of the principle that can transfer the buds of deciduous trees into the dormant state. The substance was named dormin and it was isolated from the maple. It turned out that dormin was identical with abscisin II.

The tests for abscisic acid (= abscision II = dormin) are based on the inhibition of growth, the stimulation of the shedding of leaves and fruit, the induction of the dormancy of buds and the inhibition of seed germination, to mention only the most important.

Biosynthesis

The biosynthesis of abscisic acid has still not been elucidated. Several pieces of evidence suggest that it could arise by degradation from the

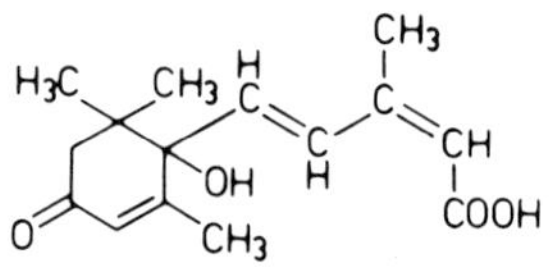

Fig. 171. Abscisic acid.

carotene zeaxanthin. Other findings, however, are compatible with the view that abscisic acid is furnished directly by the known mevalonate—isopentenyl pyrophosphate pathway—and not indirectly via a carotenoid. However, the gibberellins are also derived from the same pathway (page 110). We shall see presently that abscisic acid is an antagonist of gibberellic acid, among other compounds. A switch over of the biosynthetic route from abscisic acid to gibberellic acid or vice versa would alter the relative concentrations of the two substances and this would have consequences for morphogenesis.

Several functions of abscisic acid

The phytohormones discussed so far (indole bodies, gibberellins, and cytokinins) are invested with a positive feature. Apart from a few exceptions, they promote particular processes. Abscisic acid on the other hand is very definitely an inhibitor and thus could be characterized by a negative sign. In many cases abscisic acid acts contrary to the "positive" hormones. It is then an antagonist of the "positive" hormones. The adjustment of the equilibrium between abscisic acid and the three other phytohormones determines whether a particular morphogenic process can occur or not. If abscisic acid dominates, then the developmental process does not take place. If the three "positive" phytohormones gain the upper hand then the development in question does occur. It is hardly necessary to mention that this picture is very much oversimplified. We shall see later, for example, that equilibria are also attained between the three "positive" phytohormone groups which can be of decisive importance for the kind of morphogenesis, e.g. root- and shoot-formation (page 246).

Let us consider just one example of such an equilibrium adjustment and its morphogenetic consequences (Fig. 167). In the autumn the amount of gibberellins in the buds of our deciduous trees decreases and that of abscisic acid increases. This is primarily an effect of the short daytime. In January hardly any more gibberellin is to be found in the buds of the black currant (*Ribes nigrum*) whereas considerable amounts of an inhibitor of growth are present. The inhibitor in *Ribes* is probably, and in other deciduous trees with certainty, abscisic acid. Under the influence of the cold of winter and the increasing length of day the content of inhibitor is reduced and that of gibberellin is increased. When the content of gibberellin has finally become high enough to overcome the inhibitor, the forcing of buds in spring begins.

In Table 7 the effects of abscisic acid, IAA, gibberellins, and cytokinins on certain morphogenetic processes are contrasted with each other. The tabular summary has been considerably simplified. Nevertheless the principle is not affected by this. Abscisin counteracts the other phytohormones in quite different morphogeneses. Perhaps one and the same mechanism lies behind this diversity: abscisic acid inhibits transcription and translation, i.e. the genetic activity, as has been shown in various systems. However, the three positive phytohormone groups can promote the activity of at least certain genes. The antagonism mentioned would be intelligible on this molecular basis. However, completely convincing experimental evidence has yet to be obtained.

e. Synthetic regulators

A large number of synthetic substances is known, some of which function as growth substances, others as inhibitors. In practice they are often preferred to the naturally occurring regulators. A few of them will be mentioned.

Synthetic growth substances (Fig. 172)

α-Naphthylacetic acid is contained in many commercial preparations instead of the labile IAA. A few of its applications are root formation in cuttings, inhibition of the germination of potatoes, induction of flowering in pineapple plants, which are ripe to flower. *2,4-Dichlorophenoxyacetic acid (2, 4-D)*: in many weed killers. Dicotyledons are induced to grow abnormally, e.g. formation of roots bearing shoots, but inhibition of the main root system; serious deformation. Monocotyledons, including the cereal plants, are very much less sensitive. Evidently they can inactivate 2,4D by a mechanism which is still a matter of dispute. Thus, it is possible to eradicate the dicotyledonous weeds in a cereal crop by treatment with 2, 4-D without seriously damaging the monocotyledonous cereals. A further application of 2, 4-D: inhibition of the fall of fruit from apple, pear, orange and cotton trees.

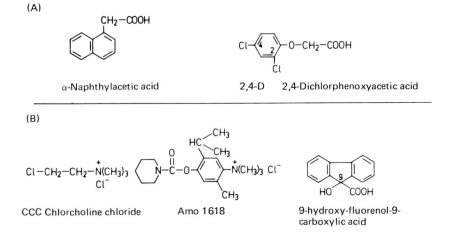

Fig. 172. Synthetic growth substances (A) and inhibitors (B).

Synthetic inhibitors (Fig. 172)

Chlorocholine chloride (CCC) and *Amo 1618* block the synthesis of gibberellins and thus inhibit growth. Cereals treated with CCC have shorter
internodes and thus are more robust. With Amo 1618 one can produce
chrysanthemums with short stalks, for example. In several cases flower
formation can be accelerated by CCC and Amo 1618.

Morphactins are a group of synthetic regulators the effect of which is
predominantly inhibitory and which possess the skeleton of 9-hydrox-
yfluorene-9-carboxylic acid (= fluorenol). The morphactins are not structural analogs of the gibberellins in spite of the fact that, superficially, their
chemical constitutions are similar. The two aromatic rings which flank the
central five-membered ring give the morphactins a disc-like structure. On
the other hand, the gibberellins with their almost saturated C skeleton exhibit an almost spherical shape. Thus, the steric configuration of the two
groups of substances is totally different.

The morphactins derive their name from the fact that they are
morphogenetically active. In the main they are inhibitors. Thus, they inhibit the germination of seeds and the growth of the seedling, the
longitudinal growth of shoots and the development of the leaf blade (Fig.
173). Apical dominance is abolished. As a result of this the side shoots can
continue to grow, leading to a broomlike form of growth. The sprouting of
rosette plants is also inhibited. In practice morphactins are used particularly in conjunction with other synthetic regulators. Thus, fluorenol and
phenoxy growth substances such as 2,4-D act synergistically and so the in-

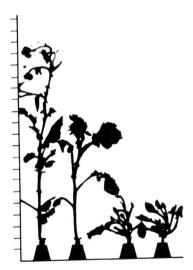

Fig. 173. The effect of chlorofluorenol, a morphactin, on the development of the
rope plant. On the left is the control. The further the other plants are to the right,
the more morphactin they have received (modified from Mohr and Ziegler 1969).

hibitory effect in combined preparations is boosted. Such combinations of fluorenol and 2, 4-D are employed in weed control in cereal crops and on pasture land. Combinations of morphactins and maleic acid hydrazide are used for "growth suppression" on pasture land. Maleic acid hydrazide inhibits especially the growth of grasses and a chlorinated fluorenol derivative that of weeds. The result of this treatment then is lawns with low growth, which obviate the necessity of repeated mowing.

f. Phytohormones and gene activity

The phytohormones intervene in the most varied developmental processes in a regulatory manner. To all appearances indole bodies, gibberellins, and cytokinins are quite unspecific inducers and abscisins just as unspecific inhibitors of morphogenesis. The three "positive" phytohormone groups can demonstrably activate gene material, and the "negative" abscisic acid represses it. All hypotheses concerning the mechanism of action of the phytohormones must do justice to these facts (unspecific action and activation of genetic material). A few of these hypotheses will be given here in simplified form. In this connection we must remember that we have considered four states of activity of genes to be possible (page 181).

(1) *General Activation or Inactivation of Genes* (Fig. 174). "Positive" phytohormones activate all activatable genes in a given tissue and phase of development, "negative" phytohormones such as abscisic acid inactivate all activatable genes in a given tissue and phase of development.

(2) *Activation or Inactivation of only One Gene by Phytohormone, Further Secondary Activations or Inactivations* (Fig. 175). A "positive" phytohormone activates only one gene locus. A "negative" phytohormone likewise inactivates only one gene locus. The gene activation or inactivation brings about a change in the metabolism of the cell. As a consequence of this further genes can now be secondarily activated or inactivated.

(3) *Stimulatory or Inhibitory Intervention in a Metabolic Reaction, Secondary Gene Activations or Inactivations* (Fig. 176). The primary mechanism of action does not consist in influencing the gene activity but rather in regulating one particular metabolic reaction. Owing to the interdependence of all metabolic reactions, a readjustment of cellular metabolism is thus brought about as in hypothesis 2 and this can then bring activations and inactivations of genes in its train.

Now let us call to mind the Jacob-Monod model (page 183). According to it, the activation of a gene implies the inactivation of a repressor and the inactivation of a gene means the activation of a repressor. If hormones are to regulate gene activities according to hypotheses 1 and 2, that must happen by appropriate modulation of repressors. One can imagine that hormones might act as effectors just like substrates and end products which alter the steric configuration, and thus the activity, of repressors.

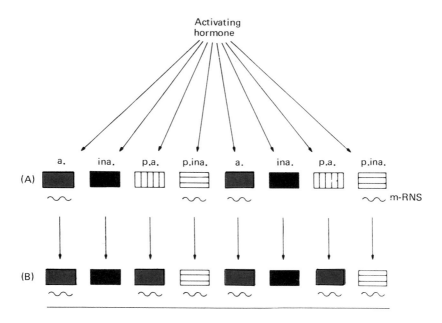

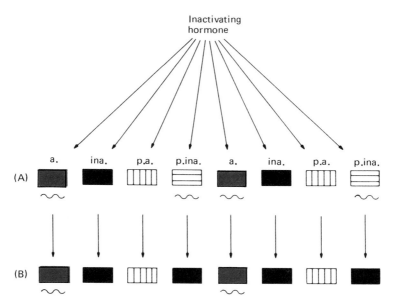

Fig. 174. A model representation of how phytohormones act: general activation or inactivation of all potentially active or inactive genes. (A) before, (B) after, the phytohormone has acted. States of activity of the genes: a = active, ina = inactive, p.a. = potentially active (activable), p. ina. = potentially inactive (repressible) (modified from Hess 1968).

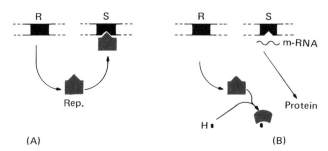

Fig. 175. Model of how a "positive" phytohormone acts: only one structural gene is activated by the inactivation of its repressor Rep. Phytohormone H acts as effector. R = regulator gene, S = structural gene (modified from Hess 1969).

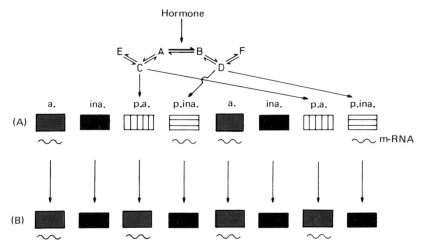

Fig. 176. Model of how phytohormones act: by influencing one, central metabolic reaction and then secondarily by furnishing or removing effectors which can activate or repress a whole spectrum of genes.

There is, however, another possibility. We have discussed that in all probability histones are repressors of genetic activity in higher organisms. Now Fellenberg has demonstrated that "positive" phytohormones, IAA, gibberellic acid, and kinetin, loosen the binding between histones and DNA. If the histone repressors were removed from DNA in this manner transcription could begin on the DNA now exposed.

If we had to choose between the three hypotheses we would settle for hypothesis 2 or 3. This is because the "positive" phytohormones can also repress genetic material occasionally. Thus, under certain conditions, IAA can repress the appearance of a peroxidase in tissue cultures of the tobacco plant. Repressions of this kind are understandable since the most varied effectors can be furnished via the reactions which are secondarily associated with them, including those which can activate particular repressors and thereby inactivate the corresponding gene loci. Above all, the hypotheses 2 and 3 do full justice to the fact that, in the living cell, no reaction can occur without influencing a large number of other systems.

The hypothesis concerning the "second messenger"

In recent years findings have, in actual fact, accumulated that indicate that many hormones exert their effect indirectly, corresponding, in principle, to our hypothesis 3. This is true not only for modulations of gene activity but also for other hormone effects, for instance, on membrane systems or allosteric enzymes. Following experiments on animal hormones by Sutherland in particular, the hypothesis of the *second messenger* was advanced. It states that hormones, the primary messengers, induce the formation of a second messenger. This second messenger is responsible for the observed "hormonal effects" rather than the hormones themselves (Fig. 176a).

Cyclic Adenosine Monophosphate (cAMP). According to our present state of knowledge the most important second messenger is cyclic 3', 5'-adenosine monophosphate (Fig. 176b). The primary messenger, i.e. the hormone, stimulates the activity of adenyl cyclase which is usually membrane-bound and which forms cAMP from ATP with the cleavage of pyrophosphate. cAMP in turn stimulates the activity of various enzymes. We will mention only one example: cAMP stimulates the activity of kinases that are able to phosphorylate certain proteins, including histones. Phosphorylation of histones reduces their inhibitory effect on transcription, at least in an *in vitro* experiment. Thus, in this way a hormone could stimulate gene activity via the second messenger cAMP.

The endogenous level of cAMP can be regulated either by activation of adenyl cyclase as described, which leads to stimulation of cAMP synthesis, or by degradation of cAMP by a phosphodiesterase, which is specific for one of the phosphate ester bonds in the molecule. The activity of the phosphodiesterase is, in turn, also subject to regulation. Thus, for example, a protein inhibitor of the phosphodiesterase has been detected in slime molds (*Acrasiales*). In some *Acrasiales* such as *Dictyostelium discoideum* cAMP is identical with acrasin, a chemotactic substance which plays an important role in the aggregation of the myxamoeba to pseudoplasmodia.

Now an example of the effectiveness of cAMP in higher plants: cAMP can replace gibberellic acid in that exemplary case already alluded to (p. 204), the synthesis of α-amylase in the aleuron of barley grains. Hence it is assumed that cAMP acts as second messenger here and that its synthesis is stimulated by gibberellic acid.

Ethylene. It seems likely that ethylene is another second messenger, typical of plant material. It can be formed in all cells, even if in varying concentrations. In most instances in higher plants the amino acid methionine is the starting material for the synthesis (Fig. 176c). Special

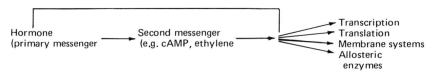

Fig. 176(A). The second messenger hypothesis (from Zenk 1970). The possibility of a direct action of the hormone without the intervention of a secondary messenger was also taken into account.

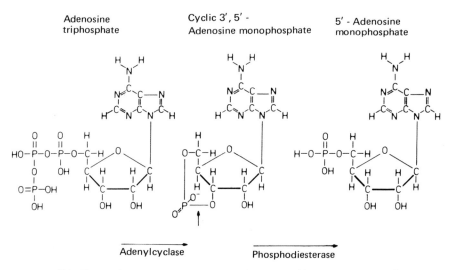

Fig. 176(B). Formation and degradation of cAMP (from Zenk 1970). The phosphodiesterase cleaves the bond shown by the arrow.

Fig. 176(C). The formation of ethylene from methionine.

degradative mechanisms are not necessary since the gaseous compound can simply be released into the atmosphere. We have already mentioned that ethylene, the formation of which is induced by IAA, seems to be the real inhibitory substance in apical dominance (page 199). The same holds true for many other kinds of morphogenesis for which one example must suffice. When growth substances such as α-naphthylacetic acid trigger flower formation in the pineapple, it is ethylene that is the immediate, active principle. In all cases like this, then, the growth substances such as IAA may be regarded as the primary messengers and ethylene as the second messenger. How ethylene exerts its effect is unknown. A direct effect on gene activity is in any case very improbable.

D. Regulation by External Factors

Up to now we have spoken only of regulation by internal factors. However, the situation within an organism is influenced quite substantially by external factors. Of the many external factors which can influence the growth and differentiation of plants, temperature and light are selected here. Moreover, we can point out only a few of the fundamentals relating to the mechanism of action of temperature and light in this section.

1. Temperature

Each metabolic reaction is characterized by a temperature coefficient which specifies how much the velocity of the reaction varies with a specified temperature shift. The temperature coefficient Q_{10} specifies by how much the reaction velocity increases for an increase of temperature of 10°C , e.g. an increase from 0° to 10°:

$$Q_{10} = \text{velocity at } 10°/\text{velocity at } \pm 0°$$

In higher plants Q_{10} has an average value of 2–3. The temperature optimum usually lies in the range 28–32°. Temperatures above 35° begin to be harmful. Since the individual reactions possess slightly different temperature coefficients, a change of temperature can favor one reaction and, at the same time, discriminate against another. In this way temperature can engage in a regulatory manner in metabolism and thus in the process of development too. Now we come to a complication. At each state of development many reactions occur, which have temperature coefficients that differ from each other to a greater or lesser extent. This leads to there being a mean temperature that is optimal for the stage of development in question. For another stage of development a quite different temperature can be optimal. Very precise studies of this point were carried out in the 1930s by Blaauw in the Netherlands. The aim was to ascertain the conditions under which bulbous plants bloomed most quickly. For this purpose bulbs were exposed to different temperatures after being harvested. It was found that each developmental process proceeds optimally at a particular temperature. The temperature optima for the individual steps are different from species to species, for hyacinths they are quite different from those for tulips, for example.

Finally, it should not be forgotten that in nature plants are not subject to the operation of single external factors but to combinations of them, e.g. combinations of particular temperature and light conditions. The flowers of the *Petunia* hybrids "Kriemhilde" are pure white if the plants are kept for 64 days in bright sunlight and at 20°. If the period of time under these conditions (bright sunlight and 20°) is shortened, anthocyanin is formed in increasing amount. After 30 days under the conditions specified only white sectors still remain in each petal. Even these can be eliminated: if the plants are kept in the shade at 30° the flowers become completely violet. Thus, in this case two external factors, light and temperature, cooperate in the formation of the character anthocyanin (Fig. 177).

It is safe to assume that gene activity is one of those metabolic processes that are influenced by temperature. This is supported by the fact that, for example, the RNA production on plant giant chromosomes (page 176) can be curbed by lower temperatures. However, experimental evidence of the same precision as that which has been obtained for the relationship between light and gene activity is not yet available.

2. Light

a. The phytochrome system

Light can intervene in metabolism via photosynthesis. However, apart from this universally given mechanism light can also regulate metabolic, and thus developmental processes, via other pigment systems and other mechanisms. A light-controlled developmental process is called *photomorphogenesis.* Correspondingly, the photoreceptors are designated as morphogenetically effective or morphogenetic pigment systems. Although it is not the only one a morphogenetic pigment system that has been intensively investigated in the last ten years is the phytochrome system.

The germination of seeds of many species is promoted by light (page 256). When the action spectrum of light for the germination of achenes of the lettuce (*Lactuca sativa*) was compiled, it was found that red light (RL) promoted germination whereas far red light (FRL) on the other hand in-

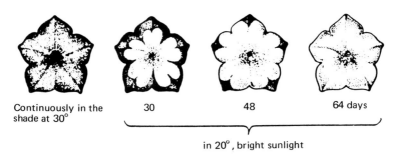

Continuously in the shade at 30°   30   48   64 days

in 20°, bright sunlight

Fig. 177. Dependence of anthocyanin synthesis in the variety "Kriemhilde" of *Petunia hybrida* on light and temperature (from Kühn 1965).

hibited it. However, Borthwick and Hendricks who have made a particularly detailed study of the germination of lettuce, advanced our knowledge a step further in the early 1950s. If achenes of the lettuce were first irradiated with RL, and then FRL, the stimulatory effect of RL was annulled. If the achenes were irradiated with RL, then FRL, and then RL again, they germinated (Fig. 178). This trick can be repeated many times. What was crucial for germination was the quality of the light which was last given: RL meant germination, FRL inhibition of germination. Similar effects have been found in other plants and with other developmental processes, e.g. in flower formation (page 298). The maximum effect of RL was found at 660 mμ and that of FRL at 730 mμ. The inference drawn from all of these data was as follows (Fig. 179). In the cells of plants a morphogenetic pigment system exists which can occur in two different states which are mutually interconvertible by light of specified wavelengths. One state has an absorption maximum at 660 mμ and is appropriately called P_{660}. It is physiologically inactive. Irradiation with RL of wavelength 660 mμ converts it into the second state. This second state has an absorption maximum at 730 mμ and, appropriately, is called P_{730}. It is physiologically active and initiates the most varied morphogeneses, such as the germination mentioned above. P_{730} is again converted into P_{660} by irradiation with FRL of wavelength 730 mμ. Owing to this oscillation between two states the system has also been known as a "reversible red-far red pigment system," a designation which today has been almost completely superseded by the shorter "phytochrome system."

| RL | RL-FRL-RL | RL-FRL-RL-FRL-RL | RL-FRL-RL-FRL-RL-FRL-RL |

| RL-FRL | RL-FRL-RL-FRL | RL-FRL-RL-FRL-RL-FRL | RL-FRL-RL-FRL-RL-FRL-RL-FRL |

Fig. 178. Proof that the phytochrome system is a factor in the germination of the achenes of the lettuce *(Lactuca sativa)*. RL = red light, FRL = far red light (from Goodwin 1965).

The phytochrome system is widely distributed in plants, even if in small concentrations. It can be extracted without too much difficulty, particularly from seedlings. Phytochrome suspended in a test-tube can also be converted from one form into the other by irradiation with RL or FRL. It has been shown that phytochrome is a chromoproteid. The pigment bound to protein in this chromoproteid is related to certain pigments of blue algae and algae. It is a phycobilin, thus, in principle, a chain of four pyrrole rings which are linked with each other by C atoms (Fig. 180).

b. Phytochrome system and gene activation

A question which is still disputed concerns the mechanism of action of the phytochrome system. In 1965, Mohr established that light is capable of activating genes via the phytochrome system. His experiments were carried out on seedlings of the white-seeded mustard (*Sinapis alba*). Such seedlings form anthocyanins after irradiation. Owing to special circumstances it is FRL and not RL that can induce anthocyanin synthesis in this case. The light-dependent anthocyanin synthesis can be stopped by actinomycin C_1, an inhibitor of transcription which is already familiar to us (Fig. 181). However, actinomycin is only fully effective if it is given before irradiation is begun or right at the beginning of the light period. If actinomycin is added later, for example 6 hours after the start of irradiation, it is capable only of delaying anthocyanin synthesis but it cannot stop it. Thus, the crucial transcription processes occur right at the beginning of the light period.

An unspecific secondary effect of actinomycin can be excluded. If this were the case then actinomycin given later, e.g. 6 hours after the beginning of irradiation, should also have had an inhibitory effect. The result of the experiment, which has been supplemented by additional data, provided the first experimental evidence that the phytochrome system is capable of activating genetic material, in this case the genes for anthocyanin synthesis.

Anthocyanin synthesis is a very complicated process in which many genes are involved (page 132). Various authors such as Zucker, Mohr, and Zenk discovered in the 1960s that light is capable of stimulating the synthesis of the phenylalanine-ammonium-lyase via the phytochrome system. Thus, light activates the genetic material for the synthesis of a key

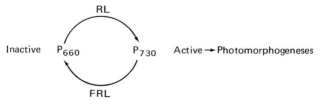

Fig. 179. Function of the phytochrome system.

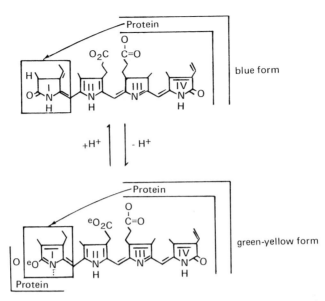

Fig. 180. A proposal for the structures of P_{660} (blue form of phytochrome) and of P_{730} (yellow-green form) and their binding to the carrier protein (from Rüdiger 1972).

enzyme of phenylpropane metabolism (page 122). Anthocyanins, too, those hybrid substances of active acetate and active cinnamic acid derivatives, can only be formed if PAL is present and active. A chain of reactions extends from PAL to the anthocyanins. Initial findings indicate that the synthesis of the regulatory enzymes is induced partly by light and partly by the appropriate substrates.

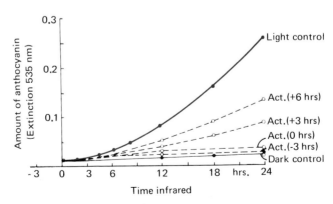

Fig. 181. Participation of the genetic material in anthocyanin synthesis in seedlings of the white-seeded mustard *(Sinapis alba)*, a type of character formation which is controlled by phytochrome. Act = addition of actinomycin C_1 (modified from Lange and Mohr 1965).

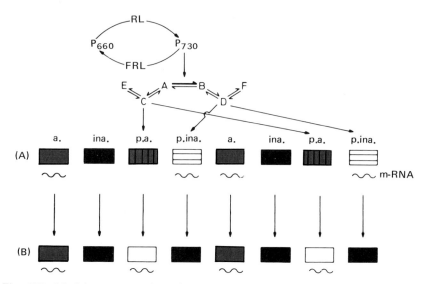

Fig. 182. Model representation of the relation between phytochrome and the genetic material (modified from Hess 1968).

Meanwhile an activation of genetic material has been demonstrated for a large number of other photomorphogeneses. Even so the mechanism of action of phytochrome has, strictly speaking, not been elucidated. It is still not known whether the chain of reactions controlled by phytochrome acts directly on the genetic material or not. According to the data presently available an indirect influence on the genetic material seems more probable (Fig. 182). The phytochrome system is cofactor in one central metabolic reaction (or perhaps several). The nature of this central reaction is unknown. It is however coupled with other reactions and, if the central reaction is influenced in any way via the phytochrome, this will have repercussions on a whole network of reactions. As a consequence changes are brought about in the concentrations of the most diverse effectors that can affect the genetic material by either activating or inhibiting it. This concept does justice to the fact that genetic material cannot only be activated but also repressed via the phytochrome system, as has been shown repeatedly. In retrospect we may notice a parallel between the external factors light and temperature and the phytohormones with respect to their mechanism of action. Both the external factors and the phytohormones can affect the state of activity of the genetic material. In both cases it is unknown whether, and to what extent, this control of gene activity has to be construed as a primary or secondary event.

Polarity And Unequal Cell Division As Fundamentals Of Differentiation

The last few chapters appear to be of purely academic interest and, moreover, rather too hypothetical. We now hasten to show that we are still in the thick of the events of differentiation.

Our fundamental problem—expressed now in a somewhat different way—was this: a cell enters the mitotic cycle. Of the two daughter cells one or indeed both do not pass through a new mitotoic cycle but begin to differentiate instead (Fig. 183). Our question is: what happened during or before the last division such that one or even both cells do not divide again, and what relationship do the regulatory mechanisms just discussed have to this event?

A. Polarity

Polarity is the term used to describe the fact that differences can arise along an axis such that one end of the axis is different from the other. Considered in detail one can distinguish between several alternatives of polar differentiation (Fig. 184).

Polarity is found in single cells as well as in multicellular units. To be sure the alga *Acetabularia* is an extreme case of a single cell with pronounced polarity. However, the oocyte in the embryo sac is also polarized, and it is polarized by its immediate surroundings, the embryo sac. The latter is polarized in turn by its position in the diploid organism. This point brings us to multicellular, polarized systems. A familiar example is a piece cut out of a willow branch (Fig. 185). Shoot-bearing roots form at its lower end but not at the upper. If the twig is rotated through 180° roots form on what was originally the "lower" end, even though this end is now uppermost. A special form of polarity is called dorsiventrality, e.g. the difference between the upper and under side of leaves.

There are hardly any cells or multicellular systems that are not polarized. One of the few exceptions are the zygotes of the brown alga of the genus *Fucus*. They do not possess a preformed polarity. As a consequence it has been possible to assess a few of the factors which can affect polarity by doing suitable experiments on them. Temperature, light, pH

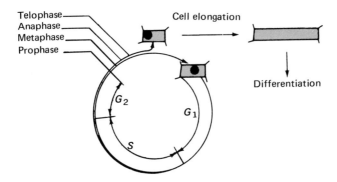

Fig. 183. The fundamental problem of differentiation.

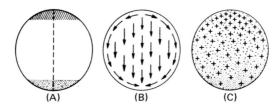

Fig. 184. Possible modes of polar differentiation. (A) Polar field polarity: a concentration of different materials at the poles of the cell. (B) Structural, directional polarity: polar directed structures pass through the interiors and/or around the periphery. (C) Gradient polarity: material gradients between the poles (from Kuhn 1965).

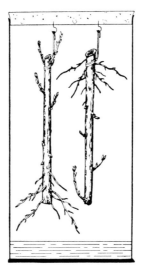

Fig. 185. A demonstration of polarity provided by the capability of a willow twig to regenerate. Normal orientation on the left, upside down on the right. Both twigs are in a moist chamber (from Walter 1962).

value, different chemical gradients and neighboring *Fucus* zygotes all play a role. In higher plants gravity is likely to be an additional factor.

As far as the material basis of polarity is concerned, one can visualize fixed structures as well as gradients of particular substances (Fig. 184).

B. Unequal Cell Division

Now let us go back to the two daughter cells, one of which does not divide any more but begins to differentiate. They arise from a parent cell which lies within a meristem such as the meristem of the shoot or root tip in higher plants. Now this parent cell is already polarized owing to its position in the meristem and possesses two poles which are different from each other. This fact alone, though, would not bring a solution to our problem any nearer. But what if the division to give these two daughter cells occurred in such a way that the new wall lay perpendicular to the axis of polarity? Two daughter cells would then result that are different from each other. A division of this kind would be unequal, not in respect to the chromosomal heredity factors but in respect to the structures and substances distributed along the axis of polarity, among which would be the cytoplasmic heredity factors. In this instance one speaks of an unequal cell division (Fig. 186).

Polarity and unequal cell division lead to daughter cells that are materially different from each other. Once cells are present that are different from each other we can also understand how different metabolic processes can be initiated or inhibited in each cell by the regulatory mechanisms that are familiar to us—including also the regulation of gene activity. This is because different regulators are available in each of these cells. The conditions within the two cells can diverge to such an extent that one can enter the mitotic cycle again whereas the other begins to differentiate.

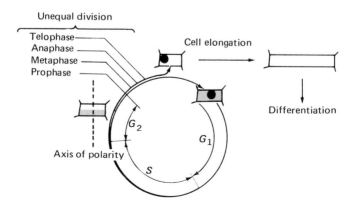

Fig. 186. Illustration of an unequal cell division.

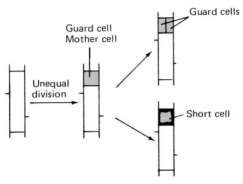

Fig. 187. Unequal cell division in the development of the stomatal apparatus and short cells in monocotyledons.

Let us now consider a few examples of unequal cell division, which illustrate the principle put forward by Bunning in particular. We shall see from the nature of the examples that the occurrence of unequal cell division is in no way limited to apical meristems.

1. Development of Stomata

The developmental history of the stomata used to be a favorite memory exercise for students. A large number of different kinds is known, all of which have been given high sounding names. However, it is a fact that a "primordial cell" always divides unequally. Usually the smaller, plasma-richer daughter cell which is endowed with a larger nucleus then passes through numerous divisions until finally the guard cell mother cell appears which divides to give the two guard cells.

The situation is at its simplest in many monocotyledons (Fig. 187). A primordial cell divides unequally into a larger and a smaller, plasma-richer daughter cell. In this case the smaller daughter cell is identical with the guard cell mother cell: it divides longitudinally to give the two guard cells. This longitudinal division need not take place, in which case "short cells" arise and they can undergo a special kind of differentiation. Thus considerable quantities of silicic acid can be stored in them.

2. Root Hair Formation

In some species every or almost every cell of the rhizodermis can develop into a root hair. In many cases, however, root hairs are formed only from quite special cells of the young rhizodermis, the trichoblasts. These trichoblasts arise by unequal cell division: a cell of the young rhizoepidermis divides into a larger and a smaller, plasma-rich cell. The fate of the larger cell is varied. It can still divide several times but it is also

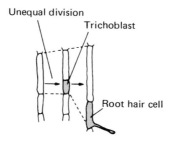

Fig. 188. Unequal cell division in root hair formation of *Phleum* (modified from Torrey 1968).

possible for it not to divide any more but to undergo considerable longitudinal growth instead, as is the case with *Phleum*. The trichoblast does not divide any more—however, instead of mitosis it can pass through several endomitoses—and its longitudinal growth is far less. Instead the root hair sprouts from it (Fig. 188). Cytochemical methods show the presence of more RNA and protein in the trichoblast then in the remaining cells of the rhizodermis. A comparative cytochemical study of some enzymes has also been made. In *Phleum* the trichoblasts are distinguished by a higher activity of acid phosphatases and peroxidases. Since the activity of other enzymes such as glucose-6-phosphatase in trichoblasts is not different from that in other rhizodermal cells, it is assumed that the two enzyme systems mentioned are related in some way to the formation of root hairs.

3. Pollen Mitosis

Haploid single cell pollen grains are immediate products of meiosis in pollen sacs. In a subsequent mitotic division, the first pollen mitosis, which occurs sooner or later, one vegetative and one generative cell arise from each pollen cell. This first pollen mitosis is unequal. The vegetative cell is larger and carries more ribonucleoprotein in its cytoplasm. Its nucleus is large and diffuse. The generative cell is very much smaller, carries less ribonucleoprotein in its cytoplasm and possesses a small but tightly packed nucleus (Fig. 189). It is often completely surrounded by the vegetative cell. The vegetative cell on the stigmata develops into the pollen tube whereas the generative cell passes through a second pollen mitosis which leads to two sperm cells or nuclei.

Sauter and Marquardt followed RNA and protein synthesis as well as changes in the stock of histones in more detail in the two cells in peonies. They found active RNA and protein synthesis in the vegetative cells. Their cytoplasm produced a strong coloration due to arginine-rich histones which were apparently localized on the ribosomes. In the generative cells practically no RNA and protein synthesis could be detected. The

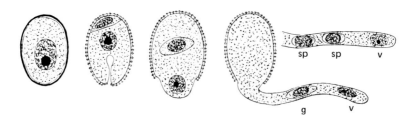

Fig. 189. Unequal cell division in the first pollen mitosis of lilies and further pollen development. g = generative, v = vegetative cell. The generative cell undergoes a second mitosis to give two sperm cells or nuclei (sp) which first occurs during outgrowth of the pollen tube (from Walter 1962).

nuclei of the generative cells showed a strong coloration due to lysine-rich histones, which was not shown by the nuclei of the vegetative cells. Perhaps these lysine-rich histones function as repressors of genetic activity and thus of RNA and protein synthesis.

Examples of unequal cell division can easily be multiplied. It must suffice simply to mention the formation of sieve tubes and accompanying cells in angiosperms, of elaters and spore mother cells in liverworts, and of hyalin and chlorophyll cells in peat mosses. After this demonstration of conspicuous cases of unequal cell division, let us return to the problem we raised at the beginning of the chapter, to cell divisions which are unequal in the sense that one of the two daughter cells can continue to divide whereas the other begins to differentiate. The first manifestation of differentiation is usually elongation of the cell in question.

Cell Elongation

A. The Phenomenon

Growth can be defined as an irreversible increase in volume. Then *elongation* would be an *irreversible increase in volume along a particular axis.* This definition is usually, but not always, applicable.

The externally visible growth of plants is based mainly on the elongation of cells which are supplied by the meristem zones of cell division. One example of the dimensions of longitudinal growth must suffice: in the entire zone of elongation of the maize root an increase of circa 20% per hour has been recorded. However, elongation is not uniform within this zone. In a major zone of elongation the increase amounts to 40% per hour whereas in the remaining zones it is much less. In absolute terms the rate of longitudinal growth lies in the range of a few μ/min, apart from exceptions such as certain grasses or pollen tubes. We have just spoken of a major zone of elongation. This brings us to a phenomenon, which was already known to the plant physiologist Julius Sachs (1832–1897) and which is paraphrased as a "major period of growth." What happens is as follows: at first each individual cell extends slightly and then maximally until longitudinal growth finally ceases. Thus each cell actually does pass

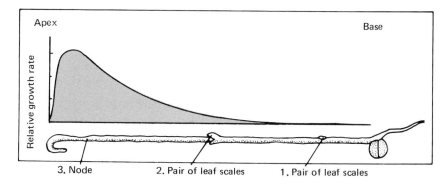

Fig. 190. "Great period of growth" of an etiolated pea seedling. The etiolated seedling only forms leaf scales rather than a completely functional leaf organ (from Torrey 1968).

through a period of growth. What is true for the individual cells is also true of the organ which they constitute. One finds that they have a major zone of elongation, that region in which the cells furnished by the meristem elongate maximally (Fig. 190).

B. The Process of Elongation Within a Cell

1. The Suction Pressure Equation of the Cell

It is common knowledge that if two sucrose solutions of different concentrations are carefully layered over each other in a glass vessel, an equalization of the concentrations by diffusion will occur. Molecules of the solvent, water, diffuse into the more concentrated sucrose solution and molecules of sucrose diffuse into the less concentrated solution until, finally, a uniform concentration has been attained. Diffusion of this kind can also occur through a semipermeable membrane. One speaks then of *osmosis.* The term "semipermeable" means that the membranes of biological systems are not equally permeable to all molecules. It can happen that a given biological membrane may allow the solvent to pass through it but not the substance dissolved in it. A membrane of this kind is designated *semipermeable.*

The plant cell is an osmotic system. Two semipermeable membranes are particularly important for its water management, the plasmalemma directly inside the cell wall and the tonoplast that surrounds the vacuole. The cytoplasm with its organelles lies between the two. It should not be forgotten that the cytoplasm certainly has its share in the water economy of the cell. One needs only to point to the potentiality for swelling of proteins. However, the content of the vacuoles is far more critical. This is because their cell sap is an often highly concentrated solution of sugars, glycosides, organic acids, and sometimes inorganic salts too, to mention only a few of the osmotically effective components of the cell sap. The semipermeable membranes of the cell allow water to pass through but not the osmotically effective substances just mentioned, or, at least, only to a slight extent. Let us now put our cell in water or in an aqueous solution of low concentration. Water molecules will then be drawn directly into the vacuoles in the process of equalizing the concentration. Hence one speaks of a suction pressure of the cell sap. The cell dilates as a result of water uptake by the vacuoles and develops a turgor pressure, a turgescence. Of course, the volume does not increase indefinitely until the concentration of the cell sap and the external medium is the same. The flexibility of the cell walls and also the pressure of the surrounding cells and tissues oppose the turgor and finally put an end to further dilation, even though the concentrations have still not been equalized. This situation is summed up in the *suction pressure equation of the cell:*

$$S_c = S_i - W$$

or, if one wishes to take into account the opposing pressure of the surrounding tissue:

$$S_c = S_i - (W + E).$$

S_c is the suction pressure of the whole cell, S_i is the suction pressure or osmotic pressure of the cell sap, W is the opposing pressure of the cell wall, and E is the external pressure of the surrounding cells. The events in the cytoplasm can be fairly safely ignored.

It is easy to convince oneself of the fact that the plant cell really is an osmotic system of this kind. This can be done by bringing about *plasmolysis*. We shall carry out our experiments on *Rhoeo discolor* from the leaves of which the lower epidermis has been removed. The lumen of the epidermal cells are simply put in water. Now the contents of the vacuole, red by dissolved anthocyanins and the cytoplasm is confined to a thin partition which is invisible in the average microscope. Now let us put the epidermal cells in a solution with a higher osmotic pressure than that of the cell sap, e.g. in glycerol. Water will then diffuse out of the vacuole, through the plasma tube and into the external medium. Owing to the loss of water the vacuole becomes smaller, the cell wall is no longer dilated by the turgor of the cell contents and the external medium penetrates through the cell wall. The plasma membrane detaches itself from the cell wall, being drawn away by the vacuole which becomes smaller and smaller (Fig. 191). Since the plasma membrane detaches itself from the cell wall, the process is known as plasmolysis. If plasmolysis is not continued for too long so that the cells are not irreversibly damaged, events can be reversed in a process known as deplasmolysis. For this to occur the epidermal cells are simply put in water. Now the contents of the vacuole which were concentrated by the preceding removal of water, has the higher osmotic pressure. Water passes through the plasma tube into the vacuole which now continues to expand until, finally, the initial state is re-established.

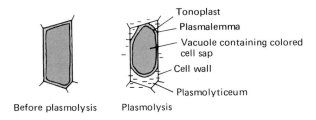

Before plasmolysis   Plasmolysis

Fig. 191. Plasmolysis. The thickness of the tube of cytoplasma has been exaggerated.

2. The Stages of Cell Elongation

We seem to have digressed somewhat from our discussion of elongation. However the connection becomes clear immediately when we mention the fact that in elongation the vacuoles or the single central space for cell sap are expanded quite considerably by water uptake. Water uptake implies an increase in the suction pressure of the cell. According to the suction pressure equation $S_c=S_i-W$, this increase can be attained by either a rise in the osmotic pressure of the cell sap S_i or a decrease in the wall pressure W.

Sometimes, but certainly not always, a correlation can be found between intensity of cell elongation and concentration of osmotically effective substances in the vacuole. Thus, we are left with the alternative: decrease of wall pressure. This lowering of wall pressure is achieved by an enhancement of the plastic flexibility of the cell wall. Crosslinks between the macromolecules of the components of the cell wall are broken, the wall is plastically extended and after elongation has occurrred new crosslinks are formed. In the cell walls of the elongation zones of the oat coleoptile different kinds of macromolecules are found (Table 8). They can all be linked to each other through functional groups in the different molecules, i.e. they can form crosslinks.

The suction pressure of the cell is increased by the enhancement of plasticity of the cell wall. This leads to the inflow of water into the vacuole and to cell elongation.

In the case of intensive cell elongation and thus severe plastic distension of the wall the different kinds of fibrils of the wall are separated from each other to such an extent that a strengthening of the cell wall is necessary. In rare cases *intussusceptive growth* occurs by which new wall material is inserted into the meshes of the fibrillar network, which have been widened by elongation. Usually, though, *appositional growth* takes place, the deposition of wall layers on the original, distended wall. As a result new networks of fibrils are, so to speak, thrown up from the interior of the cell on to the old, distended wall network. Hence, this special kind of appositional growth is known as a "multi-net growth" (Fig. 192).

Table 8. Components of the growing wall of the oat coleoptile

Component	%
Cellulose	25
Hemicelluloses	>51
Pektins	3–5
Proteins	10
Lipids	4

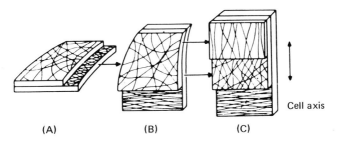

Fig. 192. Multi-net-growth of the cell wall of a cotton hair. (A) In the tip region, (B) at the intersection of the hair tip and the tubular part of the hair, (C) in the tubular part of the hair (from Cutler 1969).

The stages of cell elongation are thus:

(1) Enhancement of the plastic flexibility of the cell wall by the loosening of crosslinks.

(2) Inflow of water into the vacuole and increase in volume by cell elongation.

(3) Strengthening of the distended cell wall by "multi-net growth."

C. Regulation

1. Adjustment of the Equilibrium Between Division and Elongation

A cell comes out of the mitotic cycle and does not enter any further mitoses but begins to elongate instead (Fig. 183). Its longitudinal growth is an early step in differentiation. Which factors are influential in deciding that the cell shall not divide any more but elongate and thus begin to differentiate? Hints were obtained in experiments with tissue cultures, which are comparable with meristems in that, under certain conditions, growth with division but without differentiation can be maintained in them. In tissue cultures of tobacco callus, the following findings were made. If kinetin was added to the tissues either no effect was obtained or a slight stimulation of cell division was observed. If kinetin was combined with IAA, active division occurred. If IAA alone was present in the medium, division stopped and the cells expanded in every direction to form giant cells. Thus, the decision between division and elongation is the result of an interplay between the phytohormones:

IAA + kinetin = division, IAA = elongation.

2. Regulation by IAA

a. Polar migration

The IAA originating from the tip region of the oat coleoptile migrates downwards in a polar direction in the coleoptile. "Polar" is intended to indicate that this migration can only occur from the tip to the base and not in the opposite direction. One can convince oneself of this by a few simple experiments (Fig. 193). We first cut out a section of a coleoptile and put a small piece of agar soaked in IAA on its upper cut surface. After a while IAA can be detected in a second small piece of agar which has been placed under the coleoptile section. If the coleoptile section is turned through 180° and then placed with its morphologically upper side downward on a small piece of agar containing IAA, IAA can be detected in a piece of agar now placed on top of the morphologically lower end. Thus, gravity does not play a role in the polar migration. Only the top and bottom, morphologically speaking, are important. The IAA migrates from the morphological top to the morphological bottom, but not in the reverse direction.

The polar migration of IAA, which can also be demonstrated in other objects, is covered in the chapter on "Polarity" which we have already discussed. Its causes are unknown.

b. IAA Optima

If the elongation-promoting effect of IAA on a plant organ is plotted graphically against the IAA concentration, the curve is seen to pass through an optimum. Initially, increasing doses of IAA also lead to an increase of longitudinal growth but, finally, elongation reaches a maximum. Still higher doses of IAA then inhibit.

The optimal values for different organs of a plant are quite different. In Figure 194, mean plots for shoot, bud, and root are presented. The optimum for shoots, e.g. shoot sections of peas and the *Avena* coleoptile also,

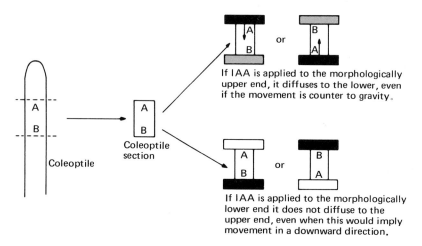

If IAA is applied to the morphologically upper end, it diffuses to the lower, even if the movement is counter to gravity.

If IAA is applied to the morphologically lower end it does not diffuse to the upper end, even when this would imply movement in a downward direction.

Fig. 193. Polar migration of IAA in a section of an *Avena* coleoptile (from Galston 1964).

lies around 10^{-5}M, for roots at 10^{-10} or 10^{-11}M, and for buds at an intermediate value. It is certainly not obvious why such optima should be found and they are casually still unexplained (Fig. 195). At any rate, the application of comparably high doses of gibberellic acid does not lead to a comparable inhibitory effect.

We have pointed out earlier that IAA is one of the factors of apical dominance (page 198). The different sensitivity of shoot and bud now make the mechanism easier to understand. For if IAA in the shoot migrates downward in a polar direction in a concentration which lies completely in the range that is optimal for the shoot, then this concentration will fall in the range that is inhibitory for a bud attached laterally to the shoot. It has already been mentioned (page 218) that IAA is effective via the second messenger ethylene in this instance.

c. Mechanism of Action of IAA

Rapid Effects of IAA

Various findings suggest that gene material can be activated under the influence of IAA. However, a certain lapse of time, a lag phase, occurs before enzymes are formed after induction by IAA. In the case of the induction of β-galactosidase of *E. coli* it amounts to 3–4 minutes, for example. In higher plants the lag phase is longer. In the case of the induction of the nitrate reductase of maize the lag phase was two hours, the shortest found so far in higher plants.

Thus, if gene material is activated, there is a lapse of time before an effect can be measured. However, it has been observed several times that elongation can take place very soon after addition of IAA. An elongation of root hairs has been detected only a few minutes after treatment with 10^{-13}m IAA. Elongation of *Avena* coleoptiles also sets in almost im-

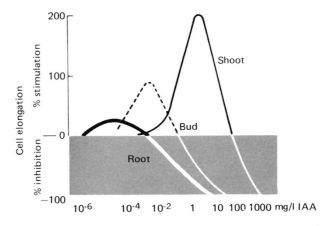

Fig. 194. IAA optima for shoot, bud, and root (from Janick et al. 1969).

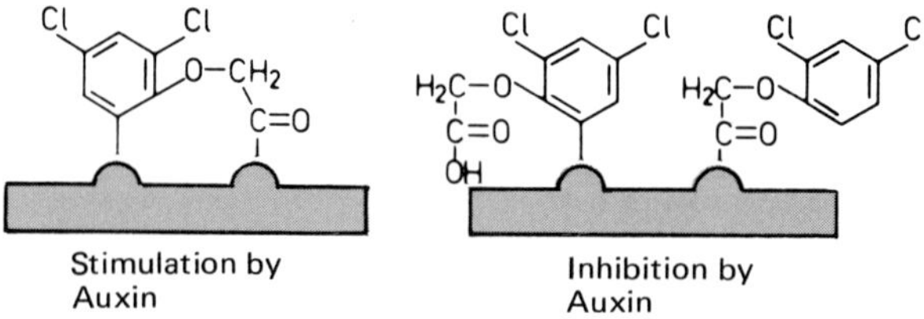

Stimulation by
Auxin

Inhibition by
Auxin

Fig. 195. Two-part attachment hypothesis of auxin action. The synthetic auxin 2,4-D is taken as an example: similar considerations apply to native auxins. The auxin can only exercise its stimulatory effect if the molecule is attached to two sites on a carrier such as a protein. The aromatic system and the side-chain are implicated, the latter probably through a covalent bond, in this two-site attachment. At high concentrations of auxin a molecule of auxin will occupy every site on the carrier. This causes the stimulatory effect to disappear; indeed, the complex may now even inhibit. Similar considerations have led to the postulation of a three-point attachment hypothesis as well as the two-point attachment hypothesis (modified from Leopold 1965).

mediately after the beginning of IAA treatment. These rapid effects of IAA cannot be explained by an induction of enzymes or other proteins which is associated with a lag phase.

A loosening of crosslinkages must also be implicated in these rapid effects of IAA. What the relationship between IAA and crosslinkages might be, is still completely unknown. A direct effect of IAA on the crosslinks can at least be excluded since the effective IAA concentration is too small, at least in some experiments. Thus, another system must intervene between IAA and crosslinkages which magnifies the IAA effect before transmitting it to the crosslinkages. One can imagine, for example, an activation of enzymes being brought about by IAA and these then lyse the crosslinkages.

The problem of the rapid effects of IAA does not become any simpler when we recall that quite different macromolecules are present in the cell wall that is being distended and thus the crosslinks are of quite different kinds. As a matter of fact, hypothetical reaction sequences between IAA and all of these components of the cell wall have been invoked to explain how IAA can sever the crosslinks in question. These speculations have led to increasing attention being paid to the proteins that are found in the cell wall. Thus a hydroxyproline-rich glycoprotein was detected in the wall substance of many lower and higher plants and was given the name *extensin* on account of its possible function in the extension of the cell wall. According to the model proposed by Lamport the carbohydrate component of extensin, which consists predominantly of galactose and arabinose, is linked to the cellulose microfibrils (Fig. 195a). By means of disulphide bridges between different extensin molecules not only the

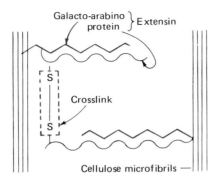

Fig. 195(A). The participation of extensin in the formation of crosslinks in the cell wall (from Lamport 1965).

glycoproteid molecules but the cellulose microfibrils attached to them become linked together. If the disulphide crosslinks are severed an extension of the cell wall can occur. However, as implied above, this is simply a model completely without any experimental foundation. Various proposals regarding a possible effect of IAA on the breaking of disulphide crosslinks are just as hypothetical.

The same also holds for the severing of crosslinks between other cell wall components. We will not concern ourselves with them but simply admit that in spite of 40 years of IAA research we still know nothing about the primary mechanism of action of IAA in elongation.

Long-lasting Longitudinal Growth

This statement is also true of long-lasting longitudinal growth induced by IAA. However, longer-lasting longitudinal growth does not seem to be possible without accompanying protein synthesis. IAA can activate gene material which can then support protein synthesis. The first evidence for this was provided in 1965 by a group of zoologists. They found that actinomycin C abolished the elongation of oat coleoptiles induced by IAA. The coleoptiles elongate in the presence of IAA and actinomycin only to the same slight extent as controls kept in water.

Since then other workers have made similar findings on other objects. IAA can evidently stimulate the processes of transcription and translation that are associated with a longer-lasting longitudinal growth. Indeed it has been found that the synthesis of an enzyme, a cellulase, is induced. Small concentrations of IAA enhance the cellulase activity in the epicotyls of young peas. The increase in activity can be blocked by inhibitors of transcription and translation. This exemplifies the fact that IAA like other phytohormones can induce the synthesis of enzymes — but unfortunately this finding does not help us very much in our analysis of longitudinal growth, because the pea epicotyls also elongate even if the synthesis of the cellulase is blocked. Thus, there was no parallel between cellulase activity and intensity of elongation.

In summary, there are effects of IAA on longitudinal growth that set in quickly and that cannot be explained by an activation of genetic material. Protein synthesis and thus genetic activity do play a role in long-lasting longitudinal growth. In this case IAA can promote transcription and translation. However, up to now no direct relationship has been found between IAA and the synthesis of an enzyme which might be of real importance for longitudinal growth. Consequently, the possibility that the stimulating effect of IAA on transcription and translation may be indirect and quite unspecific cannot be refuted. It is true to say of all of the effects of IAA on longitudinal growth that the primary mechanism of action of IAA in the regulation of cell elongation is unknown.

The Formation Of Seeds And Fruit

A. Complex Developmental Processes and Their Regulation

Cell elongation is a quite simple process of differentiation. In the next few chapters we shall consider in some detail a few complex developmental processes which consist of a large number of individual processes which interlock with each other. Longitudinal growth also participates in these developmental processes but only as one component among many.

We cannot deal here with all of the developmental processes that could be the starting points for a causal analysis, quite apart from the fact that only a very few facts are known about some morphogeneses. In view of the need to be selective it seems desirable that we be guided by the following criteria:

(1) The embryo is already invested with all of the essential organs of the plant. Consequently, the development of the embryo is a quite decisive stage of development. As it proceeds, it is tightly coupled with the formation of seeds and, in certain circumstances, of fruit and should thus be discussed in this context.

(2) In the life of the plant transitions from one phase of development to another are always particularly critical phases, the more so the more readjustment the transition requires. For if a transition of this kind occurs at the wrong time or in the wrong place the individual or indeed the whole population can be damaged. In order to avoid faulty development these kinds of transitions must therefore be subject to precise regulation.

Now transitions of decisive importance are those from dormancy to activity (and vice versa) and those from the vegetative to the reproductive phase of development. Transitions from dormancy to activity are found, for example, in the forcing of side buds and in germination. The transition from the vegetative to the reproductive phase of development in higher plants proceeds with the formation of flowers. The dormancy and activity of side buds has already been mentioned once or twice (pages 198, 202). Thus we shall concentrate here on germination and flower formation.

(3) As a multicellular organism develops the increasing size of the organism necessitates a system of transport and communication which links the individual parts of the organism with each other.

These points fix our agenda: we shall discuss, in the order shown:

(1) Seed and fruit formation including embryonal development,
(2) Germination,
(3) Differentiation and function of the vascular system,
(4) Flower formation.

B. Formation of Seeds and Fruit

Shoot and root structures are already present in the embryo of seeds. Thus the essential processes of differentiation have already been realized. In germination and subsequent processes, the organ structures present in the embryo are simply differentiated further. What interests us is how an embryo with its organ structures develops in the first place. That is why we shall now discuss a few of the details of the formation of seeds and fruit before coming to the section on germination.

1. The Process of Formation

In what follows, events will be outlined as they apply to angiosperms (Fig. 196). Meiosis starts in the microspore mother cells. Each microspore mother cell furnishes four haploid microspores. The first pollen grain mitosis takes place in each microspore leading to a vegetative and a generative cell (page 229) and then the second pollen mitosis occurs resulting in two sperm cells; often the second mitosis does not take place until the pollen grain lies on the stigma or in the style. The vegetative cell develops into the pollen tube, which thrusts forward to the embryo sac within the carpel tissue. In a quite similar manner a macrospore mother cell passes through meiosis in the nucellus of the seed structures.

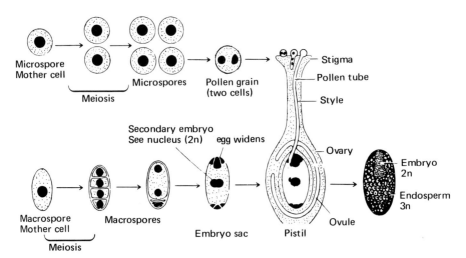

Fig. 196. Scheme of fertilization and of the initial development of endosperm and embryo (from Steward 1963).

However, three of the four haploid macrospores which result usually perish, and the fourth forms the embryo sac. In three steps its nucleus divides into eight daughter nuclei, of which one group of three appears at each end of the embryo sac and forms cells there. One of these six haploid cells in the embryo sac is the egg nucleus. The last two of the eight nuclei fuse to form the diploid, so-called secondary, embryo sac nucleus.

Fertilization is twofold. One of the two sperm cells fuses with the diploid, secondary embryo sac nucleus to form the triploid endosperm nucleus. The endosperm nucleus divides to form many nuclei each of which combines with cytoplasm to form a cell. In this way the multicellular endosperm emerges. Subsequently it can be enriched with reserve materials, vitamins, phytohormones, and other factors that are necessary for development before and, in part, also after germination. In some cases such as the coconut and pumpkin plants the endosperm is wholly or partially fluid.

The second sperm cell fuses with the egg nucleus. The diploid zygote then develops to form the embryo on which the shoot and root elements as well as the cotyledons can be recognized (Fig. 197). Thus the endosperm is the supplier of all of the necessary substances including the regulatory phytohormones. It may be completely consumed during the development of the embryo. The materials necessary for the development of the seedling are then stored at another site, e.g. in the cotyledons as is the case with our *Leguminosae*. If the endosperm in the seed is preserved to a large extent, then the substances necessary for germination are stored in it. That is the case, for example, in our *Gramineae*. The seed is defined as the dormant embryo surrounded by the more or less well-developed endosperm and the seed coat (Fig. 198). The seed coats are derived from the so-called integuments, layers which envelope the nucellus with the embryo sac.

The seed itself may be a propagation unit, i.e. in the form of seeds a species can propagate itself. However, in many cases the propagation unit is the fruit (Fig. 198). In a fruit the seeds are enveloped individually or in multiple by additional layers that are derived principally from the carpels. Inflorescence axes and leaves in the region of the inflorescence can also take part in the formation of the fruit. Passing from the outside to the inside of the fruit wall, the pericarp, can be divided into the layers of the exocarp, mesocarp, and endocarp.

Strictly speaking the development of the fruit begins with the formation of blossoms. This is because organs of the blossom are transformed into the fruit wall. This transformation is first set in motion by pollination. Thus the development of the fruit is considered to begin with pollination and the following phases are distinguished:

(1) Initial phase of fruiting which begins with pollination.
(2) Phase of division which begins with the twofold fertilization.

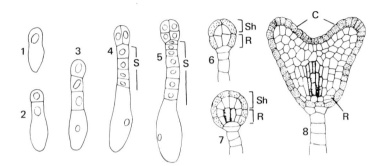

Fig. 197. Embryonal development of *Capsella bursa pastoris.* C = cotyledons, S = suspensor (uppermost cell = hypophysis), Sh = shoot (cotyledons + plumule), R = root + hypocotyl (modified from Walter 1962).

(3) Phase of cell elongation.

(4) Phase of ripening.

(5) Phase of senescence, i.e. aging, which will not concern us further here.

2. Regulation

a. Differentiation of shoot and root structures

Technique of tissue culture

The differentiation of the tissues of the fruit, the endosperm and the embryo, are finely coordinated with each other. In the rest of this chapter we shall consider a few interactions in detail. In particular, the embryo must be in close communication with the endosperm, which supplies it with all of the materials needed for its development.

Attempts have been made to determine the factors through which the endosperm engages in the development of the embryo. The experiments carried out amount, in principle, to replacing the endosperm with culture media containing defined supplements. Tissues, organs, or whole plants, e.g. even the embryo, are then cultivated on these media. Fig. 199 shows how organs (root and apical meristem) and tissues (callus from

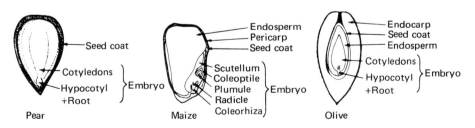

Fig. 198. A few seeds and fruits (from Janick et al. 1969).

shoot internodes) of a plant can be grown in culture. We shall give primary consideration to the culture of callus tissue. An internode is first cut out of the shoot stalk, sterilized and placed in an agar-containing culture medium. Cauliflower-like outgrowths, calluses, can then develop on the cut surfaces. Polarity can sometimes be demonstrated here too in that often, but not always, the callus develops better on the morphologically lower end.

Pieces of the callus are then transferred to fresh agar and attempts were made to keep them growing. Difficulties usually set in at this stage. Especially prior to 1940, further cultivation of the callus was successful only in exceptional cases, e.g. by Gautheret, a pioneer of the tissue culture technique. It appeared that supplements of natural or synthetic phytohormones and of vitamins are quite crucial to promoting cell division in tissue cultures. At the beginning of the 1940s, Van Overbeck then found that coconut milk, i.e. a fluid endosperm, met the demands of most organs and tissues in culture. Later experiments, particularly those of Steward showed that coconut milk is rich in different kinds of phytohormones, particularly of those with a cytokinin effect. Since then coconut milk is often added to culture media. It is obvious that there are certain problems whose solution requires that coconut milk be not added. This is especially the case when one wishes to analyze the influence of phytohormones as accurately as possible.

Induction of shoot and root structures in tissue culture

The first division of the zygote leads to two cells, one of which later

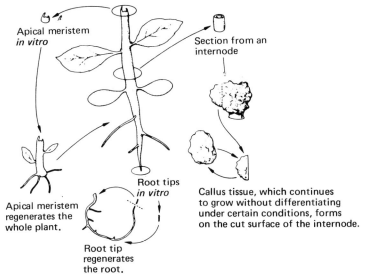

Apical meristem
in vitro

Section from an internode

Apical meristem regenerates the whole plant.

Root tips
in vitro

Root tip regenerates the root.

Callus tissue, which continues to grow without differentiating under certain conditions, forms on the cut surface of the internode.

Fig. 199. Organ and tissue cultures of different parts of a higher plant (after Torrey 1968).

develops into the suspensor and the other into the embryo (Fig. 197). The position of the cell wall in this first division is determined by the polarity of the zygote—and the polarity of the zygote is fixed, as mentioned, by the polarity of the embryo sac. We have already seen that subsequent divisions depend on the phytohormones being present in a particular ratio to each other (page 235). But divisions alone are not enough. For these divisions must occur in such a way that the essential organs of the embryo, the shoot and root structures, can develop.

Experiments with tissue cultures brought assistance at this point. It first had to be settled whether embryos can arise at all from dividing cells in tissue culture. That is indeed the case. In 1959, Reinert obtained embryos in tissue cultures of the carrot under specified conditions. Since then there have been frequent reports of the successful induction of embryo formation in tissue cultures. As already outlined the climax of this development is the formation of complete plants from single cells taken from tissue cultures (page 172). Now in the present context the whole embryo is more than we want to consider. We are interested in the more limited question as to the conditions under which the essential organ structures of the embryo, the shoot and root structures, differentiate. Here too, as in the adjustment of the equilibrium between cell division and cell elongation, the interplay between the different phytohormones is of decisive importance. In 1957, Skoog and Miller reported experiments with callus cultures of the tobacco plant (Fig. 200). IAA and kinetin were present in the culture media in a ratio that varied from experiment to experiment. At a certain ratio of IAA to kinetin growth and division of the callus occurred without differentiation. If the relative amount of kinetin is lowered, roots are formed. If the proportion of kinetin is raised, shoots develop in the callus. If kinetin is absent or present in doses that are too high, then the growth of the tissue ceases.

Of course, one can object that a callus culture cannot simply be assumed to be comparable to a developing embryo. Furthermore, similar results were also obtained with some but not all other species in tissue cultures. Nevertheless the generalization can be made with regard to the question we posed: it is very likely that the differentiation of shoot and root structures in the developing embryo is determined by a well-balanced ratio of the different phytohormones which are supplied by the endosperm.

b. Culture of embryos

After we begin to get an idea of the conditions under which the decisive step of differentiation, the development of shoot and root structures, occurs, the subsequent development of the embryo really assumes secondary importance. If we are nonetheless to give it brief consideration it is because the culture of embryos is of practical importance.

Hannig achieved the first culture of embryos in 1904. Another pioneer in this field was Laibach who, in 1925, reported the culture of embryos which arose from crosses between different species of the genus *Linum*. The culture of embryos was made considerably easier by Van Overbeck's discovery of the stimulating effect of coconut milk at the beginning of the 1940s. Coconut milk and casein hydrolysates too are usually added nowadays to culture media used for the culture of embryos. Laibach had gone ahead with embryo culture out of practical considerations. He had difficulties with the breeding of the hybrids with which he worked. These difficulties could be overcome by growing young embryos in culture instead of having to depend on seeds that were capable of germinating. Later they were planted out.

One can adopt a similar procedure with crosses of other species or genuses. Here, too, similar difficulties will turn up because in the double fertilizations gene combinations are produced which are so foreign that the frictionless interplay between them and then also between endosperm and embryo is disturbed. Thus, it often happens after a species or genus cross that the endosperm sooner or later dies or does not develop at all. In these cases the embryos would also die or at least be damaged. However, they can be isolated and grown long enough on a synthetic culture medium so that they can finally be planted out.

We now present two examples: cultivated barley *(Hordeum vulgare)* is susceptible to mildew *(Erisyphe graminis* var. *hordei)*, but *Hordeum bulbosum* is resistant. Hybrids were desired and yet after a species cross the caryopses frequently had no endosperm and the embryo was dead. However, when the embryos were isolated and cultivated in culture media, reproductive hybrids were obtained.

The true melilot *(Melilotus officinalis)* is thoroughly resistant to drought but its content of bound coumarin is troublesome to cattle and, in addition to this, the highly poisonous dicoumarol can be formed from its precursors. On the other hand, certain crosses among the related white

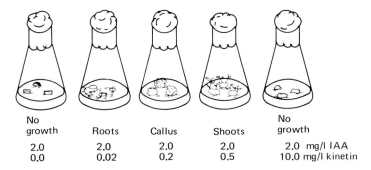

Fig. 200. Control of organ formation in tissue cultures by defined quantities of phytohormones (modified from Skoog and Miller 1957).

melilot (*M. alba*) made breeds available that are almost coumarin-free. Crosses between these breeds and *M. officinalis* foundered because first the endosperm and then the embryos died. Here, too, embryo culture helped to overcome the difficulties. Both examples may also be used to show what close interrelationships arise between so-called "applied" and so-called "pure" research.

c. Development of the fruit

The derangements that can appear in the development of the endosperm, and then also of the embryo, of hybrids of certain species and genuses demonstrate how tightly the development of the endosperm and embryo are coupled to each other. Moreover, the development of the fruit layers is also closely linked with them, especially with the development of the endosperm.

Let us trace the individual phases of fruit development (page 244). As early as 1909, Fitting had shown that fruit development begins with *pollination* and not as late as fertilization. Aqueous extracts of orchid pollen cause the ovary of orchids to swell. Later it could be demonstrated that the pollen extract contained IAA. In continuation of these experiments, formation of fruit without fertilization, parthenocarpy, was achieved in the most diverse plants by treatment with phytohormones. It depends on the species which phytohormone has the strongest effect. Parthenocarpy can be triggered in most species by IAA or by synthetic growth substances that act similarly. Among the Rosaceae the stone fruits such as the cherry, peach, and plum are exceptions. They respond to gibberellic acid, not to IAA.

Thus, changes in the prospective fruit tissue begin as early as pollination and the transfer of phytohormones associated with it. However, at the same time, the development of a partitioning layer beneath the flower, which is also under the influence of phytohormones, is inhibited, thus preventing premature fruit fall (page 199).

The next phase of fruit development starts with the double fertilization. During this phase the endosperm is formed. The embryo also grows by *dividing*. Evidently this is associated with active cell division, which takes place in the fruit tissue. Perhaps initially the regulating hormones come from the endosperm but in later stages they are derived from the embryo. In addition to IAA and gibberellins cytokinins in particular seem to be involved, as was shown by analyses of some fruits during this phase.

In the third phase vigorous *cell elongation* takes place. The regulatory phytohormones are principally IAA and gibberellins. Nitsch has shown particularly clearly in experiments on strawberries that IAA can control cell elongation. The strawberry is a syncarpous fruit in which many individual achenes sit on the flesy, swollen receptacle. If the achenes are removed before the phase of elongation, then the formation of "berries" does not take place. If only some of the achenes are removed, then inhibition of cell elongation is observed in those parts where there are no achenes (Fig. 163). If, however, the achenes are now replaced by tiny spots of paste containing growth substances, cell elongation and thus syncarp formation occur normally.

In the later stages of the elongation phase mono- and oligosaccharides are concentrated in the vacuoles. It may be that the rise in osmotic pressure associated with this acts as an additional stimulus to elongation. The phase of *ripening* is characterized by a number of biochemical changes: conversion of pectins into pectic acids by the cleavage of methyl groups (page 73), degradation of the resulting pectic acids to units of lower molecular weight; hydrolysis of different polymeric carbohydrates to oligo- and monosaccharides; development of flavoring substances; degradation of chlorophylls; in some cases, such as the apple, anthocyanin synthesis regulated by the phytochrome system and so on. However, a special characteristic of the stage of ripening is a pronounced increase in respiration which finally passes through a maximum. This stage of maximal respiratory activity is known as the climacteric. With the attainment of this maximum the fruit is "ripe."

It is not known exactly which factors control ripening. A number of hypotheses deal with systems which influence respiratory activity. However, the factor most actively discussed as being the agent controlling ripening is ethylene. If ethylene is allowed to act on fruit ripening is accelerated. Ethylene is also formed by fruits themselves, as has been shown by gas chromatographic studies and biological tests. A good specimen for biological tests are pea seedlings. Their longitudinal growth is inhibited by ethylene.

In a large number of cases ethylene production by the fruit begins even before the climacteric. That might suggest that ethylene initiates a chain of reactions that finally leads to ripening. However, it is also conceivable that it is an early by-product of a chain of reactions induced by other factors and which then, in turn, accelerates the whole process.

Germination

To define germination is not easy in that no definition seems to be free from exceptions. Usually the *forcing of the radicle* through the seed coat is taken as a criterion for the germination process. However, it must be realized that in applying this criterion, before and after the visible, morphological effect, less noticeable processes occur which also belong to germination.

In many cases the seed is capable of germinating immediately after being liberated from the parent plant. In other cases, however, the seed is capable of germinating only after a more or less long period of rest. This is known as the dormancy of seeds.

In this chapter we shall first be concerned with the causes and the breaking of dormancy and then with germination itself. At germination the seedling stage begins, which is completed with the development of the photosynthetic apparatus and thus with the transition to autotrophy. At least a few points should be mentioned about this too.

A. Dormancy

Dormancy can be due to very different causes. We will detail a few of them:

1. Incomplete Embryos

At the time of the release of the seed or fruit from the parent plant the embryos are still not fully developed. This development is made good during dormancy. An example taken from our indigenous flora is the seeds of the ash (*Fraxinus·excelsior,* cf. Fig. 201), which exhibits a period of dormancy during which the embryo first fully develops.

2. Maturation of Drying

In a number of species the full capacity to germinate is only attained after the seeds have passed through a drying process. Thus, for example, a small percentage of the seeds of maize are already capable of germinating almost immediately after being harvested. The full capacity to germinate is obtained, however, only after storage. In this connection it could be shown that the increase in the ability to germinate runs parallel to a decrease in the water content of the grains. In spite of several hypotheses it is still unclear how the full capacity to germinate is only attained by drying.

3. Impermeability to Water and/or Gases

Germination is not possible without uptake of water and gas exchange (see Conditions for Germination). Now the layers which surround the embryo are more or less impermeable to water and gases. The barrier can be furnished by the endosperm, the nucellus, the seed coat, or the fruit walls. A seed coat which is impermeable to gases is found in the ash already mentioned and the cocklebur *Xanthium strumarium*. The barriers can be made permeable by different agents, the choice of which depends on the nature of the barrier: cracking of seeds, treatment with concentrated sulfuric acid, alcohol, hydrogen cyanide, hydrogen peroxide, etc. In nature the impermeable layers are gradually broken down, particularly by the activity of microorganisms. Humidity in combination with warmth favors the process.

4. Inhibitors

All parts of a seed or a fruit including the embryo itself can contain inhibitors which block germination. Inhibitors are found in the embryo, the chemical constitution of which is unknown, e.g. in *Fraxinus excelsior* and *Xanthium strumarium*. Amygdalin, the precursor of the inhibitor hydrogen cyanide, is contained in the embryos of the Rosaceae, especially in our fruit produce containing stones and pips. When the seeds are steeped the β-glucosidase emulsin which is also present becomes active and cleaves both glucose molecules from amygdalin (Fig. 202). The aglycone, thus set free, decomposes spontaneously into benzaldehyde and hydrogen cyanide. Since the endosperm of the species mentioned is impermeable to gases the hydrogen cyanide cannot escape. It inhibits germination until the thin endosperm has rotted (also cf. Fig. 203).

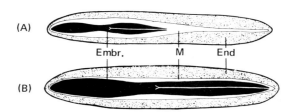

Fig. 201. Seeds of the ash *(Fraxinus excelsior)* immediately after being liberated from the parent plant (A) and after 6 months storage in moist earth (B). Embr. = embryo, M = mucous layer derived from the endosperm (End) (from Ruge 1966).

The occurrence of inhibitors in the flesh of fruit is particularly well-known. They appear to be ecologically essential. The flesh of a tomato, for instance, ought to present the best preconditions for germination when no inhibitors are present—but that would be disasterous if the conditions outside the tomato were not favorable for further development of the seedling.

The seed inhibitors localized in the fruit pulp are also sometimes known as blastocolins, a word whose choice does not appear to be justified in that it could suggest a uniformity which simply does not exist. Quite different kinds of substances appear among the inhibitors localized in the fruit pulp. Some examples are the following.

Abscisic acid is an inhibitor of germination in the fruit pulp of the hip, *trans* cinnamic is an inhibitor of germination in *Parthenium argentatum,* a shrub-forming Composite found in the southern U.S.A. and Middle America which can be used for the extraction of rubber. Caffeic acid and ferulic acid fulfil the same function in the fruit pulp of the tomato mentioned above. Coumarin and scopoletin are also discussed as inhibitors of germination. A large number of additional inhibitors of germination are known, the chemical nature of which has only been partially elucidated. However, of the majority of these substances it is true to say that the property of being able to inhibit germination has been demonstrated in the laboratory. Evidence that they act in the same way under natural conditions is difficult to obtain.

B. Conditions for Germination

As we have just seen, barrier mechanisms can be present in the embryo itself and in all of the layers that surround it. If they are removed the embryo is able to germinate. However, it germinates only if certain external conditions prevail. We shall call these external conditions that enable a seedling which is capable of germination actually to germinate conditions for germination.

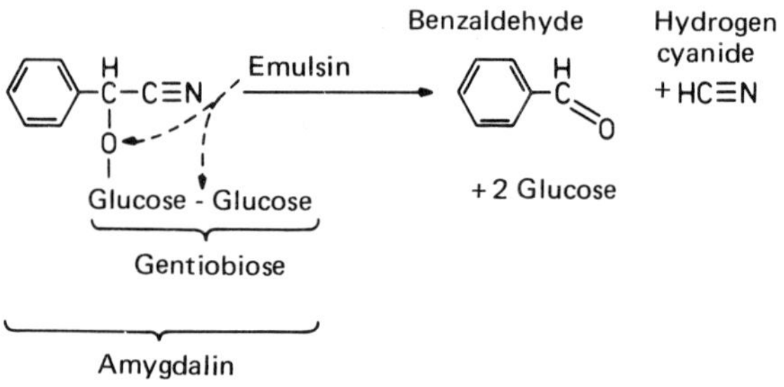

Fig. 202. Cleavage of amygdalin by the β-glucosidase emulsin.

1. Water

Water is the *conditio sine qua non*. Water uptake by the dry seed begins with swelling, a purely physical process. Hydrophilic groups, e.g. $-NH_2$, $-OH$, $-COOH$, attract the dipolar water molecules and form hydrated shells around themselves. The macromolecules that bear such hydrophilic groups, e.g. proteins and polymeric carbohydrates, thus "swell". Water uptake by swelling is followed by water uptake associated with germination. In particular, cell elongation which takes place in the course of germination is based on an intensive water intake.

As a rule of thumb it can be said that the swelling can be reversed without serious damage to the seedling by deprivation of water. This is no longer possible in the ensuing germination, which is associated with cell division and cell elongation. Otherwise, the distinctions between the physical process of swelling and the germination process, which is due to biochemical-physiological activity, are not as sharp as the definitions sometimes tend to imply. Thus, mRNA can be formed in some embryos even during swelling. In embryos of wheat the synthesis of mRNA starts only 30 minutes after the beginning of swelling.

2. Oxygen

Energy is necessary for germination. It is made available in the form of ATP, which is derived from substrate chain and respiratory chain phosphorylation. However, oxygen is a necessary prerequisite for the functioning of the respiratory chain and, thus, also for oxidative phosphorylation. For this reason the presence of oxygen is usually a condition needed for germination. Exceptions only confirm this rule: in rice and other plants too germination can proceed under water, i.e. under conditions under which the oxygen partial pressure is only very small. However, rice seedlings are equipped with a very efficient system of glycolysis which does not require any oxygen.

3. Temperature

The temperature requirements of the individual species with respect to germination clearly demonstrate the importance of different species-specific optima. This is because the temperature optima of germination correspond to the external conditions that are necessary for the development of the species in question. For example, it has been established for soil samples from the Colorado desert that winter annuals (one year plants that germinate in the autumn and blossom and bear fruit in spring or early summer of the next year) germinate preferentially at $+ 10°$ and summer annuals (one year plants that complete their whole development in one summer) at 26–30°. Thus, winter and summer annuals exhibit different temperature optima of germination that match the external conditions necessary for their further development.

The effect of low temperatures on swollen seed material presents a special case. A number of seeds require *vernalization,* storage on a moist substrate at temperatures somewhat above 0°, in order to germinate and to be able to develop without any further restraint after germination.

Low temperatures are involved in quite different processes. On the one hand they can lift barriers imposed by inhibitors and thus render the embryo capable of germinating. This particular temperature effect should actually have been included in the previous section. On the other hand germination itself can, in many cases, still take place. However, the subsequent development of the seedling is inhibited if it is not subject to lower temperatures. So-called physiological dwarfs then develop. Considered in more detail, the growth of both the epicotyl and the hypocotyl or simply the epicotyl can be inhibited. Accordingly, one speaks of either *hypocotyl* and/or *epicotyl dormancy* which must be broken by lower temperatures. In this case "dormancy" is really the wrong word since it is actually retarded development.

An example of the switching off of inhibitors in the embryo is provided by the embryo of the ash (*Fraxinus excelsior*), which, as we know, is at first incomplete. This development is made good in the summer of the next year. Only then, thus in the second winter, can the cold of winter be effective. It raises the gibberellin level in the embryo so that the inhibitors can be overcome and the seed can germinate the following spring.

Examples of Epicotyl and Hypocotyl Dormancy. It is particularly well-known of our Rosaceae that their seeds require vernalization. In the case of the apple it has been found (Fig. 203) that if the seed coat is removed germination still does not take place. If, however, the thin endosperm pellicle is also removed, the embryos germinate. (We have already mentioned that the endosperm of the Rosaceae is impermeable to gases and that, for example, hydrogen cyanide formed after water uptake cannot escape.) The seedlings of other Rosaceae as well as the apple show,

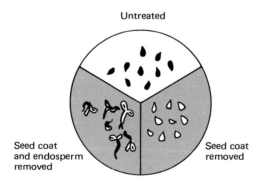

Fig. 203. Germination of apple seeds after removal of seed coat and endosperm (modified from Ruge 1966).

Fig. 204. Development of apricot seedlings after vernalization (right) and without vernalization (left) (from Ruge 1966).

however, an epicotyl and hypocotyl dormancy. They develop normally only after lower temperatures have been allowed to act on them.

Examples of Epicotyl Dormancy. In a number of species germination occurs even without the influence of lower temperatures and a root system develops, but the growth of the epicotyl remains inhibited. Plants develop that are severely compressed, such is the case with several *Rosaceae* such as the apricot (Fig. 204), a few *Liliaceae* and the shrub peony (*Paeonia suffruticosa*). After being subjected to cold the plants assume normal growth. It is precisely this group in which gibberellic acid can substitute wholly or partially for the cold. That is the reason for assuming that the cold might induce an elevation of the endogenous gibberellin level in this instance, just as in the ash.

An interesting mechanism is found among certain species of the *Liliaceae* genus *Trillium* (Fig. 205) among others. Here cold is needed twice and in between there must be a period at higher temperatures. After the effect of cold in the first winter the root system can develop. However, epicotyl dormancy is broken only in the second winter.

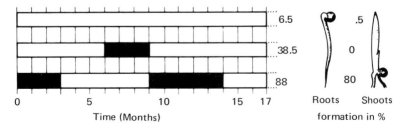

Fig. 205. Double requirement for cold of *Trillium* spec. Only a few plants germinate when no cold treatment at all has been given (above). After one cold treatment roots develop (middle). After a second cold treatment shoots also develop (below) (from Koller 1959).

4. Light

The plant grower has known for a long time that the germination of seeds of certain species can be promoted by light whereas the germination of others can be inhibited. Table 9 lists a few of the species whose germination is either promoted or inhibited by light.

It has already been mentioned (page 220) that light influences the ability to germinate via the phytochrome system. Light-stimulated and light-inhibited seeds appear to differ in the way in which their phytochrome system reacts to incident light that contains, of course, both red and far red light. In the case of the seeds that are stimulated by light P_{660} is converted into P_{730} by RL, the latter being the active state of the pigment system which can then induce germination. The seeds that are inhibited by light seem to possess a steady concentration of P_{730} in darkness which is high enough to induce germination. This point is disputed because P_{730} can gradually be converted into the more stable P_{660} in darkness. The influence of FRL predominates in dark germinators on illumination and by means of it the P_{730} present or furnished by RL, is converted into the inactive P_{660} (Fig. 206).

The further mechanism of action of light in the stimulation of germination is still unknown. In a few species it could be connected more or less directly with the production of gibberellins since the effect of light can be replaced entirely or in part by gibberellic acid in these species. But that is hardly likely to be the case for all seeds. The situation is not made easier by the fact that, in addition to the phytochrome system, other pigment systems such as the blue-light absorbing system are important.

C. Mobilization of Reserve Materials

At germination, reserve materials localized in the storage tissues (cotyledons, endosperm, and, less often, the nucellus) must be mobilized. They are the only sources of organic substances available to the seedling until it has developed its own photosynthetic apparatus. As far as this

Table 9. Several light and dark germinators (light-stimulated and light-inhibited seeds); important experimental plants italicized

Light germinators	Dark germinators
Digitalis purpurea	Amaranthus caudatus
Epilobium hirsutum	Cucurbita pepo
Lythrum salicaria	Nigella damascena
Lactuca sativa	Phacelia tanacetifolia
Nicotiana tabucum	Prenanthes purpurea
Oenothera biennis	

mobilization is concerned a few lines suffice to remind us of facts we have already discussed.

The essential reserve materials are carbohydrates, proteins and fats.

1. Mobilization of Carbohydrates. Mobilization of carbohydrates is achieved by degradation to mono- or oligosaccharides by means of the appropriate hydrolases, including, in part, phosphorylases. For example, starch degradation in barley grains under the influence of gibberellic acid (page 000).

2. Mobilization of Proteins. Mobilization of proteins involves their hydrolytic degradation to amino acids by proteases. These proteases, too, are, in part, synthesized *de novo* at germination just like the amylase of barley grains.

3. Mobilization of Fats. Mobilization of fats is brought about by hydrolysis of fats to fatty acids and glycerol by lipases (page 94) and β-oxidation of fatty acids to acetyl CoA (page 95). A part of the acetyl CoA is converted into carbohydrates via the glyoxylate cycle. The key enzymes of the glyoxylate cycle, isocitrate lyase and malate synthetase (page 96), are formed *de novo* at germination in some seeds that store fats. How their synthesis is regulated is unknown.

D. Assembly of the Photosynthetic Apparatus

Assembly of the photosynthetic apparatus completes the seedling phase of development. The development of this typical plastid structure, which we discussed earlier (page 54), can only take place in light. In darkness, it development goes awry. The phytochrome system is also engaged in the differentiation of the plastids.

What holds true for the structural elements of the plastids need not be valid for all of the enzyme systems of photosynthesis. A key enzyme in the secondary processes of photosynthesis is carboxydismutase which fixes CO_2 into ribulose-1, 5-diphosphate (page 51). The enzyme is already present before the light-dependent differentiation of the plastid structure, at least in rye seedlings. Nonetheless, illumination induces an

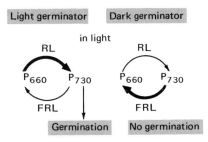

Fig. 206. Hypothesis concerning mode of action of the phytochrome system in seeds which are stimulated and inhibited by light.

active *de novo* synthesis which is regulated by the phytochrome system.

There are still other regulatory processes that are associated with the commencement of photosynthetic activity. This is because the seedling must adopt pathways that are different, in part, from those of the autotrophic, photosynthetically active plant in order to secure the materials that it needs. We may recall that pentoses and also the tetrose erythrose-4-phosphate can be furnished by the Calvin cycle (page 51) and these are needed by the plant for particular synthetic operations. The same substances can, however, also be derived from glucose-6-phosphate in the pentose phosphate cycle (page 62). The first enzyme of the pentose phosphate cycle is glucose-6-phosphate dehydrogenase. It is already present in the embryo before germination as has been shown by Feierabend in rye seedlings. During germination its activity at first increases sharply and then remains constant with the transition to photosynthesis. Evidently the seedling obtains the monosaccharides mentioned via the pentose phosphate cycle (hence the high activity of the first enzyme, glucose-6-phosphate dehydrogenase) and then switches over totally in this respect to the Calvin cycle after the photosynthetic apparatus is in operation. Now, glucose-6-phosphate dehydrogenase is no longer so important but rather the first enzyme of the Calvin cycle, carboxydismutase, becomes important. Its activity increases (Fig. 207).

E. Regulation of Germination by Phytohormones

Up to now we have gathered lots of data together and in doing so have come across the participation of phytohormones again and again. Let us

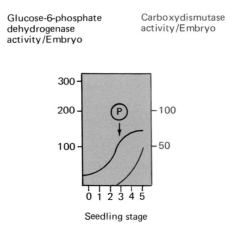

Fig. 207. Activity of glucose-6-phosphate dehydrogenase and carboxydismutase during the development of rye seedlings. P = commencement of photosynthesis (modified from Feierabend 1967).

now try to summarize the entire course of germination from the point of view of regulation by phytohormones.

First let us assume that all barriers to germination have been lifted and the seed is able to germinate. Then all of the conditions for germination must be given (water, oxygen, temperature, light). Now we can ask: how is germination induced and how can the phytohormones engage in the individual stages of the germination process? In 1968, Van Overbeck provided an answer in outline to this question as it applies to cereal grains such as a grain of barley (Fig. 208). Water penetrates the permeable seed coat and into the seed, even into the embryo within the seed. The embryo becomes active, the synthesis of mRNA for different processes begins and gibberellic acid is discharged into the aleuron. In the aleuron gibberellic acid induces the synthesis of a number of hydrolases which mobilize reserve materials. For example, one of them is α-amylase, which degrades starch. They also comprise nucleases and proteases which attack nucleic acids and proteins, respectively. As a result of the activity of the nucleases cytokinins contained in nucleic acids are set free and the proteases act to liberate amino acids including tryptophan from which IAA can be formed. Cytokinins and IAA now act on the embryo: the cytokinins induce cell division and IAA cell elongation. The embryo so induced to grow bursts its seed coat. It is helped in this by pectinases and cellulases, which it has formed if the seed coat has not already been extensively rotted by the activity of microorganisms in the soil. First the radicle appears out of the seed, then the coleoptile. As a result of gravity IAA migrates to the underside of both organs. This, in turn, leads to reactions which are geotropically different owing to the different sensitivity of the coleoptile and the radicle. The root turns downward and is positively geotropic, the coleoptile turns upward and is negatively geotropic. As soon as the coleoptile has broken through the soil its photosynthetic apparatus differentiates and with that the seedling stage is complete.

It all sounds quite plausible. However, in actual fact some of the points are still working hypotheses that need to be proved or refuted.

F. Regulation of Germination and Evolution

It is getting rather tedious to discuss phytohormones. We have learned by now that they play a role in every process of development, and also that only in very rare cases can anything definite be said about their primary mechanism of action. Indeed in their relation to germination it can be shown that the phytohormones are only part of a very complex whole. We have made a long but by no means exhaustive list of barriers to germination (page 250) and of conditions for germination (page 252). That was done not out of delight in detail but because such an accumulation of interlocking mechanisms certainly suggests that germination is a process whose regulation is of utmost importance for plants. Even if we have just emphasized the phytohormones, we have still considered only some of the complex regulatory events. In the remainder of the chapter the significance of a few other regulatory mechanisms will be discussed.

Germination under conditions that do not allow further development of the young plant signifies its death. The barriers to germination are designed in such a way that they do not allow germination under unfavorable external conditions. They are, as it were, calibrated against favorable external conditions. The gauge is provided by the barriers to germination.

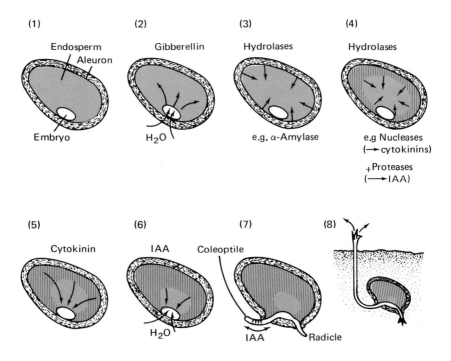

Fig. 208. Model of the course of germination taking a barley grain as an example. (1) Dormant grain. (2) Water uptake and passage of gibberellin from the embryo into the aleuron. (3) Synthesis of hydrolases in the aleuron, which then pass into the endosperm. α-amylase, which degrades starch in the endosperm, is one of the hydrolases. (4) Synthesis of hydrolases, which include nucleases and proteases, in the aleuron, which pass into the endosperm. Nucleases supply precursors for cytokinin synthesis by liberating purine nucleotides, and proteases set free tryptophan, the probable precursor of IAA. (5 and 6) Cytokinins and IAA from the endosperm stimulate the embryo to cell division and cell elongation. (7) The growing embryo breaks through the seed coat; the coleoptile grows upwards negatively geotropic and the radicle downwards positively geotropic (cf. Fig. 159). (8) The seedling breaks through the soil, is exposed to light and develops its photosynthesis apparatus (modified from Van Overbeck 1968).

Calibration against temperature

From the example of winter and summer annuals in the Colorado desert we have already learned that individual species show different temperature optima of germination. The examples could be multiplied. Thus, it is known that useful plants from cold climatic zones germinate better at lower temperatures and those from warm regions better at higher temperatures. Such data are indicative of an adaptation to existing circumstances.

Now it would be quite disastrous in our temperate zones if every seed which had a temperature optimum of germination at lower temperatures were immediately to germinate when these lower temperatures set in. That would mean that the seeds would germinate at the beginning of winter. A few species such as our winter cereals are able to withstand the subsequent cold season in the vegetative state. That is certainly not true of all species, however. At this point, then, one of the barriers to germination comes into action: only if low temperatures have been in effect throughout a certain period of time — and in nature that means the winter months — can the seed germinate. We became acquainted with this phenomenon in our discussion of vernalization.

This leads to the question as to what the gauge is, and in answer to which one can only speculate. It may be that the content of endogenous inhibitors falls when low temperatures are allowed to act for a longer period of time. In other cases, such as the ash, an increase in the content of substances that promote germination, such as the gibberellins, and that are induced by lower temperatures seems to be more important.

Chemical calibration against precipitation

We are better informed about the nature of the gauge as far as precipitation, i.e. the germination factor water, is concerned. Precipitation in particular is *the* germination factor in arid and semiarid regions. That is why Went, Evenari and other researchers carried out experiments on plants from such regions in order to investigate the chemical rain gauge.

Many annual desert plants germinate only if at least 125 mm but preferably 250–500 mm precipitation has fallen. This is practical in that the soil is then sodden enough to allow for the further development of the plant. Now the upper soil layer in which the seeds lie is just as sodden after a light precipitation of 125 mm as after a heavier one. How do the seeds measure the amount of precipitation under these circumstances?

In many cases the gauge is the inhibitors of germination, which are present in the embryo and the surrounding layers. These inhibitors of germination are washed out by rainfall and for this a certain minimal amount of precipitation is necessary. This washing out effect can easily be verified in laboratory experiments. If seeds are put on sand and sprinkled from above with at least 125 mm water they germinate. If, however, the

same amount of water is absorbed from below germination does not occur. This is because in the latter case the inhibitors cannot be washed out.

We have just selected two barriers to germination. Now time and
again several barriers to germination are found in one species. When the
barriers to germination are lifted adequate conditions for germination
must prevail. This complicated system has its advantages: all requirements for germination will not be met for all seeds of a given kind at any
one particular time. Hence the germination of seeds formed in the same
year will be drawn out over a longer time interval, possibly several years.
If now in one of these years the germinated plants die as a result of unex-
pectedly unfavorable external conditions before they form seeds, the
species nonetheless survives in the area in question.

We must now mention one further observation that can be made
repeatedly in arid regions. If the seeds of annuals have just germinated
and the external conditions are not radically unfavorable, then all plants
develop and grow. It is the more remarkable in that sometimes they grow
very close together. When that happens the individual plants are smaller,
they form fewer blossoms and seeds but they do not crowd each other out.
Similar observations can also be made in a not too thickly sown field of
corn. The "struggle for existence" and the process of selection associated
with it is in these instances concentrated in the process of germination.

Thus, the critical point in adapting to given external conditions is the
ability to germinate. This point does not make itself felt in our cultivated
plants for we have bred them precisely for rapid germination. We determine the time of sowing and thus of germination and thus render
superfluous the need to adapt as is the case with wild plants. This should
not obscure the fact that the manifestations of adaptation involved in the
germination of wild plants are realized under the pressure of a harsh
selection and further developed. In this way the regulation of germination
becomes a key factor of selection and, thus, of evolution.

The Vascular System

As long as an organism consists of only a few cells, communication and transport from one cell to another can take place without the need for systems specially designed for that purpose. If the number of cells and, thus, the size of the organism increases then special vascular systems must be developed. This need becomes the more imperative for the reason that specialization of individual cells in a multicellular organism is driven so far in quite different directions that an individual, highly differentiated cell, would be neither viable nor able to function without being supplemented by other differentiated cells of other kinds.

In animals the demands that arise in this way are met by two systems: the nervous system, which provides rapid communication between cells, and the humoral system, which is involved in communication via hormones but which is also in charge of transport of materials. The neurosecretory cells of the midbrain occupy an intermediate position. They are nerve cells in which hormones are synthesized and then conveyed by nerve fibers to the posterior lobe of the pituitary gland. In plants there is no equivalent to the nervous system; here communication *and* transport of materials are both undertaken by humoral systems. On the one hand the vascular system of plants is an organ of communication and thus of *intercellular regulation.* We need only to consider the transport of the phytohormones from the site of synthesis to the site of action. We should also remember that certain substrates can be conveyed from cell to cell. In the event that they might be brought to acceptor cells via the vascular system and trigger a substrate induction (page 184) there, then this substrate induction would provide a mechanism of intercellular regulation as well as intracellular regulation. In essential points a mechanism of this kind would show great similarity to regulation by hormones — a point which would certainly deserve examination. On the other hand the vascular system of plants is an organ of *mass transport.* Water and minerals are taken up from the soil by the root system and then conducted upward in the plant. Materials assimilated by photosynthetically active leaves are conveyed upward to the meristems and zones of elongation, as well as to younger, growing leaves. Assimilated materials are also transported downward in the stem and root and stored there.

One further property is associated with the conductive function. A part of the conducting elements, those of the xylem, make an important contribution to the *mechanical properties* of higher plants by lignification and the hardening of their walls, which that implies (Fig. 102).

A. The Elements

We cannot go into the details of the vascular system here. We will simply draw attention to the fact that two systems have to be distinguished:

(1) The *xylem* with the trachea and tracheids which are directly important for the conduction of materials. They are dead in the completely differentiated state but cannot maintain their function without living xylem parenchymal cells which surround them.

(2) The *phloem* with its sieve elements—sieve cells in gymnosperms and sieve tubes in angiosperms—which are associated with companion cells. The conducting elements are also live, though extensively degraded, in the completely differentiated state. Thus, the sieve tubes do not possess a nucleus. The energy power stations of the cell, the mitochondria, have also completely or almost completely disappeared in completely differentiated sieve tubes. It seems that the companion cells take over some of the functions which the sieve tubes can no longer carry out—an interesting field of regulation but one difficult to investigate.

B. Differentiation

Which factors are responsible for the fact that cells differentiate to form the elements of the xylem and phloem? As is often the case, important data were obtained from tissue culture experiments.

We have already mentioned that wound tissue, a callus, first forms on the cut surface of explants from higher plants. Let us now consider an explant from the xylem. It also first forms a callus out of undifferentiated cells (Fig. 209). Then pockets of cambium develop in this callus. This cambium then differentiates, xylem elements being formed on the side towards the original xylem explant and phloem elements on the other side. Now let us do the same experiment with an explant from the phloem (Fig. 209). In this case the cambium forms phloem on the side oriented towards the phloem explant and xylem elements on the other. Evidently there exists a gradient of material from the explant to the callus which specifies and directs the differentiation of the vascular tissue.

It became clear in subsequent experiments on callus tissue which substances could play a role in such gradients. The formation of vascular systems can be induced in hitherto undifferentiated tissue by grafting buds on to it as Camus showed in 1949 with the endive (*Cichorium intybus*) and Wetmore on the elder (*Syringa vulgaris*) somewhat later. If, for example, an elder bud is grafted on to callus tissue which also came from the elder, the formation of xylem elements is induced in the callus. Initially, small pockets are formed, which then enlarge (Fig. 210) and, finally, make contact with the vascular system of the bud. Induction also occurs if cellophane is inserted between the bud and the callus. Paricularly important is the fact that IAA and synthetic growth substances can be substituted for the bud: even in this instance xylem elements form in the callus. Thus, IAA is evidently one of the factors that can induce xylem formation.

Many other findings point in the same direction. We shall mention the experiments of Jacobs, which were carried out on another system at the beginning of the 1950s. Jacobs inflicted wounds on the internodes of *Coleus* and then investigated the development of xylem in the wound callus that formed. The xylem fibers that were newly formed in the callus then connect with those already present in the internodes. In these experiments it was also shown that the leaves above the site of the wound and, to a lesser extent, those below, exert a strong influence on the differentiation of xylem. Xylem formation is inhibited by removal of the leaves but recommences if IAA is applied to the cut surface of the leaf stalks after the leaves have been cut off. The role of IAA is further emphasized by a close parallel between the content of IAA which is transported from the leaves above the wound downward and from the leaves below the wound upward, and the number of xylem fibers induced in the callus (Fig. 211). In addition, these experiments also show that, in *Coleus*, the IAA transport can occur not only in a strictly polar downward

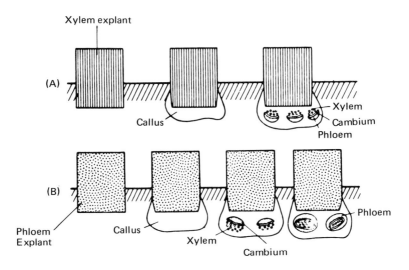

Fig. 209. Directing effects of xylem and phloem explants on the position of xylem and phloem in regeneration (from Kuhn 1965).

direction but also in the opposite direction, though to a very much less extent.

Up to now we have only discussed xylem and must now bring the second component of the vascular system, the phloem, into the picture. Both the experiments of Wetmore on the elder and those of Jacobs on *Coleus* showed that the formation of phloem can be attained if IAA and sucrose are administered together. Once again we come across phenomena that are already familiar to us: the unspecific effect of a phytohormone that can influence the most varied processes of differentiation, and the fact that combinations of the regulatory factors determine the kind of differentiation.

Homoeogenetic Induction. The experiments outlined above and many others repeatedly draw attention to the fact that differentiation that has already occurred can induce the development of new differentiation of the same kind. Thus, xylem induces the formation of new xylem in the callus, which then connects with the xylem already present (e.g. in *Cichorium, Syringa,* and *Coleus*) or orients itself with respect to it (Fig. 209). The phloem behaves similarly. This phenomenon is known as homoeogenetic induction.

Now there are objections to this concept of homoeogenetic induction, which are illustrated by an experiment carried out by Torrey (Fig. 212). In the roots of peas (*Pisum sativum*) there is a triarchic vascular bundle. As is well-known the designation triarch, tetrarch, etc., is determined by the number of radially distributed xylem strands. If root segments are grown in culture, they first regenerate a new root tip. In the presence of IAA, however, a triarchic vascular bundle is no longer formed in the cells of these new root tips but a vascular bundle with a different number of

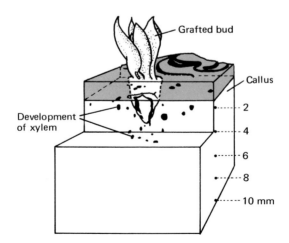

Fig. 210. Induction of xylem in a callus from elder *(Syringa vulgaris)* after an elder bud has been grafted onto it (from Torrey 1968).

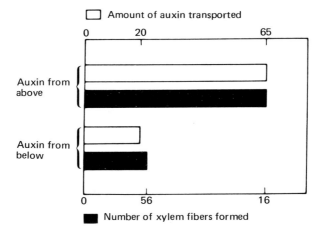

Fig. 211. Parallels between amount of auxin transported and number of xylem fibers bormed in the wound callus. Studied in shoot sections of *Coleus* (from Jacobs 1956).

xylem strands. At 10^{-5}M IAA a hexarchic vascular bundle develops and if IAA is withdrawn there follows a parallel decrease in the number of xylem strands from 6 to 5 to 4. Thus, the vascular bundles become pentarchic and tetrarchic.

At first sight such data seem to speak against a homoeogenetic induction since xylem production in the newly formed root region is evidently not determined by the xylem that was already present in the old root segment. However, it should not be forgotten that we are dealing with isolated organs under drastically altered experimental conditions. Thus, it is entirely possible that the exogenously added IAA has simply masked a

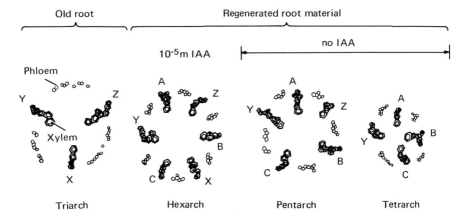

Fig. 212. Dependence of the number of xylem strands formed in regenerating root tips on the IAA concentration (modified from Torrey 1968).

concentration gradient, emanating from the old xylem strands, which was present.

Thus the last word on homoeogenetic induction has still not been said. However, we can state with certainty that phytohormones not only play a role in stimulating division in an already existing cambium (pages 198, 202), they also participate in the differentiation of the cells generated by this cambium to form elements of the xylem and phloem. As usual, the interplay of several phytohormones among themselves and with other factors determines the kind of differentiation (xylem or phloem, etc.).

C. Function

1. Transport in Both Directions

We mentioned at the beginning of the chapter that transport occurs from the top downward and also in the reverse direction in the vascular system. A simple experiment provides us with the evidence for this statement (Fig. 213). A maize plant is placed in nutrient solutions in such a way that one half of its root system is immersed in nutrient solution A and the other half in nutrient solution B. Compounds containing radioactive phosphate, P^{32}, are present in nutrient solution B. At the beginning of the

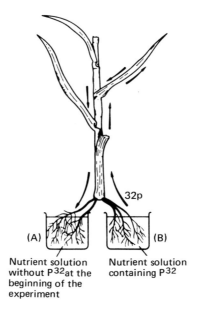

Fig. 213. Transport in both directions. Nutrient solution (A) contained no P^{32} at the beginning of the experiment; nutrient solution (B) contained P^{32}. Six hours after the beginning of the experiment the presence of P^{32} in solution (A) could be demonstrated (from Biddulph and Biddulph 1959).

experiment nutrient solution A contains no radioactivity at all. However, P^{32} can be detected in A six hours later. It had been transported in the form of particular compounds first upward in the xylem and then downward in the phloem.

2. Transport of the Xylem

a. Evidence

We have just asserted that radioactive P^{32} migrates upward in the xylem of the maize plant. Now we come to the evidence for the fact that an upward transport in the xylem can actually occur. This is illustrated by an experiment which was carried out in 1939 by Stout and Hoagland (Fig. 214). It demonstrates one of the earliest uses of a radioactive "tracer." These experiments were carried out on willow cuttings (and a few other species) that had already formed roots. The xylem and phloem were separated from each other over a defined segment of the shoot. Grease-proof paper was inserted between the two tissues. Then potassium K^{42} was administered to the plant via the nutrient soil. After 5 hours the K^{42} content of the xylem and phloem of the shoot segment mentioned was investigated. The results showed that the upward transport occurred almost exclusively in the xylem. In the segments above and below the zone of ex-

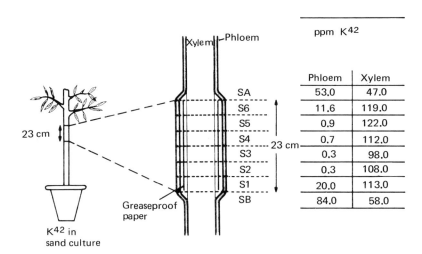

	ppm K^{42}	
	Phloem	Xylem
SA	53.0	47.0
S6	11.6	119.0
S5	0.9	122.0
S4	0.7	112.0
S3	0.3	98.0
S2	0.3	108.0
S1	20.0	113.0
SB	84.0	58.0

Fig. 214. Transport of minerals (K^{42}) in the xylem. In a willow twig (*Salix* spec) xylem and phloem were carefully separated from each other and greaseproof paper was inserted between them. K^{42} was supplied to the soil. After 5 hours the treated segment was cut into sections and in each section the xylem and phloem were analyzed separately for K^{42}. S1 to S6 = sections from the zone covered with greaseproof paper, SB = section from the zone below the greaseproof paper, SA = section from the zone above the greaseproof paper. The measured values are shown in ppm (parts per million) (modified from Stout and Hoagland 1939).

perimentation, though, K^{42} passed from the xylem into the phloem. Since then this kind of transverse transport of different substances between xylem and phloem has been demonstrated many times.

Water and dissolved mineral salts are transported primarily in the xylem. However, other substances can also be transported, e.g. sugar and several amino acids in spring. Phytohormones and synthetic growth substances are also transported in the xylem as Skoog demonstrated for IAA as long ago as 1938 (cf. Fig. 211). In many cases plant protecting substances are transported in the xylem as well. Evidence for transport in the xylem can be obtained by removal of the phloem by curling (Fig. 220). If the substance is still conducted, then it is highly probable that the xylem is the path of transport. A better technique is to examine exudates, which can be obtained by tapping the xylem of a few ligneous plants. The drawing of exudates or "bleeding sap" of this kind from the sugar maple in eastern North America is familiar and exploited commercially. In March before the trees burst into leaf such exudates contain about 3% sucrose. Bleeding sap which exudes from tree stumps or droplets due to guttation can also be investigated. Finally there is a method to suck out the contents of the xylem in a vacuum.

b. Mechanism

aa. Root pressure

We have now to enquire into the mechanism responsible for transport in the xylem. The possibility that first comes to mind is that water and the substances dissolved in it are forced upwards under pressure. Indeed, such a "pressure" does exist.

We are all familiar with the drops of liquid that can be found hanging from the tips of the leaves of lady's mantle (*Alchemilla* spec.) and grass seedlings. This represents an active secretion of water, i.e. it requires and results from the expenditure of energy by the living cells and is called guttation. It becomes particularly important when transpiration has become impossible owing to high humidity and is known to occur in the most diverse species.

Bleeding sap from wounds has also been mentioned. Drops of xylem fluid ooze out from freshly cut surfaces of tree stumps, particularly in spring. An activity of the living cell is responsible for this kind of secretion too and it is described as bleeding pressure or root pressure.

The height of the root pressure can easily be measured, simply by mounting manometers on cut surfaces, for example. If one does this, one makes the disappointing discovery that the root pressure is usually less than one atmosphere. Thus, in most cases it would be just sufficient to force a column of water in the vessels of the xylem to a height of 10 m. Even our indigenous trees are higher, to say nothing of such giant trees as *Sequoia* and *Eucalyptus*. Furthermore, the narrow lumina of the conduct-

ing zylem elements, the inner walls of which are articulated by local rein-
forcements of the wall, offer considerable resistance to the ascent of
water.

bb. Transpiration

It is clear that we must look around for another mechanism of xylem
transport. Root pressure can only be a reinforcing measure, particularly in
the spring, when very heavy demands are made on the xylem before the
leaves have burst forth and the transpiration surface of the leaves is not
yet present. This gives us our cue: transport in the xylem occurs, in the
main, as a result of suction from above rather than pressure from below.
The precondition for this suction is transpiration.

The loss of water by transpiration can take place through the cuticle
and the stomata. Correspondingly, one speaks of cuticular and stomatal
transpiration. *Cuticular transpiration* assumes some importance only in
plants with a very thin cuticle. Usually it amounts to less than 10% of the
total transpiration. Thus, *stomatal transpiration* is unquestionably the more
important process. This is true in the first place in regard to quantity, as
already mentioned. Owing to the edge effect (Fig. 215), the surface of a
leaf, which is equipped with stomata, is capable of exuding very much
more water than one would expect. In many cases this is an amount of
water equal to the weight of the plant and more per day.

Stomatal transpiration is, however, particularly important since, in
contrast to cuticular transpiration, it can be regulated by motions that
open and close the stomata. If the guard cells are completely turgescent,
then the stomata are opened. If they lose their turgescence the stomata
are closed (Fig. 216). A number of factors cooperate to regulate turgescence and, thus, opening and closing of the stomata:

(1) *Water.* An insufficient supply of water leads to loss of turgescence
of the guard cells and to closing of the stomata.

(2) *Light.* Illumination usually leads to opening of the stomata. In
many cases blue light, which is absorbed by a special pigment system, is
particularly effective. However, chlorophyll and the phytochrome system
can also be engaged by appropriate illumination.

Fig. 215. Edge effect in stomatal transpiration. Water molecules being expelled
from stomata can diffuse to the sides even in a dead calm, something that is not
possible by diffusion from a closed water surface (from Sutcliffe 1968).

(3) *Temperature.* Temperatures higher than 25° lead to closing of the stomata.

(4) *CO₂-Content.* Low CO_2 partial pressures lead to opening, high partial pressures to closing, of the stomata.

(5) *Hormonal Regulation.* Withering plants can raise their abscisic acid concentration forty fold. Abscisic acid causes the stomata to close and thus prevents further loss of water. This effect can take place extremely quickly. Maize leaves placed in an abscisic acid solution react by closing the stomata after only 3 minutes. Experiments are in progress to elucidate to what extent this abscisic acid effect is operative in practice.

These are a few of the factors which take part in regulation. Their mechanism of action is still, however, unknown. A current topic for discussion is the influence they may have on the osmotic pressure of the cell sap in the guard cells, which could be brought about by a conversion of starch into sugar or by photosynthesis in the guard cells, since they usually possess chloroplasts. Recently particular attention has been paid to the K^+ content of the guard cells. Using ingenious micromethods it has been possible in recent years to measure the K^+ content of individual guard cells. It is much higher when the stomata are open than when they are closed. Opening of the stomata is associated with transport of K^+ into the guard cells. It is certain that a still unknown but probably organic anion is simultaneously either formed in the guard cells or transported there. In any case turgescence increases and opening occurs. One attractive hypothesis attempts to link K^+ transport with the effect of abscisic acid. This involves the speculation that abscisic acid increases the permeability of the membranes to K. The result would be a K efflux from the guard cells and closing of the stomata.

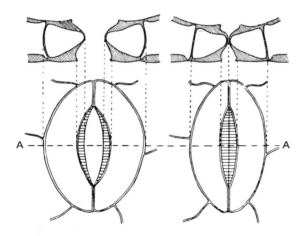

Fig. 216. Opened (left) and closed (right) stomata. Plan view (below), cross section (above) (from Walter 1962).

Neither this nor any other hypothesis can do justice to the demands of the situation and we cannot go into any further detail here. It is very likely that the regulation is multiple, obscure, and also different from one plant to the next.

cc. The diurnal acid cycle of succulents

Let us now consider in detail an obvious sign of adaptation in plants that is related to the closing movements of the stomata. The most varied morphological and anatomical devices are known in succulents which serve to reduce transpiration, including stomatal transpiration. A lowering of transpiration implies, however, not only a reduction in the loss of water vapor but also a curtailment of total gas exchange. If the stomata are completely closed — and that is the case in many succulents in the heat of the day — then almost every opportunity for gas exchange with the outside air is blocked.

This leads to problems. As desirable as the reduction of water loss is, it is just as undesirable that the uptake of the CO_2 needed for photosynthesis be inhibited. A loophole is provided in the diurnal acid cycle which is widely distributed in the plant kingdom but particularly well developed in succulents, apparently in adaptation to the conditions mentioned. At least up to now no better explanation has been found for the fact that in succulents the diurnal acid cycle has been pushed to the extreme.

As early as 1804, De Saussure demonstrated in the Indian fig that an "acidification" appeared in the plant material overnight and "deacidification" during the day. Since then similar findings have been made in many other succulents including those of the genuses *Bryophyllum, Crassula, Kalanchoe, Kleinia,* and *Sedum.* The pH value of the cell sap was always lowest in the morning and highest in the evening. Analyses showed that although other acids are present in larger quantities, the fluctuations in acid content were due, in the main, to fluctuations in the content of malate (Fig. 217). This periodic change in acid content is known as the diurnal acid cycle.

Isotope experiments demonstrated that $C^{14}O_2$ is incorporated into malate during the acidification. In 1936, Wood and Werkman discovered that bacteria can fix CO_2 in dicarboxylic acids. Now a similar kind of fixation of CO_2 is also found in higher plants and, by means of this, malate is furnished. In principle, the fixation of CO_2 can also occur in light. In light, however, it is turned off by the competing process of photosynthesis which impounds the CO_2 for its own use. Consequently, the CO_2 fixation to form malate just mentioned only assumes importance in darkness. For this reason it is spoken of, not quite correctly, as a "dark fixation" of CO_2 (C_4-dicarboxylic acid pathway, cf. page 53). How is malate formed by CO_2 fixation? Predominantly by a carboxylation of phosphoenolpyruvate (PEP), leading initially to oxalacetate (1). In the next step oxalacetate is

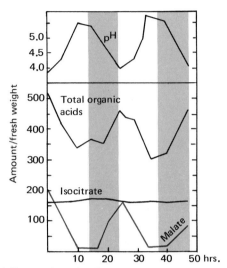

Fig. 217. Diurnal fluctuations in the pH value and in the amount of different organic acids in *Bryophyllum calycinum* (modified from Steward 1966).

hydrogenated to malate by malate dehydrogenase (MDH) (2):

1. Phosphoenolpyruvate $+ CO_2 + H_2O$ $\xrightarrow{\text{PEP-carboxylase}}$ Oxalacetate $+$ Phosphate

2. Oxalacetate $+$ NADH $+ H^+$ $\underset{\xrightarrow{\hspace{1.5cm}}}{\overset{\text{MDH}}{\rightleftharpoons}}$ Malate $+$ NAD$^+$

The required PEP is ultimately derived from the starch that is formed during photosynthesis during the day. This reminds us of the fact, which we have just mentioned, that in the light the process of photosynthesis fixes CO_2 for its own use. Now in the carboxylation of PEP the equilibrium lies completely on the right. Furthermore, PEP-carboxylase has a higher affinity for CO_2 than carboxydismutase (cf. page 51). Thus, CO_2-fixation during photosynthesis ought actually to succumb to the competing reaction. The solution of this puzzle is to be found in the fact that the activity of PEP-carboxylase, at least in succulents, is inhibited by malate. Accumulation of malate thus finally leads to inhibition of the dark fixation of CO_2 which can now be fixed in photosynthesis, the light-dependent process.

And now a complication. According to mechanism (1) *one* C atom of malate should be radioactively labeled after fixation of $C^{14}O_2$, namely C atom 4 (Fig. 218). In such experiments, however, radioactivity is also found in C atom 1. Indeed, two thirds of the radioactivity is always found in C atom 4 and one third in C atom 1. To explain this it is assumed that CO_2 is fixed twice (Fig. 218). The first fixation which is light-independent, utilizes ribulose-1, 5-diphosphate as CO_2 acceptor. Ultimately, two

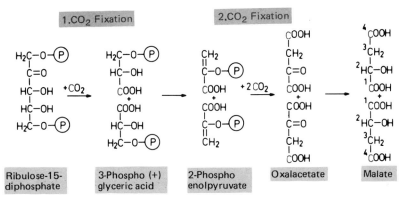

Fig. 218. CO_2 fixation occurring twice during the overnight acidification in the diurnal acid cycle (tandem hypothesis).

molecules of phosphoenolpyruvate arise from this acceptor, one having radioactivity in C atom 1 and one without. This reaction thus furnishes the first third of the radioactivity. Labeled CO_2 is now fixed in each of the two molecules of phosphoenolpyruvate according to reaction (1), i.e. in a second reaction which is light-independent. This second reaction accounts for the other two thirds of the radioactivity in C atom 4.

The deacidification by day is the result of an oxidative decarboxylation brought about by the malic enzyme:

$$\text{Malate} + \text{NADP}^+ \xrightarrow{\text{malic enzyme}} CO_2 + \text{Pyruvate} + \text{NADPH} + \text{H}^+$$

Since the malate that has accumulated overnight inhibits further fixation of CO_2 into PEP, the CO_2 set free by the malic enzyme can be fixed in photosynthesis.

Let us now look again at the whole sequence of events from the standpoint of the *adaptation of succulents*. During the night respiration predominates. The CO_2 so formed is, however, not lost to the atmosphere but stored in the form of malate after being fixed in probably two light-independent reactions. During the day photosynthesis occurs. At the same time the external temperatures rise leading to greater and greater losses of water by stomatal transpiration. Then transpiration is choked by complete or partial closure of the stomata. This also shuts off the CO_2 supply from outside. However, the plant now draws on its CO_2 reservoir, malate: malate is broken down to form CO_2, which can then be used in photosynthetic CO_2 fixation even when the stomata are closed.

dd. Cohesion theory

After our digression into the diurnal acid cycle of the succulents, let us return to the mechanism of water transport in the xylem. We have already pointed out that the root pressure can only be of secondary impor-

tance. We then identified the real driving force for the transport in the
xylem as transpiration. This is because the cells of the leaf parenchyma
that have transported water through the internal holes and then through
the stomata to the outside, make good their water deficit from the fine
ramifications of the xylem in the leaf parenchyma. In this way suction is
exerted on the water filaments, which fill the elements of the xylem from
the root region up to the leaf parenchyma. Rupture of these water filaments is prevented by high cohesive forces between the molecules of
water in the capillaries of the xylem. Thus, if at the upper end of one of
these water filaments suction is exerted as a result of transpiration, the entire column of water moves upwards. One speaks here of a cohesion
theory of the rise of water in the xylem (the designation "theory" is, in
contrast to many other occasions, justified here). Furthermore,
astonishingly high cohesive forces can be measured in plants, in certain
regions of the fern sporangium up to 250 atm for example.

Added to this is the adhesion of water molecules to the molecules of
the wall lining of the xylem which prevents the water filaments from
being detached from the vessel walls, thus making their rupture more
difficult. The sheath of living xylem parenchyma around the conducting,
dead xylem elements serves an additional purpose. It prevents the infiltra-
tion of air into the conducting elements, which are subject to considerable
suction at the time of intensive transpiration.

We are sceptical and would like to have at least a few pieces of evidence to support the cohesion theory of water ascent. Assuming that
transpiration is the driving force, the following experiment (Fig. 219),
which was carried out before the end of the last century, shows that
cohesive forces actually do permit liquids to be raised in capillaries. A
piece of plaster that has been soaked in water is placed on top of a capillary
that is filled with water. Plaster and capillary are placed in a dish containing mercury. Transpiration, which today can easily be brought about by
using a hair dryer, for example, causes mercury to be drawn upwards in
the capillary. Later, the plaster was replaced by a twig (Fig. 219), with the
same result. Thus, water is sucked up through the elements of the xylem
just as it is through the glass capillary beneath it.

If the cohesion theory holds, the suction must begin at the tips of the
plant. There the movement of water should actually be initiated. The evidence for this was obtained in 1936 by Huber. The content of the xylem
was electrically heated locally and the migration of the hot fluid was
followed using thermoneedles, which were stuck into the conduction
systems under investigation. It was shown that at the beginning of heat-
ing and, thus, of transpiration in the morning, a movement of water was
first discerned in the tips of branches and then later in the trunks of trees.

The suction could also be measured. Each individual conducting element in the xylem is contracted somewhat when transpiration is intensified owing to the adhesion of the column of water to the wall lining of

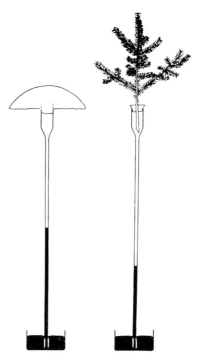

Fig. 219. An experiment to demonstrate the cohesion theory of transport in the xylem. Transpiration from a block of plaster or a twig causes mercury to be sucked up into a glass tube (from Walter 1962).

the xylem. That, of course, cannot be measured. However, the sum of the cross-sectional contraction of all of the individual xylem elements can be measured: the diameter of the trunk of ligneous plants becomes noticeably less in the midday heat owing to intense transpiration.

3. Transport in the Phloem

a. Evidence

It was demonstrated in principle by Malpighi as early as 1679 that a downward transport of substances can take place in the phloem. He removed a ring of bark from different trees over their entire circumference and observed as a result a bulging of the bark above the ring (Fig. 220). In the 19th century the conducting elements of the phloem were then discovered and it was established that the downward directed flow of substances occurred in it.

Subsequently, ringing experiments of this kind have been frequently carried out. However, as in the case of the xylem, isotopes were used here too as soon as the tracer technique was sufficiently well developed. One such experiment is reproduced in Figure 221. The experiment is a com-

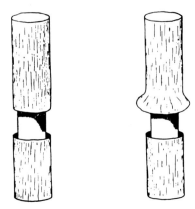

Fig. 220. A ringing experiment. A girdle of bark is removed from a woody stem. Several weeks later the bark about the girdle swells because the materials to be transported downward accumulate here (from Richardson 1968).

bination of ringing and isotope application. It can be seen that ringing blocks the downward transport of assimilated materials in the phloem but not the upward transport of phosphorus compounds in the xylem. Thus,

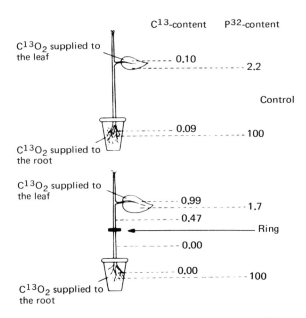

Fig. 221. Combination of a ringing and tracer experiment. $C^{13}O_2$ is supplied to a leaf and P^{32} to the root. In the control, C^{13} labeled assimilated materials move downward and P^{32} upward. If the phloem is removed by ringing, the upward transport of P^{32} is not blocked, only the downward transport of C^{13} assimilates (modified from Rabideau and Burr 1945).

phloem and xylem are not only separated from each other by their positions in the cross-section of the shoot, they can also function quite independently of each other.

Now we must enquire into the nature of the substances that are transported in the phloem. Information of this kind can be obtained from autoradiographic investigations of the phloem after administration of labeled materials. Fluorescent dyes have also been used frequently. In addition biologists use a technique here which puts every biochemical or biophysical technique in the shade as far as sensitivity and exactness are concerned: the *aphid technique.* Aphids remove necessary nutrients from plants. To do this they puncture the sieve tubes. Microscopic inspection has shown that the proboscis always pierces only one sieve tube. Several hours after the puncture the aphid under investigation is anaesthetized, e.g. with CO_2, and cut off from its proboscis. Exudate from the punctured sieve tube now emerges from the cut surface and this can be collected in micropipettes and then analyzed chromatographically (Fig. 222). Aphids which are frequently used are *Tuberolachnus salignus,* which live off willow trees (*Salix* spec.) or *Acyrthosiphon pisum,* which feeds on the broad bean *Vicia faba,* among other plants. Shield-lice (*Coccidae*) can also be used instead of aphids.

Using the method that has just been outlined, the substances transported in the phloem could be determined. Quantitatively, carbohydrates predominate, the most important being sucrose (Fig. 223). In addition there are small amounts of oligosaccharides such as raffinose, stachyose, and verbascose. The first three compounds are well-known to us; verbascose is stachyose to which an additional galactose unit has been added. As for the rest sieve tube sap contains, in addition to the phosphates of various hexoses, amino acids, amides, nucleotides, nucleic acids, virus particles, phytohormones, including the controversial flowering hormone (page 291), and inorganic ions.

It has been shown with regard to inorganic ions that by no means can each and every ion be transported in the phloem. Thus, for example, calcium and borate can apparently be transported only in the xylem (Fig. 224).

Fig. 222. The aphid technique. On the left a sucking aphid, on the right the anaesthetized aphid has been cut off. Sieve tube sap streams out of the proboscis, which has punctured only one sieve tube, and can be collected.

b. Mechanism

The mechanism of transport in the sieve tubes has still not been elucidated. Nonetheless most of the presently known findings argue in favor of the correctness of the *mass flow hypothesis* put forward by Munch in 1926. According to it, convection or *mass flow* is responsible for the transport in the sieve tubes just as for transport in the xylem or in the blood vessel system of animals. The driving force of this mass flow in the sieve tubes is a concentration gradient of osmotically active substances decreasing in the direction of transport.

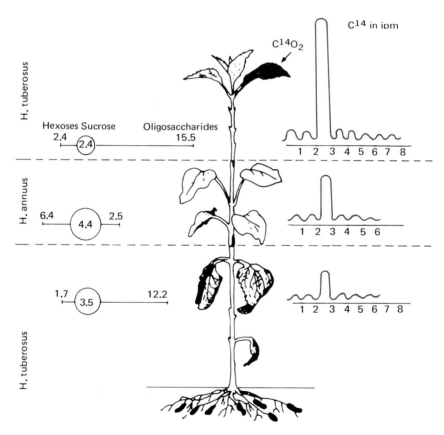

Fig. 223. Sucrose as the most important carbohydrate transported in the phloem. Double graft between *Helianthus tuberosus* and *H. annuus* (cf. left hand margin). On the left is shown the relative content of different carbohydrates in the stem of the separate graft segments. On the right, the relative radioactivity of different carbohydrates from the separate graft segments. (1) fructose, (2) glucose, (3) sucrose, (4-8) oligosaccharides whose molecular weight increases to the right. Sucrose is the most important carbohydrate to be conducted irrespective of the plant species (modified from Kursanov 1963).

Let us make the principle of the mass flow hypothesis clear by resorting to a model put forward by Munch (Fig. 225). Two inverted cells A and B, which are covered with a semipermeable membrane, are joined to each other by a capillary R. Cell A contains a 10% sucrose solution to which Congo Red was added in order to make the flux of material visible. Cell B contains water. Owing to its high osmotic pressure cell A takes up water from the surrounding vessel through the semipermeable membrane. The excess pressure which develops in cell A drives the colored sucrose solution through the capillary R and across to cell B. Indeed water is forced out through the semipermeable membrane at the lower end of cell B.

Now Munch equated the cell A with the sites of assimilation, i.e. primarily the green deciduous leaves with their high concentration of assimilated materials. The capillary represents the sieve tubes and the cell B the sites of storage, the trunk or even underground storage organs. At the sites of storage the substances that have been assimilated are deposited in high molecular, often solid, form so that the concentration of osmotically active substances should really be low there. A mass flow transports substances between these two sites, which have different concentrations of osmotically active substances. Even the experiments with aphids showed that a mass flow actually can take place in the sieve tubes. In this case a discharge of assimilated materials from the proboscis can continue for days and this can be explained only with difficulty as being

Fig. 224. Conduction of borate in the xylem. The root system of a leaf cutting of *Nicotiana rustica* is divided. The right half is maintained in a nutrient solution containing borate, the left half in a nutrient solution without borate (modified from Ziegler 1963).

due to something other than the flow of material. However, there are quite conclusive criteria, which Ziegler put forward in 1962. A flow of mass requires:

(1) an osmotic gradient which decreases in the direction of flux. In this point the original concept of Munch (osmotic gradient falling from the leaves to the roots) could not be confirmed. However, in the sieve tubes a gradient of osmotically active substances falling in the direction of flow has been found many times and this gradient would be quite sufficient to make mass flow feasible.

(2) semipermeability of the phloem compared with the surrounding tissue which must give up or take up water (in Figure 225 this is represented by the vessels filled with water surrounding the cells A and B). This requirement is fulfilled.

(3) continuous efficiency of the conducting system for the streaming fluid. On this point tempers run high since an impediment to streaming in the sieve tubes could actually be provided in the form of the sieve plates. Various additional hypotheses attempt to overcome this controversial point.

(4) Up to now we have postulated what conditions must hold. Now we must point out something that may *not* happen if the mass flow hypothesis is to have validity, a *bidirectional transport in one and the same sieve tube*.

Bidirectional transport in one and the same plant is completely compatible with the mass flow hypothesis and has also been demonstrated. Thus, assimilated materials from leaves can be conducted both upward into the shoot meristem and also downwards in the direction of the root. However, bidirectional transport in one and the same sieve tube would contradict the reality of the mass flow hypothesis.

It is precisely this that Eschrich appeared to have demonstrated in 1967 with the aphid technique on *Vicia faba* (Fig. 226). Fluorescein, a

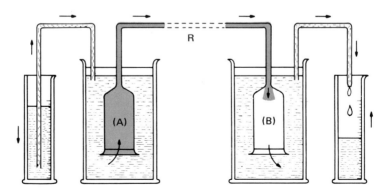

Fig. 225. Model experiment of Münch to demonstrate mass flow (modified from Ziegler 1963).

fluorescent dye, was applied to a leaf at the foot of a plant and C^{14}-labeled urea to a leaf at the top of the plant. Aphids were then introduced between the two leaves. The honey dew of each of the aphids was examined. The result was that in the honey dew of one aphid both substances, fluorescein and C^{14}-urea, were detected. Since each aphid punctures only a single sieve tube this result seems to argue strongly for a bidirectional flow and against the correctness of the mass flow hypothesis.

However, Eschrich provided an explanation that is compatible with the mass flow hypothesis (Fig. 227). In one sieve tube fluorescein streams upwards and in another C^{14}-urea downward. The two sieve tubes make contact with each other through transverse connections. Via the latter the substance in one sieve tube can pass over into the other and both substances are then swept along by mass flow in both directions.

Active Transport

Transport in both the xylem and the phloem is a long distance transport. We have just mentioned that the sieve plates still raise problems for the mass flow hypothesis. According to some hypotheses transport over the short distance through the sieve plates is not accomplished by mass flow as in the sieve tubes but by an active transport. This active transport is over a short distance and its essential characteristic is that it *requires* the *expenditure of energy.*

An active transport can convey substances over short distances against a concentration gradient. It is particularly important when it is necessary to pass through membranes. We need only to be remainded of the ATP-driven sodium and potassium pumps that are indispensable to the bioelectric effects in the nervous system.

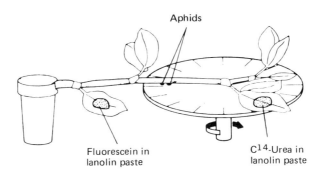

Fig. 226. Apparent two-directional transport in one sieve tube. A *Vicia faba* plant was laid horizontal and fluorescein was supplied to a lower leaf and C^{14}-urea to an upper. Aphids were applied between the two leaves and their honey dew (cf. Fig. 222) was collected in a collector rotating beneath them. Both substances could be detected in the honey dew of one aphid. Since one aphid punctures only one sieve tube this result at first suggested a two-directional transport in one sieve tube (from Eschrich 1967).

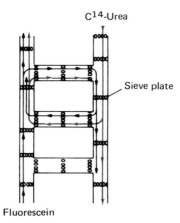

Fig. 227. Explanation of the result of the experiment shown in Fig. 226 without assuming a two-directional transport: the materials escaped into neighboring sieve tubes via transverse connections and are there carried along by mass flow in the opposite direction (from Eschrich 1967).

Several models have been developed to account for the mechanism of active transport through membranes (Fig. 228). In part, they are based on the existence of carriers (carrier hypotheses) which are charged with a particular substance, which they then transport through the membrane.

It is likely that active transport also plays a crucial role in the uptake of ions by the root. The ions, which are dissolved in the water of the soil, first come into contact with the cell wall. Passage through the cell wall appears not to cause any further difficulties. According to a current hypothesis the water-saturated cell wall is an "apparently free diffusion space." Within it ions can move by virtue of diffusion.

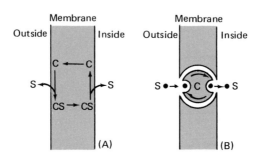

Fig. 228. Models of active transport. (A) A carrier C, which is located in the membrane, is charged with the substance S to be transported and carries it from the outside to the inside of the membrane. (B) A rotating system C is located in the membrane, which accepts the substance S and conveys it to the inside by rotation (from Luttge 1968).

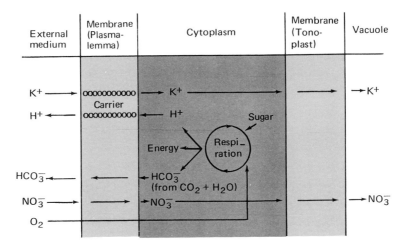

Fig. 229. Theory of active ion uptake according to Lundegardh (modified from Finck 1969).

The plasmalemma, on the contrary, is an obstacle that must be taken more seriously. In 1932, Lundegardh put forward a hypothesis of anion respiration which has hardened to a *theory of active ion uptake* in the meantime (Fig. 229). As a result of respiration CO_2 is formed which readily gives rise to H^+ and HCO_3^- in an aqueous medium. H^+ Ions are channeled to the outside of the plasmalemma by an active transport mechanism involving the participation of carriers. There they are exchanged for other cations, which have diffused through the cell wall and have now reached the plasmalemma. The carrier is charged with these ions and transports them in the opposite direction through the plasmalemma into the inside of the cell. There they can be employed in metabolism, transferred to neighboring cells via the plasmodesmata or via a suitable active transport system or, finally, secreted into vacuoles. Depending on the situation prevailing, transport through the tonoplasts surrounding the vacuoles is due to pure osmosis or, again, an active mechanism.

That all sounds very impressive. However, the fact should not be hidden that there are other hypotheses regarding the uptake of ions by the root. Most of the facts, though, do argue in favor of the fundamental validity of the theory of active ion uptake.

Flower Formation

A. Definitions

The first essential phase in the life of a plant was the development of the embryo during which all of the essential organs of the vegetative plant were molded. A second crucial phase was germination, the transition from the seedling, which is dependent upon the reserve materials of the parent plant, to the autonomous, photosynthetically active young plant. This transition is safe-guarded by many control mechanisms.

Few important new phenomena are encountered in the subsequent vegetative development. In the main those structures that were already present in the thriving seedling are further developed. We have considered in some detail a system of the kind that is developed particularly intensively during the vegetative phase of growth, namely, the vascular system.

Flower formation represents a further transition that, like germination, is of crucial importance and is therefore subjected to the most varied controls. *Flower formation signifies a transition from the vegetative to the reproductive phase of development.* The shoot meristem is now induced to develop sepals, petals, stamens, and carpels instead of leaves. This transition can only occur at a particular time in the life of the plant, which, within certain limits, is determined genetically: the plant has to have reached the stage of *ripeness-to-flower.* When that occurs varies very much from species to species. Once a plant is ripe to flower, it can be induced to form flowers. In this process two stages must be distinguished from each other, the induction of flowering and the differentiation of flowers and inflorescences.

By *induction of flowering* we mean all those processes that cause the cells of the shoot meristem to switch from the course of development that they have hitherto pursued and now form the leaf organs of the flower instead of leaves. Once this pivotal event has been initiated, the switches are thrown so that the *processes of differentiation leading to flowers or inflorescences* follow (Fig. 230). The induction of flowering is of particular interest to the physiologist and the plant grower. This is because differentiation follows the induction of flowering almost automatically. Control of flower

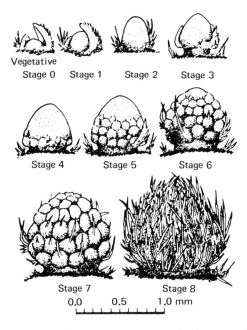

Fig. 230. The apical meristem of *Xanthium strumarium* in the vegetative state and in 8 different stages of flower differentiation (modified from Salisbury 1963).

induction thus implies control of flower formation. For that reason we shall give pride of place to the induction of flowering.

Now a few technical terms. During flower induction plants must be exposed to certain external conditions for a defined period of time. The time interval is known as the *induction period* and the efficacious external conditions as the *inductive conditions*. The natural, inductive, external conditions can, in many cases, be replaced by treatment with certain chemicals. External conditions under which plants remain in the vegetative state are called *noninductive*.

In the following sections we shall consider in some detail the two most important inductive external conditions, defined conditions of temperature and light. First we shall collect facts together and then try to fit the individual pieces of data into an overall picture which is still necessarily hypothetical.

B. Temperature and Flower Induction: Vernalization

It is sufficiently well known that our winter cereals thrive and bear corn only after they have been subjected to the cold of winter. In 1918, Gassner established that the cold can be effective even during germination. He spoke of a cold requirement, which can also be met by an experimental cold treatment. Experimental cold treatments aimed at flower induction have been carried out on very many species in the years and decades since then.

We have already discussed the effect of low temperatures on the development of plants several times, during germination (page 253), in the breaking of epicotyl dormancy (page 254) and in the dormancy of buds or its breaking (page 203). All of these phenomena, including sometimes the effect of cold on the induction of flowering, are usually comprised in the expression "vernalization." In the following we intend *vernalization* to mean only those processes which are triggered by cold treatment and which stimulate flower induction to a greater or lesser degree. The temperatures which are effective usually lie between a few degrees above zero and 15°.

In addition to our winter cereals many other winter annuals (page 253) and biennials (two-year plants) are dependent on vernalization. In many cases the biennials form a rosette of leaves which is attached to the soil in the first year with which they pass the winter. In the second year when the days become long enough shoots sprout forth and flowers are formed.

Let us now compile facts derived from a few more closely investigated plants.

1. Petkus Rye

In recent decades, Gregory and Purvis in particular have worked on the Petkus rye. Summer and winter varieties of it are known. The differences with regard to the cold requirement have been shown to be genetically controlled. The summer rye flowers in our latitudes without any clearly recognizable dependence on external factors. The winter rye first requires cold treatment, which is provided in nature by the winter following sowing, and then long days, which are presented to it in the following summer, in order to be able to flower. We shall consider the relationships between length of day and flower formation in more detail later (page 297).

We have already mentioned that seedlings can be vernalized. However, still earlier *stages of development* respond to cold treatment: vernalization can be undertaken just 5 days after pollination of the parent plant, by cooling in ice, for example. At this point in time the embryo consists of only a few cells. This finding is important for the reason that the number of cells is increased quite considerably before the plant reaches the flower forming stage. Nevertheless there is in no way a "dilution" of the single flowering impulse. One cannot help thinking of the participation of a system that can reduplicate identically such as the nucleic acids—a point we shall come across many times in our discussion of flower formation.

A further question concerns the *site* at which low temperature treatment acts. Information on this point was obtained from cultures grown on artificial nutrient media. Not only can isolated embryos be vernalized on defined media but vernalization can also be successfully carried out on isolated shoot tips. Plants capable of flowering can regenerate from shoot tips that have been subjected to low temperature. In the case of Petkus rye the apical meristem is the receptor site for low temperature treatment. Now of course this is not the case with all plants as we shall see.

The longer rye is vernalized the more quickly it flowers. Only after vernalization has lasted for about 20 days is a further extension of vernalization not accompanied by a shortening of the time taken to flower (Fig. 231). Thus, apparently during vernalization processes occur which proceed step by step and finally lead to a specific end product. The more of this end product that is present the more quickly the plant is induced to flower after completion of vernalization.

If vernalization is followed by treatment at high temperatures (in the case of Petkus rye at about 40° for not more than a maximum of two days) then the effect of vernalization is abolished to a large extent. This is known as *devernalization*. The shorter the preceding period of vernalization, the more complete is the devernalization. In this case too a stepwise reaction sequence leading to an end product can be inferred. If the period of vernalization was long enough and consequently sufficient end product had accumulated, then flower formation ensues rapidly. It should be noted that the high temperatures do not damage the rye. Devernalized plants can subsequently be vernalized again.

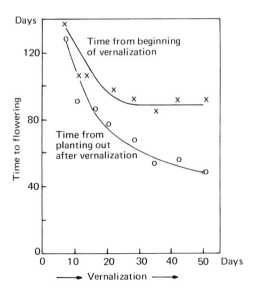

Fig. 231. Vernalization of Petkus winter rye (from Purvis and Gregory 1937).

2. Henbane (Hyoscyamus Niger)

Melchers chose henbane for his experiments. *Hyoscyamus* is familiar as a component of the witches' ointments of the Middle Ages which can give rise to certain kinds of hallucinations. Inspite of its dubious past the species turned out to be extremely useful in investigations of vernalization and also of photoperiodism.

An annual and a biennial breed of *Hyoscyamus niger* are known. The annual flowers in the same year in which it is sown. In the first year the biennial forms only a rosette of leaves that is attached to the soil and with which it passes the winter. After vernalization brought about by the cold of winter it forms flowers in the second year provided that the days are long enough (Fig. 232). Long days prevail in our latitudes in the summer. The difference between the annual and biennial breeds has also been shown to be genetically determined. Of the many facts discovered first by

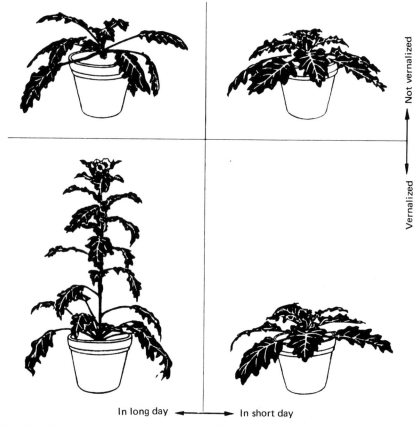

Fig. 232. The biennial variety of henbane *(Hyoscyamus niger)* and its formation of flowers under defined conditions of temperature and light (from Ruge 1966).

Melchers and then Lang we shall mention only a few that are relevant and helpful in this connection. The first important finding was that, in the case of the biennials, the low temperature must be presented first and then the long day afterwards. The same external conditions presented in the reverse order do not act inductively. Thus, in *Hyoscyamus* there are linked in series in this, and only in this order: processes which are dependent on low temperatures and processes which are dependent on long days.

Further important findings were made in *graft experiments* which were carried out on *Hyoscyamus* by Melchers in a manner similar to that applied in earlier experiments on beets by Vochting in the 1930s. The biennial breed of *Hyoscyamus* can be induced to flower even under noninductive external conditions by a flowering graft from the annual breed (Fig. 233). Results such as this led to the hypothesis that there is a *flowering hormone* or *florigen* that migrates from the flowering donor into the vegetative recipient and causes it to flower.

This assumption was made the more reasonable by the fact that a mutual growing together of the two tissues was not always necessary. Sometimes a close contact of the tissues did, no doubt, make the migration of the florigen from the donor to the recipient possible. Apparently agar too can transmit the florigen.

Now the graft experiments were successful not only within a species but also in the combination of different species. Thus, a biennial *Hyoscyamus* can also be brought to flower under noninductive conditions by a graft from a flowering tobacco twig (*Nicotiana tabacum*). Thus the florigen is not species specific. However, other findings do point to differences between the flowering hormones of some species. Details of attempts to isolate the flowering hormone will be discussed later (page 301).

We had just called to mind the effect of low temperature on the breaking of the dormancy of seeds, epicotyls, and buds. In all of these instances treatment with gibberellins can substitute for the effect of low temperature, at least in a number of plants which have been examined. Thus it seems reasonable to ask the question whether it might not be possible to administer gibberellins instead of cold treatment to bring about vernalization. That is indeed the case. Since 1956, Lang has induced flowering in different species that require vernalization such as carrots (*Daucus carota,* Fig. 234) by treatment with gibberellic acid without employing lower temperatures. Additional similar findings followed. Now this makes us sceptical and arouses the suspicion that gibberellic acid cannot be the true agent but rather a skeleton key which, as it were, passes by chance in the lock of vernalization. An objection to this possibility is that changes in the content of endogenous gibberellins could be demonstrated in vernalized plants as compared to nonvernalized plants. This is true too

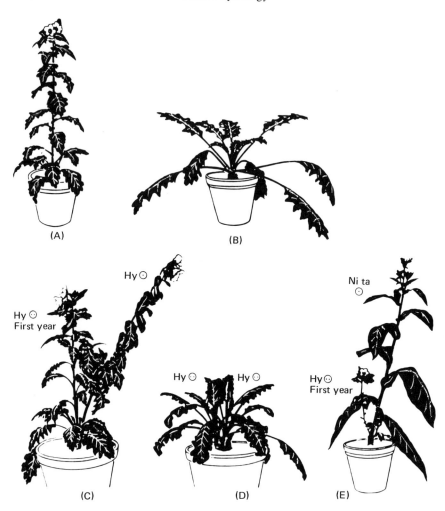

Fig. 233. Grafting experiments on *Hyoscyamus*. (A) Annual, (B) biennial in the first year, (C) annual (right) grafted on biennial in the first year, (D) control: biennial graft on biennial in first year; (E) *Nicotiana tabacum* graft on biennial in the first year (from Kuhn 1965).

in *Hyoscyamus* (Fig. 235). Thus, in all probability, gibberellins also participate in vernalization under natural conditions.

3. Streptocarpus Wendlandii

In discussing Petkus rye and henbane we have dealt with classical plants, as it were, which may be considered representative of very many other species. Let us now bring a further species into the picture, which ought to remind us that nature is inexhaustible in introducing new

Fig. 234. Replacement of cold treatment by gibberellic acid in the carrot *(Daucus carota)*. Not vernalized (left), treated with gibberellic acid (center), and vernalized (right) modified from Lang 1957).

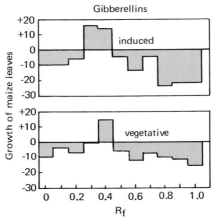

Fig. 235. The gibberellin content of cold-induced and noninduced plants of the biennial variety of *Hyoscyamus*. The cold treated plants contain somewhat more gibberellins. Autoradiography was carried out as in Fig. 167 (modified from Leopold 1964).

variants. In addition, we shall be able to further enlarge our collection of facts.

Streptocarpus wendlandii is a species of the family *Gesneriaceae,* which is native to the mountains of East to South Africa. To judge from its vegetative habits it is an oddity: one of the original two cotyledons dies early. The other, on the other hand, develops into a leaf train over a meter long which is the only leaf organ of the vegetative plant. The growth of this leaf is catered for by a meristem which is localized at the edge between the leaf surface and the hypocotyl.

Streptocarpus is induced to flower by being exposed for about 8 weeks to low temperatures (+ 10°) and short days. After this induction two things happen:

(1) The leaf begins to grow vigorously. During the 8 weeks of induction growth had been completely inhibited.

(2) Out of the same meristematic complex which is responsible for the vegetative growth an inflorescence develops which can be followed by others in serial arrangement.

In Petkus rye and other species such as *Hyoscyamus* the stimulus of low temperature is picked up by the shoot tips but in *Streptocarpus,* on the other hand, by the leaf. Oehlkers verified that in elegant experiments in 1955. The *Streptocarpus* leaf can be cut up into slips which easily form roots and grow into new plants. Slips from vegetative plants form only leaves at their base, never inflorescences without induction. Slips from induced leaves fall into several groups: some form only leaves, some first leaves and then inflorescences and some form inflorescences immediately (Fig. 236). Accordingly, this last group should show the highest content of florigen. Leaves were then cut into slips after different periods of flower induction. Slips which formed flowers immediately showed a

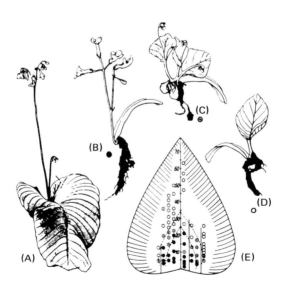

Fig. 236. Experiments with leaf cuttings from *Streptocarpus wendlandii.* (A) Flowering plant. (B) Cutting, which forms inflorescences immediately. (C) Cutting, which first forms leaves and then inflorescences. (D) Cutting, which forms only leaves. (E) Distribution of the cuttings B, C, and ,D over the leaf blade (from Kuhn 1965).

greater tendency to have been derived from the middle of the leaf after a short period of induction. With increasing time of induction the zone giving rise to slips which formed flowers immediately was displaced more and more towards the base of the leaf. This result demonstrated two things. Firstly, florigen is formed in the leaf rather than in the meristem under the influence of the inductive conditions. Secondly, florigen migrates from the middle of the leaf to its base where the inflorescences then usually develop.

Wellensiek was able to show on another plant requiring vernalization, the silver leaf (*Lunaria rediviva*), that the stimulus of the cold was also assimilated by the leaf. In the same experiments indications were obtained that the assimilation of the cold stimulus might be associated with the existence of cells capable of division. This shows a parallel with the assimilation of the stimulus of cold by the shoot meristem. There, too, cells capable of division are indeed present.

But now to return to *Streptocarpus*. It has been shown that 2-thiouracil, a structural analogue of uracil (page 22), is incorporated into RNA in *Streptocarpus*. If the plant is now treated with 2-thiouracil during the induction of flowering, the formation of flowers is completely blocked or, at least, seriously delayed, whereas the growth of the leaf is not impaired. Thus, flower formation is *selectively inhibited* (Fig. 237, cf. see below).

The result of the experiment can be interpreted as follows: under the conditions of induction genes involved in flower induction are activated.

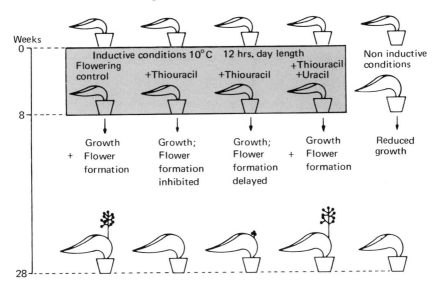

Fig. 237. Selective inhibition of flower formation in *Streptocarpus wendlandii* by 2-thiouracil. Similar results are obtained after treatment with ethionine, a structural analog of methionine (modified from Hess 1968).

They form mRNA. Incorporation of 2-thiouracil leads to a spurious mRNA and thereby abolishes the effect of flower induction. Other genes are responsible for the growth of the leaf. The activity of these genes is temporarily inhibited during flower induction: the leaf does not grow and also cannot be induced to grow by administration of phytohormones. Thus, during flower induction the genes for the growth of the leaf do not form any mRNA. This, in turn, means that thiouracil treatment does not lead to any spurious mRNA for leaf growth. It has already been demonstrated (page 24) that there really are RNA systems that are important for vegetative growth. After the completion of induction, genes for the growth of the leaf then become active again. If thiouracil is first supplied at this point both the growth of the leaf and the differentiation of the inflorescences are disturbed. However, the induction of flowering can no longer be abolished by treatment at such a late stage: all plants flower sooner or later — unless such a high dose of thiouracil has been chosen that the whole plant dies.

These and other confirmatory findings left no doubt that genes for flower induction become active in *Streptocarpus* during the induction period. Apart from that we also have here an example of differential gene activity. Before induction the genes for the growth of the leaf are active, during induction the genes for flower induction are active and those for leaf growth temporarily repressed, and after induction the genes for leaf growth become active again and, in addition, those for the differentiation of inflorescences.

Since these first experiments, carried out in 1959, similar experiments with antimetabolites of transcription and translation have been carried out on very many other species that are dependent on temperature and length of day, with similar results. Thus, there is adequate demonstration that the genetic material participates in the processes of flower induction.

4. A Hypothesis Concerning Vernalization

Let us now try to assemble the facts compiled thus far into a hypothesis concerning the events occurring in vernalization (Fig. 238). Two hypotheses that resemble each other in important points were put forward by Gregory and Purvis and by Melchers and Lang.

First of all a genetic background can be taken for granted. We shall leave it at this rather vague expression, for we do not know what the individual processes are in which these genes engage. In front of this genetic backdrop the following events occur. A substance A is converted at low temperature and in one or more steps into a substance B. B is unstable. At higher temperatures it can be converted either back into A or into a by-product D. This interpretation does justice to the eventuality of a devernalization. If the low temperatures persist B is converted in one or more steps into a substance C. C is the stable end product of vernalization.

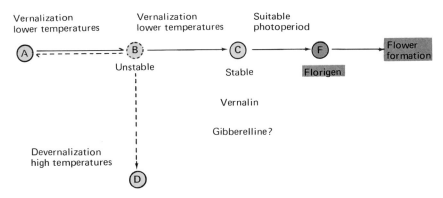

Fig. 238. Hypothesis concerning vernalization.

Accordingly, one can call it *Vernalin*. Now this complies with the fact that during vernalization an end product gradually accumulates over several reaction steps. As far as the chemistry of C is concerned the possibility of its being a gibberellin is discussed. The main reason for saying this is that gibberellins can substitute for the effect of low temperature.

However the formation of vernalin is not enough to bring about vernalization. In addition, a suitable length of day is necessary: for most plants requiring vernalization, such as winter rye and henbane, a long day, and in the case of *Streptocarpus,* for example, a short day. In the appropriate photoperiod (page 298) vernalin is then converted into a substance F, the actual flowering hormone or florigen. An alternative might be that vernalin itself is not converted into florigen but that it regulates the synthesis of florigen from other precursors. Florigen then induces meristematic cells, usually those of the shoot meristem, to adopt the new direction of reproductive development. With that flower induction comes to an end and the processes of flower differentiation commence.

In our discussion of vernalization we have more than once mentioned the dependence of flower formation on the length of day. A dependence of this kind is shown not only by species requiring vernalization but by very many others too.

C. Length of Day and Flower Induction: Photoperiodism

In 1920, in the neighborhood of Washington, U.S.A., Garner and Allard endeavored to bring a variety of the tobacco plant with particularly large leaves to flower. The variety was given the name of "Maryland Mammoth." In Washington it flowered so late in the year that the formation of seeds in the open fields would have occurred in winter. *Nicotiana tabacum* "Maryland Mammoth" had to be brought into the greenhouse if the seeds and, thus, the variety were to survive. After many fruitless experiments Garner and Allard solved the puzzle: "Maryland Mammoth" flowered only if the plant had been maintained for a certain time under conditions of short days and long nights. Thus, the phenomenon of photoperiodism was presented to the scientists in an acute form. If in what follows photoperiodism as it affects flower induction is placed in the foreground, we nevertheless ought not to forget that there is, in addition, a large number of other manifestations of photoperiodism.

1. Long and Short Day Plants, Neutral Day Plants

Once having been alterted to the fact, scientists were soon able to draw up long lists of plants that have different requirements for the length of day or night. Leaving aside special cases, one can distinguish between long day plants, neutral day plants, and short day plants. For both long and short day plants there is a critical length of day which is species specific and usually lies between 10 and 14 hours. Long day plants require a period of illumination exceeding the critical day length (more than 10–14 hours), neutral day plants show no clearly recognizable dependence on the length of day and short day plants demand illumination for a time shorter than the critical day length (usually less than 10–14 hours), before they switch over to flower formation. Close relatives can belong to quite different groups. Thus, *Nicotiana tabacum* "Maryland Mammoth" is a short day plant, *N. sylvestris* a long day plant. As we shall see a critical day length corresponds to a night length that is just as "critical." For this reason it has even been proposed that the short day plants be renamed long night plants and the long day plants be renamed short night plants. A few examples of each group are presented in Table 10.

2. Analysis of Photoperiodism in Flower Induction

a. The leaf as the site of light uptake

Now that we are familiar with the phenomenon, let us consider a few data relating to its causation. First of all it was shown that the leaf was the

Table 10. Several short day plants and long day plants (SDP and LDP); important experimental plants are italicized

SDP	LDP
Cannabis sativa	Allium cepa
Chrysanthemum indicum	Avena sativa
Dahlia variabilis	Beta vulgaris
Helianthus tuberosus	Daucus carota
Kalanchoe bloßfeldiana	*Hyoscyamus niger*
Nicotiana tabacum	Nicotiana sylvestris
Perilla ocymoides	Lactuca sativa
Soja hispida	Papaver somniferum
Xanthium strumarium	Vicia faba

organ that received the light stimulus. If, for example, only a single leaf of
N. sylvestris is grafted on to a plant of *Nicotiana tabacum* "Maryland Mammoth" that is kept under long day conditions, *N. tabacum* was induced to
flower (Fig. 239). Another experiment, which was carried out by Harder
on *Kalanchoe blossfeldiana,* a well-known red flowering plant of the
Crassulaceae from Madagascar, will also be mentioned (Fig. 240).
Kalanchoe is a short day plant which flowers only if it does not receive
more than 10–12 hours of light per day for a certain period of time. Now it
suffices to keep just one leaf under these inductive conditions of day
length, as by putting a little bag over it to keep it dark, for the shoot above
this leaf to be induced to flower.

b. Experiments to demonstrate a flowering hormone

The light stimulus is received by the leaf but the pivotal event takes
place in the shoot meristem. This finding alone suggests the existence of a
flowering hormone even in plants that are dependent on day length,
which migrates from the leaves to the apical meristem. Another experi-

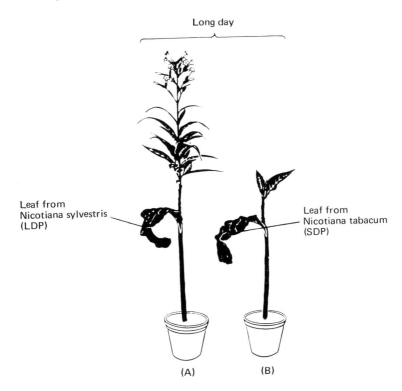

Fig. 239. Flower induction in *Nicotiana tabacum* "Maryland Mammoth," a short day
plant, in the long day by grafting a leaf of the long day plant *N. sylvestris.* (A). Control:
leaf from *N. tabacum* M. M. grafted onto *N. tabacum* M. M. (B) (from Kuhn 1965).

ment, which points in the same direction, will now be outlined (Fig. 241). Even the cotyledons of *Pharbitis nil*, one of the *Convolvulaceae*, can be induced. *Pharbitis* is a short day plant. Now Zeevaart kept the plants for different lengths of time in darkness to induce them and then removed the cotyledons on bringing them into the light. If the cotyledons are removed after 14 hours of darkness, none of the plants were able to flower. Fourteen hours of darkness, i.e. a single inductive cycle, are normally quite sufficient to induce flowering. Controls that had retained both cotyledons, were able to flower normally after 14 hours induction.

If, however, the cotyledons are first removed after 18 hours of dark induction, the plants are able to flower. A further extension of the dark period is not able to further enhance the flowering effect. The fact that only 5 blossoms were formed per plant instead of 7 as in the controls is due to damage sustained during removal of the two single leaf organs, the cotyledons, which were fully developed at the time of induction. The result of the experiment can hardly be interpreted otherwise than by a migration of a flowering hormone from the cotyledons into the shoot meristem. After 14 hours of darkness the flowering hormone has already been formed in the cotyledons but has not yet migrated out of them. On the other hand, after 18 hours sufficient florigen has left the cotyledons to guarantee a full flowering effect.

Finally, here again graft experiments suggest the existence of a

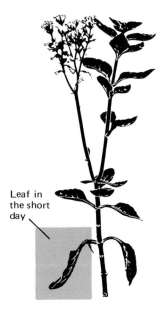

Leaf in
the short
day

Fig. 240. Uptake of the light stimulus by the leaf. The vegetative part of the plant of *Kalanchoe blossfeldiana* begins to flower if one single leaf is maintained under short day conditions (modified from Kuhn 1965).

flowering hormone. As Tschailachjan, among others, showed, plants that are dependent upon the length of day can also be induced to form flowers during photoperiods that are not inductive if they are combined with a flowering graft partner. We have already mentioned a graft of this kind between tobacco and henbane (page 291). This example also shows us that these kinds of grafts can be successfully carried out between different species as well as within a species. Thus, the flowering hormone is not species specific — that was also shown by suitable grafting in the case of plants that require vernalization. In addition, long day plants can be induced to flower by short day plants and vice versa. Thus, the flowering hormone in long and short day plants should be identical. Its migration occurs in the phloem. Now an induced plant can be grafted on to a noninduced plant and cause it to flower. Then the plant that was induced to flower by grafting can be grafted on to a second noninduced plant and cause the latter to form flowers. This trick can be repeated again and again. The important point is that all of the plants in this series are induced to flower to their full intensity. Thus, dilution of the florigen does not occur. This again suggests the participation of processes involving identical reduplication in flower induction, without its being possible up to now to say anything more specific than that.

In view of apparently convincing indications of the existence of a flowering hormone it is reasonable to ask whether this substance has

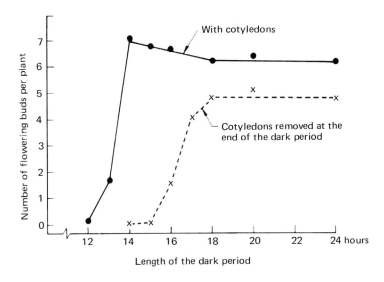

Fig. 241. Evidence for the migration of the flowering hormone from the leaf to the apical meristem in *Pharbitis nil*. The test for migration consisted of the removal of the cotyledons at different times after the beginning of the inductive dark period. *Pharbitis* is a short day plant in which even the cotyledons can be induced (modified from Zeevaart as presented in Beermann et al. 1966).

already been isolated and characterized. To this question one must give the almost embarassing answer that the plants have outwitted even experienced biochemists throughout the world, despite all their efforts. Up to now no one has succeeded, in definitive and reproducible experiments, in extracting the flowering hormone and using it to induce plants to flower under noninductive conditions. Only in the case of the cocklebur *Xanthium strumarium* have a few, apparently positive findings occasionally been published, the last in 1969. Perhaps a turning point is in the offing here.

c. Effect of light breaks: significance of the dark period

If short day plants are subject to a period of darkness, the duration of which would normally be sufficient to induce flowering, and this is then interrupted by light in the form of a flash or a brief time exposure, then the plants remain vegetative (Fig. 242). Conversely, if a dark period, which is normally too long to permit the induction of flowering in long day plants, is interrupted, the long day plants flower as if they had been exposed to a long day.

These experiments with light breaks show us first of all that crucially important processes occur in plants which are dependent upon the length of day even in the dark period. As already mentioned, that is why the designations long night plants and short night plants would be just as appropriate as short day plants and long day plants. Finally, we must also mention that these experiments with light breaks, which pure scientists tackle, bring benefits in hard cash. If, for example, one wishes to induce long day plants to flower in winter it is not necessary to expose them continuously to light. Instead the dark period can be interrupted for a short time with a light break and electricity is saved.

d. Participation of the phytochrome system

We have just learned that interruption of the dark period of short day plants can prevent the induction of flowering. These light breaks involve

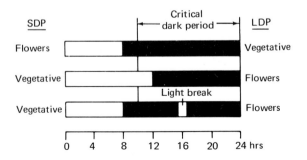

Fig. 242. Effect of light breaks during the dark period on flower formation in long day plants (LDP) and short day plants (SDP).

ordinary white light. It now becomes interesting to know the wavelength of the active component of the white light. To this end, Borthwick and Hendricks and their colleagues carried out experiments with *Xanthium strumarium,* the short day plant already mentioned.

The active component is red light. If red light is given in a light break, flower formation does not occur (Fig. 243). Furthermore, it could also be shown that, if exposure to red light is followed by exposure to far red light, the plants flowered. Thus, far red light had overcome the inhibitory effect of red light. This kind of alternation between red and far red light could be repeated. If we now recall similarly designed experiments on the germination of lettuce seeds which had similar results (page 221), then we can identify the common factor: the phytochrome system is also engaged in the induction of flowering. The next question we must ask concerns the manner in which phytochrome participates. Borthwick and Hendricks have developed the following hypothesis for short day plants (cf. Fig. 244). Irradiation with red light leads to the formation of P_{730}, the active phytochrome. Its activity consists in an inhibition of the induction of flowering via an unknown mechanism. In daylight more red light is present than far red. Consequently, in the daytime the concentration of P_{730} is adjusted to a level that is high enough to prevent the induction of flowering. At night P_{730} is converted back again into P_{660} in a process which to all appearances seems to be enzymatically controlled. A large number of findings have proved that this conversion of P_{730} into P_{660} actually can occur without irradiation by far red light. As a consequence of this conversion during the night the content of inhibitory P_{730} falls: the synthesis of the flowering hormone, blocked up to now, can now begin.

We must not hide the fact that this hypothesis, even in the case of some short day plants, can only be made plausible by invoking additional

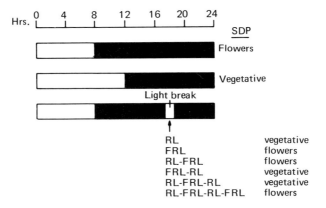

Fig. 243. The participation of the phytochrome system in flower induction. Experiments on the short day plant (SDP) *Xanthium strumarium.* RL = red light, FRL = far red light (modified from Galston 1964).

hypotheses. Attempts to adapt them to the circumstances found with long day plants have been relatively unconvincing up to now.

e. Gibberellins and flowering hormones

Gibberellins are able to substitute for the effect of low temperature on many plants that require vernalization. It seemed reasonable to test gibberellins on plants that are dependent on day length, too. Let us now briefly summarize the results relating to both kinds of effect:

(1) Most species which require low temperature treatment can be induced to flower by treatment with gibberellins.

(2) If after cold treatment plants still require long days then usually treatment with gibberellins can only substitute for the effect of cold and not for that of the long day.

(3) In the case of long day plants that have no requirements for cold, gibberellin treatment can substitute for the long day. That is especially true of those long day plants in which the transition to flower formation is associated with the sprouting of rosettes.

(4) With short day plants, the short day cannot be replaced by gibberellin treatment.

Thus, gibberellins can substitute for either the effect of cold or the long day as shown in point 2 but seldom both at the same time. Usually the short day cannot be replaced by gibberellins. We can also exclude with certainty the possibility that the gibberellins might be identical with the much sought after flowering hormone. The reasons for this are:

(1) according to the results of grafting experiments long day plants and short day plants possess the same flowering hormone (page 301). However, gibberellins are able only to substitute for the long day but not for the short day;

(2) even in long day plants gibberellins do not always act to trigger flowering. It happens particularly in rosette plants. Long day plants with normally extended internodes, so-called caulescent long day plants, cannot be induced to flower by gibberellins.

Thus, gibberellins play a role in flower formation but are not flowering hormones. We shall provide further confirmation for this statement by sketching experiments which were carried out in 1962 by Zeevaart and Lang on the *Bryophyllum daigremontianum*. As far as its requirements for a photoperiod are concerned the bryophyllum is a special case. The plant requires first long day, then short day in order to be able to flower; thus, it is a long day-short day plant. Treatment with gibberellins can substitute for the long day: gibberellin treatment under short day conditions induces the plant to flower (Table 11). The short day, on the other hand, cannot be replaced by gibberellins. Thus, if gibberellins are administered to the plants under long day conditions, they remain vegetative. All of these findings are consistent with the statements we made above.

Table 11. The relationship between gibberellic acid and florigen in the long day-short day plant *Bryophyllum daigremontianum*. LD = long day, SD = short day, GA = gibberellic acid, → from … to, − = no flowers, + = flower formation

| Treatment | Flower formation | Florigen donor | Graft experiments | |
			Florigen recipient	Flower formation
LD	−	LD → SD	SD	+
LD+GA	−	SD+GA	SD	+
SD	−	SD (control)	SD	−
SD+GA	+	LD → SD →		
		LD	LD	+
LD → SD	+	SD+GA → LD	LD	+
		LD (control)	LD	−

Let us now consider an extension of these findings by grafting experiments. Plants are again treated under short day conditions with gibberellic acid. They begin to flower. These plants are then used in grafting experiments as "flowering hormone donors." Their graft partners, the "flowering hormone recipients," are now induced to flower under long day conditions. That finding is the crucially important one: gibberellins cannot substitute for the short day in *Bryophyllum* but the "flowering hormone donor," on the other hand, which has been induced to flower with the aid of gibberellins can. This demonstrates once again that gibberellins evidently participate in the synthesis of the flowering hormone but are not identical with it.

f. Participation of the genetic material

The administration of antimetabolites of transcription and translation inhibits the induction of flowering in plants that are dependent on the length of day just as it does in those requiring vernalization. The inhibitors produce their effect partly in the leaf, partly in the shoot meristem. As far as their activity in the leaf is concerned, it may be assumed that they impair the activity of the genes that are involved in the synthesis of the flowering hormone. At least in part their effectiveness in the shoot meristem has to do with subsequent flower differentiation rather than with flower induction: the activity of the genes for flower differentiation is interferred with.

Furthermore, older findings of Salisbury and Bonner on *Xanthium strumarium* and more recent ones (1969) of Krekule on *Chrysanthemum rubrum,* another short day plant, suggest that antimetabolites such as 5-fluorodeoxyuridine might inhibit DNA reduplication in the shoot meristem, which appears to be necessary for successful induction of flowers. Here again a parallel with vernalization is revealed, since successful vernalization also appears to be associated with cells that are capable of dividing (page 301). The ability to divide also implies the ability to reduplicate DNA. It hardly makes sense at the moment to try to interpret these findings.

g. A hypothesis concerning the photoperiodic induction of a short day plant

Let us now make an attempt to unite in a hypothesis the data that have been brought together in the preceding sections. Of course, this is something that can only be done for the better studied short day plants (Fig. 244). Ordinary daylight contains more red than far red light. Consequently, in the daytime a constant concentration of P_{730} is maintained in the leaf and this is high enough to bring about a chemically unknown type of inhibition of the synthesis of flowering hormone. In darkness, P_{730} is gradually converted into P_{660}. If the darkness lasts long enough and the critical dark period exceeded then so little P_{730} is present that the synthesis of flowering hormone can begin. It is controlled by appropriate genes and can be inhibited by antimetabolites of transcription and translation. Gibberellins are involved in the synthesis of florigen in a manner that is still unknown.

The more often the plants are exposed to an inductive dark period, the more the flowering hormone that is formed. After a number of light-dark cycles, which varies from species to species, sufficient flowering hormone is present. It then migrates in the phloem through the leaf stalk into the shoot and upward to the apical meristem. There florigen activates the genes for flower differentiation. This gene activation can also be inhibited by appropriate antimetabolites. With the bringing into action of the genes for flower differentiation the process of flower induction is already over. That the scales have been tipped becomes microscopically visible (Fig. 230).

In summary, it must be repeated that this is a hypothesis. In order to

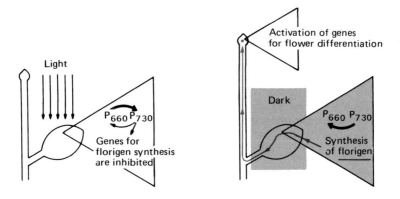

Fig. 244. Hypothesis concerning photoperiodic induction of a short day plant.

be able to judge the situation we need only to bear in mind that the existence of florigen, a point central to this scheme, has still in no way been rigorously proved.

3. Photoperiodism in Flower Induction as a Sign of Adaptation

In our discussion of germination, we had established that the different barriers to, and conditions for, germination developed as adaptations to quite well-defined external conditions that varied from case to case. The same seems to be true of flower formation, an equally critical transition in the life of the plant. This is because the formation of flowers and seeds at the wrong time can be just as disastrous as germination under unfavorable conditions. An adaptation to external realities is particularly obvious in the dependence of flower formation on the photoperiod.

First it can be stated that plants in higher latitudes are chiefly long day plants. Thus they are induced by increasing day length and can still complete flowering and fruit formation before winter begins. Short day plants would be out of place here. This is because they would be induced when the days became shorter in the autumn and would then inevitably reach their reproductive phase in the winter. On the other hand, it makes sense that plants in very low latitudes, i.e. tropical plants, should be short day plants if they are not neutral day plants. This is because long day plants would not be induced to flower on the equator. Up to now we have spoken only of high and very low latitudes. In the intermediate geographical latitude of 35–40°, however, both long day and short day plants abound. In particular, data have been compiled for our cultivated plants which come from these latitudes. In the high latitudes it was the harsh winter and in the low latitudes the constant, short day length, which caused an appropriate adaptation. In the intermediate latitudes another kind of regulation often prevails: the dry period.

If we simplify matters, we can say that a dry period is found either in summer or in winter in the intermediate latitudes. In 1957, Junges investigated the connections between the origin of a number of cultivated plants, the climate in the place of origin, and the photoperiodic behavior. It was found that cultivated plants from regions with winter drought such as certain regions of China, India, and Central America are short day plants whereas cultivated plants from regions with summer drought such as certain regions of Central Asia, the Near East and the Mediterranean area are, on the contrary, long day plants. The sense of this "arrangement" probably is to be seen as follows. The permanent organ of the plant, which endures the period of drought whenever it falls, is the seed. Thus, to put it anthropomorphically, the plants in regions with dry winters hurry to pass over to flowering and fruit-bearing as the days become shorter. For plants in regions with dry summers the lengthening days are the alarm signal to begin sexual propagation as fast as possible.

Man has planted the surface of the entire earth with certain cultivated plants, even in geographical latitudes that make flowering at the right time impossible. However, knowledge of the photoperiodic bases of flower induction makes it an easy task to induce useful plants to propagate by exposing them either to light or darkness. Conversely, he can, if desired — think of a lettuce running to seed — prevent the formation of flowers completely at will.

4. Light and Circadian Rhythms

The sections on the induction of flowering will have appeared to some readers as rather too hypothetical inspite of all the facts that we know. This is because there are considerably more data which we don't know. It is just in this connection with photoperiodism that we cannot help mentioning one more subject, the central relationships of which are unknown inspite of a wealth of pieces of data. We intend to discuss briefly circadian rhythms.

a. The phenomenon of the physiological clock

We have discussed that plants can require long or short days in order to be able to flower. In the discussion one question was passed over: how do plants actually notice whether they are being maintained under long day or short day conditions? How do plants measure the length of day? Thus, we are asking what is their internal clock.

First we must distinguish between two different types of clock:

(1) *The Hour-glass.* A certain process is set in motion by a stimulus and proceeds until it is finished, just as the sand in an hour-glass trickles from the upper to the lower half. Thus, time is measured by the expiration of a particular process. A clock of this kind is easily imaginable in a living organism.

(2) *The Oscillating Clock.* The real "physiological" or "biological" clock comes under this heading and has been studied in detail in plants, particularly by Bunning. Consider an example: the petals of *Kalanchoe blossfeldiana* open in the morning and close again in the evening. Now it is possible to maintain *Kalanchoe* for some time in complete darkness too. If the plants had first been kept in the normal light-dark cycle, the movement of the petals also continues in constant darkness with only minor damping (Fig. 245).

Behind the visible phenomenon of petal movement there must be a chronometric mechanism, which goes through similar oscillations, i.e. an oscillator. Once this physiological clock is set, it continues in this fixed rhythm. In the example the timing was determined by the light-dark alternation first presented to the plant. The timing factors are called the pace setters.

Physiological clocks like this are widely distributed in living creatures of all kinds. They are coupled to the most varied rhythms. As far as 24 hour rhythms are concerned, one used to speak of "endogenous diurnal rhythms." It was soon noticed, however, that these rhythms usually show periods of 18–21 hours, not exactly 24, under constant external conditions. For this reason the designation *circadian rhythms* (circa = about, dies = day) is preferred nowadays.

b. Photoperiodism in flower induction and the physiological clock

Let us return to our special problem, photoperiodism in flower formation. How is time measured in this instance, by an hour-glass or by an oscillating clock, the physiological clock in its narrower sense?

According to presently available data the answer seems to be simple: by an hour-glass. A particular process is set in motion by the light-dark alternation and the plant measures time by the course this process takes. A process of this kind is apparently already present. By this is meant the conversion of P_{730} to P_{660} that begins with the transition from light to darkness. Under certain conditions this process of conversion might indeed be the hour-glass which is used by plants to measure the duration of darkness.

However, this hour-glass mechanism can function only if the hour-glass is turned over again after it has run down: there must be a continual alternation between light and dark. During the period of darkness P_{730} is converted into P_{660} and in the light P_{730} is regenerated, the clock is wound up again. If the plants are brought into constant conditions, such as constant darkness, then this chronometric mechanism should run down.

However, that is just what does *not* happen. Even in constant darkness the plants pass through periodically recurring phases in which they show a varied sensitivity to light. First let us look at an example of evidence in short day plants (Fig. 246). As already mentioned (Fig. 242), interruption of the dark period of short day plants such as *Kalanchoe blossfeldiana* by

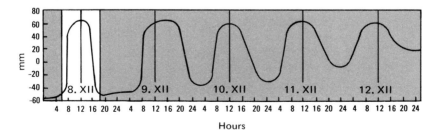

Fig. 245. Circadian opening (rising of the curve) and closing (falling of the curve) of the petals of *Kalanchoe blossfeldiana*. After initial setting by light-dark alternation, the succession of movements can be maintained in constant darkness (modified from Bunning 1967).

light breaks can prevent flowering. After being subjected initially to circadian light-dark cycles *Kalanchoe* was brought into darkness for a period that was considerably prolonged and the darkness was interrupted at defined intervals with light breaks of two hours' duration. Thus, the darkness was probed with light breaks for possible differences in sensitivity to light. As the effect on flower formation showed periodically recurring phases of different sensitivity to light are indeed present, even in constant darkness.

And now a similar experiment with a long day plant (Fig. 247). Flower formation in long day plants such as *Hyoscyamus niger* is stimulated by light breaks (Fig. 242). After being subjected to circadian light-dark cycles *Hyoscyamus* was put into prolonged darkness and the darkness was again probed with light breaks of two hours duration for phases of different light sensitivity. That such phases periodically recur, even in constant darkness, was revealed by the effect on flower formation.

After being subjected initially to circadian light-dark cycles plants maintain the same periodicity with respect to their sensitivity to light for 2–3 days under conditions of constant light or darkness. Up to now we have always spoken of photoperiodism but only now do we see how justified we were. For phases of equal sensitivity to light actually do recur periodically. Behind these appearances of photoperiodism in the induction of flowering there is an oscillating, physiological clock.

c. Hypothesis concerning the nature of the physiological clock

We do not know the nature of physiological clocks. However, results in recent years do, at least, allow hypotheses to be put forward for animals as well as plants. One of these hypotheses, which is based on experiments

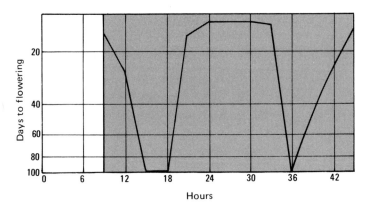

Fig. 246. Photoperiodism in flower formation of a short day plant *(Kalanchoe blossfeldiana)*. In separate experiments a prolonged period of darkness was probed each time with a two hour light break. As the effect on flower formation shows, phases of different sensitivity to light recur periodically (modified from Bunning 1967).

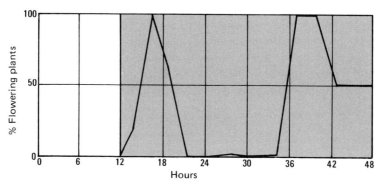

Fig. 247. Photoperiodism in flower formation of a long day plant *(Hyoscyamus niger)*. In separate experiments, a prolonged period of darkness was probed each time with a two hour light break. As the effect on flower formation shows, phases of different sensitivity to light recur periodically (modified from Bunning 1967).

by Engelsmaa on cucumber seedlings, will be presented in simplified form. In the case of the clock investigated light was again the pace setter, or, more precisely, the alternation between light and dark.

Let us call to mind the key enzyme of phenylpropane metabolism, phenylalanine-ammonium-lyase or PAL (page 122). Its synthesis can be induced by light in a variety of plants, including cucumber seedlings. One of the products resulting from PAL activity, *p*-coumaric acid, can repress the synthesis of the enzyme. The nature of this repression is complex. In one sense it appears to be end product repression in accordance with the Jacob-Monod model, but there is also circumstantial evidence for the participation of a newly formed protein. However it may be, the accumula-

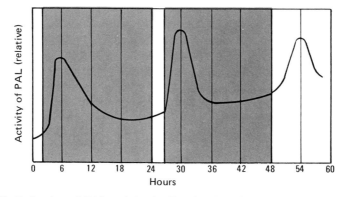

Fig. 248. Induction of PAL activity by illumination. Cucumber seedlings are first grown in the dark. Then, at O hour, they were exposed to a light period. The activity of PAL shot up as a result. If one wishes to induce a similar increase in activity later, one must increase the duration of the light period. This is because the light must counteract increasing end product repression by *p*-coumaric acid. The periods of darkness are shown in red (modified from Engelsma 1968).

tion of *p*-coumaric acid in the tissues of the cucumber seedlings leads to the inhibition of the synthesis of PAL.

If cucumber seedlings are first grown in the dark and then exposed to light, the activity of PAL increases rapidly (Fig. 248). After passing through a maximum the activity gradually begins to fall if the seedlings are placed in darkness again. The reason for this is repression of PAL synthesis by the *p*-coumaric acid, which has accumulated in the meantime.

If one wishes to produce an increase in activity by renewed exposure to light, which is just as high as that obtained after the first light exposure, then the time of exposure must be prolonged in order to overcome the end product repression by *p*-coumaric acid (Fig. 248, after 24 hours). The same is true of a third attempt to boost PAL activity by exposure to light. The exposure time must be prolonged still further (Fig. 248, after 48 hours). Thus, the system becomes less light sensitive the more *p*-coumaric acid has accumulated as a consequence of increases in PAL activity.

Up to now we have been dealing with an hour-glass which is set running by exposure to light. Now one can visualize how this kind of system might be turned into an oscillating clock. For that to happen two conditions must be fulfilled:

(1) A single induction by light must serve to promote the genes for PAL synthesis to a permanently active state which can be inactivated only temporarily by repression.

(2) Provision must be made for the removal of the inhibitory principle, *p*-coumaric acid — by consumption or transport.

Under these conditions the initial repression by *p*-coumaric acid would be relaxed with increasing consumption of the substance and gene activity, having been induced once and for all by light, would be able to reassert itself without renewed exposure to light. After a time there would again be an increase in the activity of PAL even in darkness, *p*-coumaric acid would accumulate, there would be a decrease in activity of PAL due to end product repression by *p*-coumaric acid, *p*-coumaric acid would be consumed etc. In this way a physiological clock would be established, at least in principle. The point that is of particular interest to us is that phases of different sensitivity to light would recur periodically just as we have seen in photoperiodism. This is because a high content of *p*-coumaric acid always implies a limited light sensitivity of the system for PAL synthesis (cf. above).

This hypothesis certainly rests on a weak foundation. But let us not lose sight of the crucially important point: the fact that one can even *attempt* an explanation at the molecular biological level of so complex a phenomenon as the physiological clock is a better indication than many others of the revolution which has occurred in modern botany.

Bibliography

Books for reference and suggested reading are listed below. *Introductory text; **more advanced text.
Review articles (for advanced students), which are published periodically, include Fortschritte der Botanik, and Annual Reviews (particularly of Biochemistry, Genetics and Plant Physiology).

General

* Baron, W. M. M.: Organization in Plants. E. Arnold Ltd, London 1967, 2. ed.
* Bennet, Th. P.: Elements of Protein Synthesis. An Instructional Model. Freeman, San Francisco 1969.
** Bielka, H.: Molekulare Biologie der Zelle. Gustav Fischer, Stuttgart 1969.
* Bogen, H.: Knaurs Buch der modernen Biologie. Droemer-Knaur, Munchen, Zurich 1965.
** Bonner, J., und E. Varner: Plant Biochemistry. Academic Press, New York and London 1965.
* Bresch, C., und R. Hausmann: Klassische und Molekulare Genetik. Springer, Berlin-Heidelberg-New York 1972, 3. ed.
* Buddecke, E.: Grundriß der Biochemie. W. de Gruyter, Berlin 1970.
Clowes, F. A., und B. E. Juniper: Plant Cells. Blackwell, Oxford und Edinburgh 1968.
* DeBusk, A. G.: Molecular Genetics. Macmillan, New York and London 1968.
* Devlin, R. M.: Plant Physiology. Van Nostrand Reinhold Co., New York, Cincinnati, Toronto, London, Melbourne 1969, 2. ed.
** Dickerson, R. E., und I. Geis: The Structure and Action of Proteins. Harper and Row, New York, Evanston und London 1969.
* Durand, M., und P. Favard: Die Zelle. Vieweg, Braunschweig 1970. Frey-Wyssling, A., und K. Muhlethaler: Ultrastractural Plant Cytology. Elsevier, Amsterdam 1965
* Galston, A. W.: Physiologie der grünen Pflanze. Franckh, Stuttgart 1964.
* Goldsby, R. A.: Cells and Energy. Macmillan, New York and London 1967.
* Günther, E.: Grundriß der Genetik. Gustav Fischer, Stuttgart 1971, 2. ed.
Hayes, W.: The Genetics of Bacteria and their Viruses. Blackwell, Oxford and Edinburgh 1968, 2. ed.
** Hess, D.: Biochemische Genetik. Springer, Berlin-Heidelberg-New York 1968.
*Hess, D.: Genetik, Herder, Freiburg, Basel, Wien 1974 3. ed.
* Karlson, P.: Kurzes Lehrbuch der Biochemie. G. Thieme, Stuttgart 1972, 8. ed.

* Kimball, J. W.: Cell Biology. Addison-Wesley-Publ. Co., Reading, Menlo Park and London 1970.

* Kuhn, A.: Grundriß der Vererbungslehre. Quelle und Meyer, Heidelberg 1965, 4. ed.

** Ledbetter, M. C., and K. R. Porter: Introduction to the Fine Structure of Plant Cells. Springer, Berlin, Heidelberg, New York 1970.

* Lehninger, A.: Bioenergetik. G. Thieme, Stuttgart 1969.

* Lenhoff, E. S.: Tools of Biology. Macmillan, New York 1966.

** Mohr, H.: Lehrbuch der Pflanzenphysiologie. Springer, Berlin-Heidelberg-New York 1971, 2. ed.

* Nagl, W.: Chromosomen. Goldmann, Munchen 1972.

* Novikoff, A. B., and E. Holtzman: Cells and Organelles. Holt, Rinehart and Winston, New York 1970.

* Nultsch, W.: Allgemeine Botanik. G. Thieme, Stuttgart 1968, 3. ed.

** Price, C. A.: Molecular Approaches to Plant Physiology. McGraw Hill, New York, 1970.

Ruhland, W.: Handbuch der Pflanzenphysiologie, Bd. I-XVIII. Springer, Berlin-Heidelberg-New York 1955—1967.

* Schlegel, H. G.: Allgemeine Mikrobiologie. G. Thieme, Stuttgart 1969.

** Sitte, P.: Bau und Feinbau der Pflanzenzelle. Gustav Fischer, Stuttgart 1965.

* Stent, G. S.: Molecular Genetics. Freeman, San Francisco 1971.

Steward, F. C.: Plant Physiology, vol. I—VI C. Academic Press, New York and London 1960-1972.

* Stiles, W., and E. C. Cocking: An Introduction to the Principles of Plant Physiology. Methuen, London 1969, 3. ed.

Steward, F. C.: Plant Physiology, vol. I—V B. Academic Press, New York and London 1960—1969.

* Strasburgers: Lehrbuch der Botanik fur Hochschulen. Gustav Fischer, Stuttgart 1967, 29. ed.

* Träger, L.: Einfuhrung in die Molekularbiologie. Gustav Fischer, Stuttgart 1969.

* Vogel, G., und H. Angermann: dtv-Atlas zur Biologie, Bd. 1 und 2. Deutscher Taschenbuch Verlag, Munchen 1967/1968.

* Walter, H.: Grundlagen des Pflanzenlebens. Eugen Ulmer, Stuttgart 1962, 4. ed.

** Wagner, R. P., und H. K. Mitchell: Genetics and Metabolism. John Wiley, New York, London and Sydney 1965, 2. ed.

* Watson, J. D.: Molecular Biology of the Gene. W. A. Benjamin, New York and Amsterdam 1970 2. ed.

* Wieland, Th., und G. Pfleiderer: Molekularbiologie. Umschau-Verlag, Frankfurt 1969, 3. ed.

* Wissen im Uberblick. Band "Das Leben". Herder, Freiberg i. Br. 1970.

Metabolism

** Alston, R. E., and B. C. Turner: Biochemical Systematics. Prentice Hall, Englewood Cliffs 1963.

* Beadle, G., und M. Beadle: Die Sprache des Lebens. S. Fischer, Frankfurt 1969.

 * Brewbaker, J. L.: Angewandte Genetik. Gustav Fischer, Stuttgart 1967.
** Bu'Lock, J. B.: Biosynthese von Naturstoffen. Bearbeitet und ubersetzt von H. Grisebach und W. Barz. Bayerischer Landwirtschaftsverlag, Munchen 1970.
** Davies, D. D., J. Giovanelli und T. AP Rees: Plant Biochemistry. Blackwell, Oxford 1964.
Freudenberg, K., und A. C. Neish: Constitution and Biosynthesis of Lignin. Springer, Berlin-Heidelberg-New York 1968.
Goodwin, T. W.: Chemistry and Biochemistry of Plant Pigments. Academic Press, London and New York 1965.
Goodwin, T. W.: Biochemistry of Chloroplasts. vol. I und II. Academic Press, London and New York 1966/1967.
Goodwin, T. W.: Porphyrins and related Compounds. Academic Press, London and New York 1968.
Goodwin, T. W.; Natural Substances formed biologically from Mevalonic Acid. Academic Press, London and New York 1970.
Harborne, J. B.: Biochemistry of Phenolic Compounds. Academic Press, London and New York 1964.
Harborne, J. B.: Comparative Biochemistry of Flavonoids. Academic Press, London and New York 1967.
Haslam, E.: Chemistry of vegetable Tannins. Acedemic Press, London and New York 1964.
Hawkes, J. G.: Chemotaxonomy and Serotaxonomy. Academic Press, London and New York 1968.
Hegnauer, R.: Chemotaxonomie der Pflanzen, Bd. 1—5. Birkhauser, Basel and Stuttgart 1962—1969.
 * Heywood, V. H.: Plant Taxonomy. E. Arnold, London 1967.
Heywood, V. H.: Modern Methods in Plant Taxonomy. Academic Press, London and New York 1968.
Ikan, R.: Natural Products. A. Laboratory Guide. Academic Press, London and New York 1969.
Karrer, W.: Konstitution und Vorkommen der organischen Pflanzenstoffe. Birkhauser, Basel and Stuttgart 1958.
Kirk, J. T. O., und R. A. E. Tilney-Basset: The Plastids. Freeman, San Francisco and Folkestone 1967.
** Luckner, M.: Der Sekundarstoffwechsel in Pflanze und Tier. Gustav Fischer, Stuttgart 1970.
Metzner, H.: Progress in Photosynthesis Research, vol. I—III. IUBS-Publication, Tubingen 1969.
** Moritz, O.: Einfuhrung in die Pharmazeutische Biologie. Neubearbeitet von O. Moritz und D. Frohne. Gustav Fischer, Stuttgart 1967, 4. ed.
Mothes, K., und H. R. Schutte: Biosynthese der Alkaloide. VEB Deutscher Verlag der Wissenschaften, Berlin 1969.
Paech, K.: Biochemie und Physiologie der sekundaren Pflanzenstoffe. Springer, Berlin-Gottingen-Heidelberg 1950.
Pridham, J. B.: Terpenoids in Plants. Academic Press, London and New York 1967.
Reinhold, L., und Y. Liwschitz: Progress in Phytochemistry, vol. 1. Interscience (John Wiley), London, New York, Sydney 1968.

* Richter, G.: Stoffwechselphysiologie der Pflanzen. G. Thieme, Stuttgart 1969.

** Robinson, T.: The Biochemistry of Alkaloids. Springer, Berlin-Heidelberg-New York 1968.

Roy-Burman, P.: Analogues of Nucleic Acid Components. Springer, Berlin-Heidelberg-New York 1970.

Schnepf, E.: Sekretion und Exkretion bei Pflanzen. Protoplasmatologia VIII 8. Springer, Wien and New York 1969.

** Steinegger, E., und R. Hansel: Lehrbuch der Pharmakognosie. Springer, Berlin-Heidelberg-New York 1968, 2. ed.

* Sullivan, N.: Die Botschaft der Gene. Suhrkamp, Frankfurt 1969.

Swain, T.: Chemical Plant Taxonomy. Academic Press, London and New York 1963.

Swain, T.: Comparative Phytochemistry. Academic Press, London and New York 1966.

Swan, G. A.: An Introduction to the Alkaloids, Blackwell, Oxford and Edinburgh 1967.

** Swanson, C. P., T. Merz und W. J. Young: Zytogenetik. Gustav Fischer, Stuttgart 1970.

* Tribe, M., and P. Whittaker: Chloroplasts and Mitochondria. Arnold, London 1972.

Vernon, L. P., und G. R. Seely: The Chlorophylls. Academic Press, New York and London 1966.

Ycas, M.: The Biological Code. North Holland Publishing Co., Amsterdam and London 1969.

* Zähner, H.: Biologie der Antibiotica. Springer, Berlin-Heidelberg-New York 1965.

Development (including water conservation and transport)

Beermann, W. et al.: Cell Differentiation and Morphogenesis. North Holland Publishing Co., Amsterdam 1966.

Bernier, G.: Cellular and molecular aspects of floral induction. Longman, London 1970.

* Black, M., und J. Edelman: Plant Growth. Heinemann, London 1970.

** Bonner, J.: The Molecular Biology of Development. Clarendon Press, Oxford 1965.

Bonner, J., und P. Ts'o: The Nucleohistones. Holden Day Inc., San Francisco, London, Amsterdam 1964.

** Bünning, E.: Entwicklungs- und Bewegungsphysiologie der Pflanze. Springer, Berlin-Gottingen-Heidelberg 1953, 3. ed.

** Bünning, E.: The Physiological Clock. Springer, New York 1967, 2. ed.

Busch, H.: Histones and other Nuclear Proteins. Academic Press, New York and London 1965.

Butenko, R. G.: Plant Tissue Culure and Plant Morphogenesis. Israel Program for Scientific Translation, Jerusalem 1968.

* Butterfass, Th.: Wachstums- und Entwicklungsphysiologie der Pflanze. Quelle und Meyer, Heidelberg 1970.

Carr, D. J.: Plant Growth Substances 1970. Springer, Berlin-Heidelberg-New York 1972.

** Cutter, E.: Plant Anatomy: Experiment and Interpretation. Part I and II. Arnold, London 1969 and 1971.

DeReuck, A. V. S., und J. Knight: Histones. Churchill Ltd., London 1966.

* Ebert, J. D., and I. M. Sussex: Interacting Systems in Development. Holt, Rinehart und Winston, New York 1970.

Evans, L. T.: The Induction of Flowering. Macmillan of Australia, North Melbourne 1969.

* Finck, A.: Pflanzenernahrung in Stichworten. Ferdinand Hirt, Kiel 1969.

* Galston, A. W., and P. J. Davies: Control Mechanisms in Plant Development. Prentice Hall, Englewood Cliffs 1970.

John, B., und K. R. Lewis: The Chromosome Cycle. Protoplasmatologia VI B. Springer, Wien and New York 1969.

Kaldewey, H., and Y. Vardar: Hormonal Regulation in Plant Growth and Development. Verlag Chemie, Weinheim 1972.

** Kühn, A.: Entwicklungsphysiologie. Springer, Berlin-Heidelberg-New York 1965, 2. ed.

** Leopold, A. C.: Plant Growth and Development. McGraw Hill, New York 1964.

Lüttge, U.: Aktiver Transport (Kurzstreckentransport bei Pflanzen). Protoplasmatologia VIII 7 b. Springer, Wein und New York 1969.

Milborrow, B. V. (ed.): Biosynthesis and its Control in Plants. Academic Press, London and New York 1973.

Mitrakos, K., and W. Shropshire: Phytochrome. Academic Press, London and New York 1972.

Mohr, G., und H. Ziegler: Symposium uber Morphaktine. Gustav Fischer, Stuttgart 1969.

** Mohr, H.: Lectures on Photomorphogenesis. Springer, Berlin-Heidelberg-New York 1972.

** Mohr, H., und P. Sitte: Molekulare Grundlagen der Entwicklung. BLV, Munchen, Bern, Wien 1971.

Nikolaeva, M. G.: Physiology of Deep Dormancy in Seeds. Israel Program for Scientific Translation, Jerusalem 1969.

** Ruge, U.: Angewandte Pflanzenphysiologie. Eugen Ulmer, Stuttgart 1966.

* Richardson, M.: Translocation in Plants. E. Arnold, London 1968.

** Salisbury, F. B.: The Flowering Process. Pergamon Press, Oxford-London-New York-Paris 1963.

** Sinnot, E. W.: Plant Morphogenesis. McGraw Hill, New York, Toronto, London 1960.

Sitte, P.: Probleme der biologischen Reduplikation. Springer, Berlin-Heidelberg-New York 1966.

Slatyer, R. O.: Plant-Water Relationships. Academic Press, London and New York 1967.

** Stahl, F. W.: Mechanismen der Vererbung. Gustav Fischer, Stuttgart 1969.

** Steward, F. C., and A. D. Krikorian: Plants, Chemicals and Growth. Academic Press, New York 1971.

Street, H. E. (ed.): Plant Tissue and Cell Culture. Blackwell, Oxford 1973.

* Street, H. E., and H. Opik: The Physiology of Flowering Plants. Arnold, London 1970.

* Sutcliffe, J.: Plants and Water. E. Arnold, London 1968.

Sweeney, B. M.: Rhythmic Phenomena in Plants. Academic Press, London and New York 1969.

* Torrey, J. G.: Development in Flowering Plants. Macmillan, New York and London 1968, 3. ed.

* Waddington, C. H.: Principles of Development and Differentiation. Macmillan, New York and London 1966.

Wardlaw, C. W.: Morphogenesis in Plants. Methuen, London 1968, 2. ed.

** Wareing, P. F., and I. D. J. Phillips: The Control of Growth and Differentiation in Plants. Pergamon Press, Oxford 1970.

** Wilkins, M. B.: Physiology of Plant Growth and Development. McGraw Hill, London 1969.

Ziegler, H.: Symposium Stofftransport. Gustav Fischer, Stuttgart 1968.

Sources of Illustrations

The sources of those illustrations which have been taken from the literature are given below. With a few exceptions these illustrations have been adapted considerably to the needs of the text; half-tone engravings have been reproduced in practically every case as line drawings. All other illustrations are original.

Baron, W. M. M.: Organization in Plants. E. Arnold, London 1967, 2. ed.: 33, 34, 35.

Beermann, W.: Jahrbuch der Max-Planck-Gesellschaft 1966, 69: 144.

Beermann, W., et al.: Cell Differentiation and Morphogenesis (Symposium). North-Holland Publishing Co., Amsterdam 1966: 241.

Bennett, Th. P.: Elements of Protein Synthesis. An instructional Model. Freeman, San Francisco 1969: 4,12.

Biddulph, S., und O. Biddulph: Scientific American 200, 44, 1959: 213.

Bielka, H.: Molekulare Biologie der Zelle. Gustav Fischer, Stuttgart 1969: 134, 155.

Bonner, J., und J. Varner: Plant Biochemistry. Academic Press, New York and London 1965: 18.

Bopp, M.: Naturwiss. Rdsch. 16, 349, 1963: 140.

Bünning, E.: The Physiological Clock. Springer, New York 1967, 2. ed.: 245, 246, 247.

Chroboczek, H., und J. H. Cherry: Biochem. biophys. Res. Commun. 20, 774, 1965: 146.

Clowes, F. A. L., und B. E. Juniper: Plant Cells. Blackwell, Oxford und Edinburgh 1968: 49.

Cutter, E. G.: Plant Anatomy: Experiment and Interpretation. Part I. Cells and Tissues. E. Arnold. London 1969: 192.

Dure, L., und L. Waters: Science 147, 410, 1965: 19.

Engelsma, G.: Umschau 1968, 727: 248.

Eschrich, W.: Planta 73, 37, 1967: 226, 227.

Feierabend, J.: Umschau 1967, 494: 207.

Finck, A.: Pflanzenernahrung in Stichworten. Ferdinand Hirt, Kiel 1969; 229.

Freudenberg, K., und A. C. Neisch: Constitution and Biosynthesis of Lignin. Springer, Berlin-Heidelberg-New York 1968: 100, 102.

Galston, A. W.: Physiologie der grunen Pflanze. Franckh, Stuttgart 1964: 193, 243.

Goldsby, R. A.: Cells and Energy. MacMillan, New York and London 1968, 2. ed.: 29, 30, 62.

Goodwin, T. A.: Chemistry and Biochemistry of Plant Pigments. Academic Press, London and New York 1965: 178.

Hess, D.: Z. Pflanzenphysiol. 56, 295, 1967: 148.

Hess, D.: Biochemische Genetik. Springer, Berlin-Heidelberg-New York 1968: 9, 75, 76, 85, 136, 142, 149, 151, 170, 174, 175, 176, 182, 237.

Hess, D.: in Wissen im Uberblick, Band "Das Leben". Herder, Freiberg i. Br. 1970: 137.

Hotta, Y., and H. Stern: Proc. Nat. Acad. Sci. (Wash.) 49, 648, 1963: 150.

Jacobs, W. P.: Amer. J. Bot. 90, 163, 1956: 211.

Janick, J., R. W. Schery, F. W. Woods und V. W. Ruttan: Plant Science. Freeman, San Francisco 1969: 1, 194, 198.

Karlson, P.: Kurzes Lehrbuch der Biochemie. G. Thieme, Stuttgart 1970, 7. Aufl.: 47,48.

Kaudewitz, F.: in 9. Kolloquium der Ges. f. Physiol. Chemie in Mosbach, 104, 1958:5.

Key, J. L., und J. Ingle: Proc. nat. Acad. Sci. (Wash.) 52, 1382, 1966: 17.

Kimball, J. W.: Cell Biology. Addison-Wesley Publ. Co., Reading, Menlo Park und London 1970:11.

Koller, D.: Scientific American 200, 75, 1959: 205.

Kreutz, W.: Umschau 1966, 806: 37 a und b; in H. Metzner, Progress in Photosynthesis Research vol I, Tubingen 1969: 37 c and d.

Kühn, A.: Grundriβ der Genetik. Quelle and Meyer, Heidelberg 1965, 4 Aufl.: 143, 177.

Kühn, A.: Entwicklungsphysiologie. Springer, Berlin-Heidelberg-New York 1965, 2. Aufl.: 184, 209, 233, 236, 239, 240.

Kursanov, A. L.: in Advances in Botanical Research vol I, 209. Academic Press, London and New York 1963: 223.

Lamport, D. T. A.: in R. D. Preston, Advances in Botanical Research vol. 2, S. 151. Academic Press, London and New York 1965: 195a.

Lang, A.: Proc. Nat. Acad. Sci. (Wash.) 43, 709, 1957: 234.

Lange, H., und H. Mohr: Planta 67, 107, 1965: 181.

Lehninger, A.: Bioenergetik. G. Thieme, Stuttgart 1969: 26, 61, 63, 81.

Leopold, A. C.: Plant Growth and Development. McGraw-Hill, New York, San Francisco, Toronto, London 1964: 195, 235.

Levine, R. P.: Scientific American 221, 58, 1969: 32.

Lingens, F.: Hohenheimer Reden und Abhandlungen 26, 24, 1969: 158.

Lüttge, U.: Aktiver Transport (Kurzstreckentransport) in Pflanzen. Springer, Wien and New York 1969: 228.

Lynen, F.: Jahrbuch der Max-Planck-Gesellschaft 1969, 46: 21, 65.

Mohr, G., und H. Ziegler: Symposium Morphaktine. Gustav Fischer, Stuttgart 1969: 173.

Nagl, W.: J. Cell Sci. 6, 87, 1970: 145.

Nultsch, W.: Allgemeine Botanik. G. Thieme, Stuttgart 1968, 3. ed.: 139.

Overbeck, J. van: Scientific American 219, 75, 1968: 208.

Plant Research '66. MSU/AEC Plan Res. Laboratory; East Lansing/Michigan 1966: 168.

Purvis, O. N., und F. G. Gregory: Ann. Botany 1, 569, 1937: 231.

Rabideau, G. S., und G. O. Burr: Amer. J. Bot. 32, 349, 1945: 221.

Richardson, M.: Translocation in Plants. E. Arnold, London 1968: 220.

Rüdiger, W., in K. Mitrakos und W. Shropshire: Phytochrome. Academic Press, London and New York 1972.

Ruge, U.: Angewandte Pflanzenphysiologie. Eugen Ulmer, Stuttgart 1966: 201, 203, 204, 232.

Ruhland, W.: Handbuch der Pflanzenphysiologie, Bd. XIV, 1162. Springer, Berlin, Gottingen, Heidelberg 1961:166.

Salisbury, F. B.: The Flowering Process. Pergamon Press, Oxford, London, New York, Paris 1963: 230.

Skoog, F., und C. O. Miller: Biological Action of Growth Substances. Cambridge University Press, Cambridge 1957: 200.

Stahl, F. W.: Mechanismen der Vererbung. Gustav Fischer, Stuttgart 1969: 135.

Steward, F. C.: Scientific American 209, 104, 1963: 196.

Steward, F. C.: Plant Physiology, vol. IV A, 1965: 147; vol. IV B, 1966: 217. Academic Press, New York and London.

Steward, F. C.: Pflanzenleben. Bibliographisches Institut, Mannheim and Zurich 1969: 159.

Steward, F. C., M. O. Mapes, A. E. Kent and R. D. Holsten: Science 143, 20 (1964).

Stout, P. R., und D. R. Hoagland: Amer. J. Bot. 26, 320, 1939: 214.

Strasburgers: Lehrbuch der Botanik fur Hochschulen. Gustav Fischer, Stuttgart 1967, 29. ed.: 141.

Sutcliffe, J.: Plants and Water. E. Arnold, London 1968: 215.

Taylor, J. H., P. S., Woods und W. L. Hughes: Proc. nat. Acad. Sci. (Wash.) 43, 122, 1957: 138.

Torrey, J. G.: Development in Flowering Plants. Macmillan, New York and London 1968, 3. ed. 163, 164, 188, 190, 210, 212.

Walter, H.: Grundlagen des Pflanzenlebens. Eugen Ulmer, Stuttgart 1962, 4. ed.: 161, 185, 189, 197, 216, 219.

Wilkins, M. B.: Physiology of Plant Growth and Development. McGraw-Hill, London 1969: 31, 167.

Zachau, H. G., D. Dutting, und H. Feldmann: Angew. Chemie 78, 392, 1966: 10.

Zenk, M. H.: Ber. dt. Bot. Ges. 83, 325 (1970): 176a.

Ziegler, H.: Naturwiss. 50, 177, 1963: 224, 225.

Index

Leading references are given in bold type; astericks refer to pages carrying illustrations.

Abscisic acid, 105, 210*, 252
  assays for, 210
  biosynthesis, 210
  chemical constitution, 210
  functions, 211
  historical, 210
Abscisins, 101, 203*
Acetal, 63*, 64
Acetaldehyde, 78*
Acetate-malonate pathway, 118, **120***, 129, 133, 137*
Acetate-mevalonate pathway, 104, 118, 137*
Acetyl-CoA, 79*, 82*, 90, 91, 94, 95, 96, 97*, 103* (*see also* active acetate)
Aconitase, 80
cis-Aconitate, 80
Actidione, 22
Actinomycin C_1 (-D), **22**, 23*, 24, 176, 222, 223*
Active acetate, 79*, 80*, 81*, 82* (*see also* acetyl CoA)
Active ion uptake, **285***
Active transport, **283**, 284*
Acyl carrier protein, 92, 93*
Acyl-CoA, 120*
Adaptive enzyme formation, 184
Adenine, 3*, 157, 205
Adhesion, 276
ADPG, 59, 69
Aesculetin, 125*
Aglycone, **63**, 64, 109, 131
Alanine, 141, 142*, 143*
Aldolase, 52*, 75, 76*
Aldol condensation, 52
Alkaloids, 115, **144**, 145*
  place in metabolism, 159
Allosteric effects, 192
Amaryllidaceae alkaloids, 145*, **150**
Amination, reductive, 140

Amines, biogenic, 144, 146, 147, 150, 155
Amino acids, 138, 144
  C skeleton, 142*
  families, 143*
Amino-acyl tRNA, **15**, 16*, 18*
Amino-acyl tRNA synthetase, **15**, 16
δ-Aminolaevulinic acid, 160*
Amo 1618, 212*
Amygdalin, 251, 252*
Amylases, 70, 179, 204
  α-amylase, 68*, 70, 204*, 205, 259
  β-amylase, 68*, 70
Anabasin, 147*, 149*
Anthocyanidins, 117*, 118, 130*, **131**, 132*
  biosynthesis, **133**, 134*
Anthocyanins, 8*, 131, 158, 186, 220*, 223*, 249
Anthranilic acid, 119*
Antibiotics, **22**
Anticodon, 13*, 16*, **17**
Antimetabolites, **20**, 179
  flower formation, 294, 295*, 305
Aphid technique, **279***, 282, 283*
Apical dominance
  cytokinins, 207
  gibberellins, 203
  IAA, 198
  morphactins, 213
Apigenin, 130*
Appositional growth, 235*
Arabinose, 58*
Arbutin, 117*, 128*, 129
Arginine, 142*, 143*, 188
Arginine operon, 186
Asparagine, 140, 156
Aspartic acid, 54, 141*, 142*, 143*, 148*, 156, 193*, 196, 200
Aspartokinase, 193*
ATP, 85*, 86*

Autoradiography, 50, 51*, 167, 169*
Auxins, 197 (see also IAA)
Avena curvature test, 196*, 197
Avena section test, 197*

Balata, 100*, 115
Balbiani ring, 175
Base pairing, 5, 6*, 13
Belladine, 151*
Benzaldehyde, 251, 252*
Benzoic acid, 150*, 200
Benzylaminopurine, 205, 206*
Benzylisoquinoline alkaloids, 153
Betacyanidins, 153
Betacyanins, 144, 145*, 152, 153*
Betalamic acid, 153*
Betanidine, 153*
Betaxanthine, 145*, 152, 153*
Bile acids, 107
Biological oxidation, 74*, 85
Biotin, 90, 92*
Blastocolins, 253
5-Bromouracil, 20, 21*
Bufadienolide, 109*
Building blocks, analysis of, 132
Butein, 130*

1-6 C-Acylation, 120*
Cadaverine, 146*, 147*
Caffeic acid, 122*, 123*, 125*, 129, 253
Caffeine, 156, 157*, 158
Callus, 246, 247*, 264, 265*, 266* (see also) wound callus
Calvin cycle, 51, 52*, 75, 258
Cambium,
    gibberellins, 202
    IAA, 198
Camphor, 100*, 105*
Carboxydismutase, 51, 52*, 56, 257, 258*
Cardenolides, 109*
Carotene,
    α-, 112*, 113, 114*
    β-, 112*, 113, 114*
    γ-, 114*
    δ-, 114*
    ζ-, 112, 113, 114*
Carotenes, 100*, 112*
Carotenoids, 100*, 111, 136
    biosynthesis, 113
    chemical constitution, 111
Carotenoid acids, 113*
Carrier hypotheses, 284*
Catalases, 41, 83, 179

Catechols, 130*, 131
C₄-dicarboxylic acid pathway, 53
Cell division, 163
    cytokinins, 206, 207, 235
    equilibrium with elongation, 235
    fruit development, 249
    gibberellins, 201, 202
    IAA, 198, 235
    unequal, 225, 226
Cell elongation, 198, 204*, 231
    course, 232
    definition, 231
    fruit development, 198*, 249
    gibberellins, 201
    IAA, 198, 236
    regulation, 235
Cell hemins, 41, 101
Cellobiases, 72
Cellobiose, 64, 65*
Cellulases, 72, 259
Cellulose, 71*
Chalcones, 129, 130*, 134*
Chemosystematics, 158
Chinchona alkaloids, 156
Chloramphenicol, 22
Chlorcholine chloride, 212*, 213
Chlorogenic acid, 122
Chlorophyllides, 160, 161*
Chlorophylls, 38, 46*, 101
    absorption spectrum, 39*
    active chlorophyll, 45*
    ionization, 40*
    localization, 54, 55*, 56
    structure, 40*
    synthesis, 160*
Chloroplasts, 54, 55*
Cholesterol, 106*, 108
Chorismic acid, 119*
Chromatin, 29*, 188
Chromosomes, 1, 164, 188
    semiconservative replication, 167, 168, 169*
    single strand model, 169
Chymochromic pigments, 136
Cinnamic acid, 119, 122*, 123*, 252
Cinnamic acid CoA, 133*, 135
Cinnamic acids, 117*, 119, 121, 122*, 128, 131, 134
Cinnamic acid start hypothesis, 135*
Cinnamyl alcohol, 117*
Circadian rhythms, 308
Citral, 100*, 105*
Citrate, 80, 82*
Citric acid cycle, 74*, 75, 80, 82*

Citronellol, 105*
Climacteric, 249
Cocaine, 150*
Cocoa nut milk, 206, 246, 247
Code, genetic, 11, 12*
Codeine, 154*
Codogen, 13*,
Codon, 11, 12*, 13*
Coenzyme A, 79*, 80*, 81, 82*
Cohesion, 276
Cohesion theory, 275, 277*
Colchicine, 144, 145*, 150, 152*, 168
Compartmentalization, 98, 104
    chloroplasts, 56
    mitochondria, 87
Competitive inhibition, 20, 21*, 81, 82*,
    191*, 192
Complex formation (chelates), 137
Condensing enzyme, 80, 82*
Coniferin, 126*, 128
Coniferyl alcohol, 126*, 127
Convergence, 158
Copigments, 137
o-Coumaric acid, 124*
p-Coumaric acid, 119, 122*, 123*, 124*,
    125*, 126*, 127, 128*, 129, 311, 312
Coumarin, 122, 123*, 252
    bound, 124, 248
Coumarins, 117*, 122, 124
p-Coumaryl alcohol, 126*, 127
Cristae, mitochondrial, 86, 87*
Crocetin, 113*
Crosslinks, 234, 238, 239*
Cyanidin, 132*, 135*, 136
Cycloartenol, 106*, 108, 110
Cyclopentanoperhydrophenanthrene
    see Steran
Cysteine, 140, 142*, 143*
Cytochromes, 41, 42*, 46*, 84*, 87*
Cytochrome oxidase, 83, 84*, 87*
Cytokinins, 101, 157, 205, 259, 260*
    assays for, 205
    biosynthesis, 157, 206
    chemical constitution, 205
    functions, 207
    historical, 205
    tRNA, 209
Cytosine, 3*

2,4-D, 200, 206, 212*, 213, 214
Dark fixation of CO$_2$, 273
Dark germinators, 256
Decarboxylation
    α-ketoglutarate, 81, 82*

pyruvate, 75, 78, 79*, 81*
5-Dehydroquinic acid, 118, 119*
5-Dehydroshikimic acid, 118, 119*, 120
Delphinidin, 132*, 136
Demissidin, 110*
Deoxyribonucleic acid
    see DNA
2-Deoxyribose, 3*
Deplasmolysis, 234
Depside, 122
Development, definition, 163
Devernalization, 289, 297
2,4-Dichlorophenoxyacetic acid
    see 2,4-D
Dicoumarol, 248
Differential gene activity, 172, 294, 296*
    definition, 175
Differentiation
    definition, 163
    shoot and root, 245, 246, 247*
Digitalis glycosides
    see Heart glycosides
Digitoxigenin, 109*
Dihydroflavanol, 134*, 135
Dihydroxyacetone phosphate, 51*, 52*,
    75, 76*, 77*, 94*
Dihydroxyphenylalanine
    see DOPA
Dimethylallyl pyrophosphate, 103*, 104,
    115
1,3-Diphosphoglyceric acid, 75, 77*, 78
Direct oxidases, 83
Disaccharides, 64, 65*
Diterpenes, 100*, 110
Diurnal acid cycle, 273
DNA
    autocatalytic function, 10, 165
    crown galls, 171
    heterocatalytic function, 10
    synthesis in a cell-free system, 165, 166*
    Watson-Crick model, 5, 6*
DNA ligases, 167
DNA polymerases, 165
DNA synthesis cycle
    see Mitotic cycle
DOPA, 153*, 154*
Dormancy, 250 (see also Germination, bar-
    riers to)
    of buds 202, 204*, 211
Dormin, 210
Double labelling technique (nucleic acids),
    178*
Dwarf mutants, 200, 202*
Dwarfs, physiological, 201, 255*

Ecdysone, 107, 204
Edge effect, 271*
Effectors, 184, 185, 186*
Electron transport, **36**, 37*, 42*, 83
  cyclic, 40*, 46*, 48
  non cyclic, 46*, 48
Electron transport chains
  see Electron transport
Embryonal development, 243*, 244*
Embryos, 242, 243*
  culture 245*, **247**
Emerson effect, **44**
Emulsin, 251, 252
End product activation, 194*
End product inhibition, 192*, 193*, 196*
Energy balance
  biological oxidation, 85*
  fermentations, 78
  glycolysis, 76, 77, 78
  α-oxidation (fatty acids), 96
  β-oxidation (fatty acids), 96
Enolase, 76, 77*
Enol pyruvate, 76, 77*
Enzyme activity, regulation, **191**,
Enzyme synthesis
  cytokinins, 208
  gibberellins, 204
  IAA, 200, 239
Epicotyl dormancy, 254, 255*
Epimerases, 59
  4-epimerase, 59, 60*, 61*
  ribulose-3-epimerase, 59, 60*
Ergoline skeleton, 155*
Ergot alkaloids, 155*, 158
Erythrose-4-phosphate, 52*, 118, 119*,
  194*
Esterase, pH 7.5, 32, 33*
Ethanol, 78*
Ethionine, 20, 21*, 297*
Ethylene, 199, 200, 218, 249
Excretions, 102
Extensin, 72, 239*
External factors, regulation, 218

FAD, 42*, **43**, 46*, 81, 82*, 95*, 138, 139
Farnesol, 100*, 105
Farnesyl pyrophosphate, 103*, 104, 105,
  106*, 110
Fats, **87** (see also neutral fats)
Fatty acids, 88*, 94, 96*, 97*
  biosynthesis, **90**, 92*
  chemical constitution, **88**, 91*
  unsaturated, 91*, 93
Fatty acid synthetase, **92**, 93*

Fermentation
  alcoholic, 78*
  latic acid, 78*
Ferredoxin, 43, 46*
Ferredoxin-NADP reductase, 46*
Ferredoxin reducing substance (-FRS),
  43, 46*
Fertilization, 243*, 248
Ferulic acid, 122*, 123*, 125*, 126*, 127,
  129, 252
Flavan, 130*
Flavan-3,4-diols, 118, 130*, **131**, 134
Flavan-3-ols, 130*, 131
Flavan derivatives, 117*, 118, 120*, **129**,
  135
  biosynthesis, **132**
  chemical constitution, **129**
Flavanones, 117*, 118, 130*, 134*
Flavine adenine dinucleotide
  see FAD
Flavine mononucleotide
  see FMN
Flavone, 130*
Flavones, 117*, **130***, 134*
Flavonols, 117*, 118, **130***, 134*, 198
Flavoproteins, **41**, 42*, 43, 46*, 84*, 85,
  87*, 138
Florigen
  see flowering hormone
Flower differentiation, 286, 287*
Flower formation, 204*, **286**
  definition, 286
Flower induction
  definition, 286
  gibberellins, **304**
  length of day, **297**
  temperature, **287**
Flowering hormone
  extraction, 302
  gibberellins, **304**
  graft experiments, 291, 299*, 301, 305
  migration of 294, 300, 301*
    (see also graft experiments)
  vernalin, 297*
Flower pigmentation, **136**
5-Fluordeoxyuridine, 21*, 171*, 192, 305
Flourenol, 212*, 213*, 214
5-Fluoruracil, 21*, 24
FMN, 41, 92*
Fructosans, 67*
Fructose, 57, 58*, 280*
Fructose-1,6-diphosphate, 52*, 75, 76*
Fructose-6-phosphate, 52*, 53, 60, 62, 75,
  76*

Fruit, 244
  development, 244, 248
  ripening, 249
Fruit fall, 199, 204*, 248
Fumarate, 81, 82*
Fumarate hydratase, 81, 82*

Galactinol, 66
Galactose, 58*, 59, 66
β-Galactosidase, 184, 237
Galacturonic acid, 60, 73
Gallic acid, 117*, 127*, 128
Gallotannins, 128
GDPG, 59, 72
Gene activity
  differential, **172**, **175**
  phytohormones, 214
  primary, 174, 175
  regulation, 183
  states of activity of genes, **181**
Gentiobiose, 64, 65*, 113*
Geotropism, 195*, 261*
Geraniol, 100*, 105*
Geranyl-geranyl pyrophosphate, 103*,
  113, 114*
Geranyl pyrophosphate, 103*, 104, 105
Germination, 204*, **250**
  cytokinins, 207
  definition, 250
  evolution, 259
  gibberellins, 203, 256
  light (phytochrome), **220**, **256***
  morphactins, 213
  phytohormones, **258**
  temperature, **253**
Germination, barriers to, **250**, 256
  impermeability, 251
  incomplete embryos, 250
  inhibitors, 251
  maturation, 250
Germination, conditions for, **252**, 261
  light, 256
  oxygen, 253
  temperature, 253, 261
  water, 253, 261
Giant chromosomes, 175, 220
  plant, 176*, 177*
  RNA synthesis, 175
  structure, 175*
Gibban, 111*
Gibberellic acid, 111*, 201*, 204, 211,
  255, 291, 305
Gibberellins, 100*, 101, **110**, 111*, 200

assays for, 200
biosynthesis, 111*
chemical constitution, 200
flower formation, 291, 293*, **304**
function, 201
historical, 201
Gluconic acid-6-phosphate, 61, 62*
Glucose, 57, 58*, 63*, 75, 76*, 104, 280*
Glucose-1-phosphate, 59*
Glucose-6-phosphate, 52*, 53, 58*, 60, 61,
  62*, 75, 76*
Glucose-6-phosphate dehydrogenase, 258*
β-Glucosidases, 124, 127*, 128, 251, 252*
Glucoside, 63*
Glucuronic acid, 73
Glutamic acid, 140*, 141*, 142*, 143*,
  196
Glutamic acid dehydrogenase, 140*
Glutamine, 140, 141*, 156
Glutamine synthetase, 140, 141*
Glyceric acid-1, 3-diphosphate
  see 1,3-Diphosphoglycerate
Glycerol, 88, 90*, 95*, 148*
Glycerol phosphate, 94*
Glycine, 142*, 143*, 156, 157*, 160*
Glycolysis, 74*, 75, 76*, 77*, 118
Glycosidases, 64, 66, 70
Glycosides, **63***, 118, 122, 129, 131
  N-glycosides, 64
  O-glycosides, 64
Glyoxalate, 97*
Glyoxalate cycle, 94, **96**, 97*, 98
Glyoxysomes, 98
Growth, definition, 163
Growth substances, 197
  synthetic, 212
GTP, 81
Guanine, 3*, 156
Guttapercha, 100*, 115, 116*
Guttation, 305

Head-tail addition, 103*, 104
Heart glycosides, 109*
Hellebrigenin, 109*
Hemicelluloses, 72
Hemiterpenes, 100*
Heterochromatin, 190
Hexoses, 57, 58*, 97*
Histidine, 143*
Histones, 1,
  phytohormones, **216**
  repression, **188**, 216, 230
Homoeogenetic induction, 266

HS-CoA
  see Coenzyme A
HS-Enzymes, 75, 93*, 140
Hybrid enzyme, 33*
Hybridization of nucleic acids, 25
  DNA/DNA, 26, 158
  DNA/RNA, 26, 27*
Hydrogen cyanide, 251, 252*, 254
Hydrolases, 70, 259, 260*
Hydroquinone, 117*, 128*, 129
p-Hydroxybenzoic acid, 117*, 128*, 129, 198
5-Hydroxy-ferulic acid, 122*, 123*
Hydroxylupanin, 147*
p-Hydroxyphenylpyruvate, 119*
4-Hydroxyproline, 142*, 143*
Hyoscyamine, 150*
Hyperchromic effect, 26*
Hypocotyl dormancy, 254

IAA, 123, 170, 188, **196**
  assays for, 196*, 197*
  biosynthesis, 119*, 197
  chemical constitution, 196
  cell elongation, 235
  degradation, 198
  enzyme synthesis, 238
  functions, 198
  gene activity, 238
  historical, 196
  mechanism of action, 237
  multiple effects, 195*
  optima, 236, 237*
  phloem differentiation, 266
  polar migration, 236*, 265
  two-point attachment hypothesis, 238*
  xylem differentiation, 265, 266*, 267*
IAA oxidases, 123, 131, 198
Indicaxanthin, 152, 153*
Indole alkaloids, 145*, **154**
Indole derivatives, 101
Indole-3-acetic acid
  see IAA
β-Indolylacetic acid
  see IAA
Inhibitors, synthetic, 212, 213
Inosine, 16*
Inosine-5'-phosphate, 156, 157*
Inositol, 187*
Intermediates
  natural, 121
  obligatory, 121
Intussusceptive growth, 235

Inulin 67*
Inversion
  see Epimerases
Invertase, 66, 191
α-Ionone ring, 122*
β-Ionone ring, 112*, 113
IPA, 16*, 157, 205, 206*
IPP
  see Isopentenyl pyrophosphate
Isoamylase, 70
Isocitratase, 97*
Isocitrate, 80, 82*, 97*
Isocitrate dehydrogenase, 80, 82*
Isocitrate lyase
  see Isocitratase
Isoenzyme pattern, 178, 179*, 180*
Isoenzymes, **32**, 178, 179*, 180*, 200
  fine regulation in branched biosynthetic
    systems, 33, **34***, 193*, 194*
Isoleucine, 142*, 143*, 192*, 193
Isomaltose, 70
Isomerases, 59
  phosphogluco-isomerase, 52*, 60, 62, 76*
  phosphoribo-isomerase, 60
  phosphotriose-isomerase, 51, 52*, 76*, 77*
  (-triosephosphate isomerase)
Isopentenyl adenine
  see IPA
Isopentenyl pyrophosphate, 99, 103*, 144, 155, 211
Isoprene, 99
Isoprenoids, 99
Isoprene rule, 99
Isoquinoline alkaloids, 145*, 153
Isosteric effects, 191

Jacob-Monod model, **183**

Kaempferol, 130*, 131, 198
α-Ketoglutarate, 79, 80, 82*, 140*, 141*, 142*
α-Ketoglutarate dehydrogenase, 80, 82*
Kinases, 57
  hexokinase, 58*, 76*
  phosphofructokinase, 75, 76*
  phosphoglycerokinase, 77*
  pyruvate kinase, 76, 77*
Kinetin, 205, 206*, 208*, 235

Lac-operon, 183, **184**

Lactate, 78*
Lactose, 185
Lanosterol, 106*, 108, 110
Latex, 115
Latex tubes, 115
Leaf, fall of, 204*, 211
  abscisic acid, 200, 211
  ethylene, 200
  IAA, 199*
Leucine, 142*, 143*
Leucoanthocyanins, 131
Light, 220
  germination, 220, 256
Light germinator, 256
Lignin, 117*, **124**, 125*
Limonene, 100*, 105*
Linoleic acid, 91*, 93
Linolenic acid, 91*, 93
Lipases, 94, 95*, 96
Lipoic acid amide, 79*, 80*, 81
Long day plants, 290*, 298, 307, 311*
Longitudinal growth
  see Cell elongation
LSA
  see Lipoic acid amide
Lupanin, 147*
Lupin alkaloids
  see Quinolizidin alkaloids
Lupinin, 147*
Lutein, 112*
Luteolin, 130*
Lycopene, 112*, 113, 114*
Lycophyll, 112*
Lysergic acid, 155*
Lysine, 142*, 143*, 144, 145*, 146*,
  147*, 188, 193*

Malate, 54, 82*, 97*, 273, 274*, 275*
Malate dehydrogenase, 81, 82*
Malate synthetase, 97*
Maleic acid hydrazide, 214
Malonate, 81, 83*, 91, 92*, 192
Malonyl-CoA, **90**, 91, 92*, 94, 120*,
  133*, 135*
Maltase, 70
Maltose, 64, 65*
Malvidin, 132*, 136
Mannans, 71
Mannose, 72
Mass flow, 280, 282*
Mass flow hypothesis, **280**
Matrix
  (chloroplast), 56

(mitochondria), 86
Melting of DNA, 25, 26*
Membranochrome pigments, 136
Menthol, 100*, 105*
Menthone, 100*, 105*
Meselson-Stahl experiment, 167, 168*
Metabolism, survey, 161*
Methionine, 20, 21*, 108, 142*, 143*,
  193*
5-Methylcytosine, 3*
Mevalonic acid, 103*, 104, 155, 206, 211
Mistletoe proteins, 189
Mitochondria, **86**, 87*
Mitotic cycle, **163**, 164, 181
Molecular genetics (concept), **11***
Monosaccharide, **57**
Monoterpenes, 100*, **104**, 105*
Morphactins, 212*, 213*, 214
Morphine, 153, 154*
Morphine alkaloids, 153, 154*
Multienzyme complexes, **78**, 79, 81, 90,
  92, 93*
Multi-net-growth, 235
Mutases, 58
  phosphoglucomutase, 58
  phosphoglyceromutase, 76, 77*
Myoinositol, 66
Myricetin, 130*, 131

NAD$^+$, 42*, **43**
NADP$^+$, 42*, **43**
α-Naphthylacetic acid, 200, 206, 212*
Neurosporene, 113, 114*
Neutral day plants, 298
Neutral fat, 90*
  biosynthesis, 94*
  degradation, 94, 95*
  place in metabolism, 98
Nicotiana alkaloids, 145*, **147**
Nicotinamide, 42*, 43
Nicotinamide-adenine dinucleotide
  see NAD$^+$
Nicotinamide-adenine dinucleotide phos-
  phate
  see NADP$^+$
Nicotine, 147*, 149*
Nicotinic acid, 144, 148*, 149*
Nitrate, 138, 139*, 143
Nitrate bacteria, 138
Nitrate reductase, 138, 139*, 186, 208,
  209, 237
Nitrite, 138, 139*
Nitrite bacteria, 138

Norbelladin, 151*,
Norlaudanosolin, 153, 154*
Nornicotine, 147*, 148, 149
Nucleases, 259, 260*
Nucleic acids, chemical constitution 1, 3
    (see also DNA, RNA)
Nucleosides, 3*, 4*
Nucleotides, 3*, 4*

Oleic acid, 91*, 93
Oligosaccharides, 63, 64
One-electron transition, 41
One gene–one enzyme hypothesis, 30
One gene–one polypeptide, 30
Operator gene, 183, 184*, 186*
Operon, 184*, 186*
Organ cultures, 245*
Ornithine, 144, 145*, 146*, 147, 148,
    149*, 150*
Osmosis, 232
Osmotic pressure, 233
Oxalacetate, 54, 80, 81, 82*, 92*, 97*, 98,
    140, 141*, 142*, 274, 275*
Oxalsuccinate, 80, 82*
α-Oxidation, 96*
β-Oxidation
    cinnamic acids, 128*, 129
    fatty acids, 95*
Oxidation, terminal, 74*, 75

Pace setter, 308
PAL
    see Phenylalanine-ammonium lyase
Palmitic acid, 91*
Pantheteine, 80*, 92, 93*
Papaverine, 153, 154*
Parthenocarpy
    gibberellins, 202, 248
    IAA, 200, 248
Pectin acids, 72*, 73, 249
Pectinases, 72*, 73, 259
Pectin esterases, 72*, 73
Pectins, 72*, 73, 249
Pectin substances, 60, 72*
Pentose phosphate cycle, 52*, 53, 62*,
    118, 258
Pentoses, 3, 57, 58*
P-Enzyme, 69
Peonidin, 132*, 133*, 136
Period of growth, 232
Peroxidases, 41, 83, 90, 128, 179, 180*,
    198, 229

Petunidin, 132*, 136
Phenolases, 83, 128
Phenol carboxylic acids, 117*, 120, 127*,
    128
Phenols, 117
    biosynthesis (general), 118
    biosynthesis (particular), 121
    chemical constitution, 117
    place of metabolism, 137*
    simple, 117*, 128
Phenylalanine, 23*, 118, 119*, 122, 123*,
    142*, 143, 144, 145*, 150*, 151*,
    152*, 194*
Phenylalanine-ammonium lyase, 122, 123*
    light, 222, 223
    physiological clock, 311*, 312
Phenylpropane derivatives, 117*, 119,
    126, 133*
Phenylpyruvate, 119*
Phloem, 127*, 264
    differentiation, 266
    transport, 277
Phosphatidic acids, 94*
2-Phosphoenolpyruvate, 53*, 54, 76, 77*,
    97*, 98, 118, 119*, 194*, 274, 275*
6-Phosphogluconic acid, 53
3-Phosphoglyceraldehyde, 51*, 52*, 75,
    76*, 77*
2-Phosphoglyceric acid, 76, 77*
3-Phosphoglyceric acid, 49*, 50*, 51, 52*,
    54, 76, 77*
Phosphorylases, 69
Phosphorylation, 57, 58*
Phosphotriose dehydrogenase, 75, 77*, 78
Photolysis, 36, 47
Photomorphogenesis, 220, 222, 224
Photoperiodism, in relation to flower induc-
    tion, 297
    as a sign of adaptation, 307
    hypothesis, 306*
    light breaks, 302*
    physiological clock, 309
    phytochrome, 302, 303
    site of light uptake, 299
Photophosphorylation, 47
Photosynthesis, 35, 37*
    CO$_2$ acceptor, 49*, 53
    first light reaction, 47
    pigment system I, 44, 46*
    pigment system II, 45, 46*
    primary processes, 35, 36, 46*, 50
    quantum yield, 48
    secondary processes, 35, 49, 52*, 53
    second light reaction, 47

Photosynthetic apparatus, assembly of, 257, 258
Phototropism, 195*
Phycobiliproteins, 161, 162
Phycocyanins, 162
Phycoerythrins, 162
Physiological clock, 308
 flower induction, 309
 hypothesis, 310
Physostigmin, 155*
Phytase, 187*
Phytic acid, 187*
Phytin, 187*
Phytochrome system, 101, 162, 249
 flower formation, 302*, 303*, 309
 gene activation, 222, 223*, 224*
 germination, **220**, 221*, 256, 257*
 structure, 223*
Phytoen, 113, 114*
Phytofluen, 113, 114*
Phytohormones, 182, **194**
 comparison, 204*
 gene activity, 214
 multiple effects, 195*
Phytol, 39, 40*, 100*, 110
Phytosterol, 108
α-Pinene, 100*, 105*
β-Pinene, 100*, 105*
Piperidine, 146*
Piperitenone, 105*
Plasmochrome pigments, 136
Plasmolysis, 233*
Plastoquinone, **41**, 42*, 46*
Polarity
 differentiation, **225**, 236, 246
 nucleic acids, 5
Pollen development
 induction of thymidine kinase, 185
 unequal division, 229
Pollination, 242, 243*, 248
Polynucleotides
 see nucleic acids
Polyribosomes, **15**, 17, 19*
Polysaccharides, 63, **67**
Polyterpenes, 100*, **115**
Poly U, 12
Porphobilinogen, 160*
Porphyrins, 40*, 101, **160**
 place in metabolism, 162
Precipitation (germination), 261
Precursors, 121
Pregnenolone, 109*, 110
Prephenic acid, 119*

Primary gene activity, 174, 175, 178
Proline, 142*, 143*, 152, 153*
Propionic acid, 96, 97*
Proteases, 259, 260*
Protein pattern, 178, 179*, 180*
Proteins, structure
 primary, 30, 31*
 quaternary, 31*, 32
 secondary, 31*
 tertiary, 31*
Protoalkaloids, 144
Protocatechuic acid, 117*, 127*, 128, 129
Protochlorophyllides, 160*, 161
Protopectin, 73
Protoporphytin, 160*
Provitamins A, 112
Puff, 175, 176*, 177*
Pulegone, 105*
Purine alkaloids, 145*, **156**
Puromycin, **23***, 24
Putrescine, 146*
Pyridine, 146
Pyridine nucleotide cycle, 123*
Pyridine nucleotides, 42*, 43
Pyridoxal phosphate, 141*
Pyridoxamine phosphate, 141*
Pyrrole, 39, 40*, 160*, 161
Pyrrolidine, 146*
Pyruvate, 74*, 75, 76, 77*, 78, 141, 142*, 143*, 275

Q-Enzyme, 69
Quercetin, 130*, 131
Quinic acid, 118, 119*, 120, 122
Quinine, 155*, 156
Quinoline alkaloids, 145*, 155*, 156
Quinolinic acid, 148*
Quinolizidine alkaloids, 145*, 146, 147*

Raffinose, 66, 278
Redox systems, 37*, **38**, 42*, 83, 84*
Regeneration, 172
 from single cells, 174*
Regulation, **181**
 enzymatic, 182, 191
 enzyme activity, 182, 191
 external factors, 182, **218**
 gene activity, 182, 183
 intercellular, 182, **194**
 internal factors, 182, 183
 phytohormones, 182, **194**
 point of departure, 182

transcription, 182, **183**
translation, 182, **191**
Regulator genes, 183, 184*, 186*
Repressors, 183, 184*, 186*, 193
Reserpine, 155*
Reserve materials, mobilization of, 256
Respiratory chain, 75, **82**, 84*, 86
Respiratory group, 87*
Reticulin, 154*
Ribitol, 41, 42*
Ribonucleic acids
  see RNA
Ribose, 3*
Ribose-5-phosphate, 52*, 60, 62, 157*
Ribosomes, **14**, 16, 18*
Ribulose, 59
Ribulose-1,5-diphosphate, 49*, 52*, 53, 274, 275*
Ribulose-5-phosphate, 52*, 60*, 62*
Ringing, 277, 278*
RNA
  building blocks, **1**, 3*
  and histones, 190*
m-RNA
  degradation, 191
  evidence for, **24**, 25*
  flower formation, 294, 295
  half life time, 187
  long lived, **27**, 28*
  puff, 175, 176*
  phase-specific, 176, 177, 178*
t-RNA, 15, **16***, 157, 209
RNA polymerase, **12**, 188
Root formation, 188, 195*, 198
Root hairs, 228, 229*
Root pressure, 270
Rubber, 100*, **115**, 116*
Rubber plants, 115*

S, Assimilation of, 139
S-Adenosyl-methionine, 73, 108, 129, 133*
Scopoletin, 123, 125*, 253
Secretions, 102
Secondary plant substances, **100**, 161*
Second messenger, 217
Seeds, 244*
  formation, 241
  germination, 250
Selective inhibition, 22, 294, 296
Semiconservative DNA replication, **165**
Semipermeable membrane, 232
Senescence, 204*, 207

Separation zone (leaf, fruit fall), 199*, 249
Serine, 142*, 143*
Sesquiterpene, 100*, 105
Shikimic acid, 118, 119*, 120
Shikimic acid pathway, **118**, 119*, 129, 137*
Short day plants, 298, 307, 310*
Sinapic acid, 122*, 123*, 126*, 127, 152*, 155
Sinapyl alcohol, 126*, 127
β-Sitosterol, 106*, 108
Sparteine, 147*
Squalene, 100*, 106*, 110
Stachyose, 66, 278
Starch, **68**, 204
  amylopectin, 68, 69*
  amylose, 68*
Starch transglucosylases, 69
Stearic acid, 91*, 93
Steran, 107*
Steroid alkaloids, **110**, 144
Steroid hormones, 107
Steroids, 100*, 107
Sterols, **108**
Stigmasterol, 106*, 108
Stomata
  development, 227
  movements, 271, 272*
Streptomycin, 22
Stroma, 56
Structural analogs, **20**, 21*, 22, 23*
Structural genes, **183**, 184*, 186*
Strychnine, 155*, 156
Substitution hypothesis, 135*
Substrate induction, **184**, 193
Substrate level phosphorylation, 78
Succinate, 81, 82*, 83*, 97*
Succinate dehydrogenase, 81, 82*, 87*, 192
Succinyl-CoA, 81, 82*, 160*
Succinyl-CoA synthetase, 81, 82*
Suction pressure, 233
Suction pressure equation, 232, 233
Sucrose, 64, 65*, 67, 266, 278, 280, 281
Sugar esters, 118, 122, 196
Sugar nucleotides, 59*
Sulfate, 139
  active, 139*
Surface enlargement
  chloroplasts, 56
  mitochondria, 87
Swelling, 254
Syringic acid, 127*

Syringin, 126*, 128
Systematics, biochemical, 157

Tail-tail addition, 103*, 104, 106*, 113, 114*
Temperature, **219**
  anthocyanins, **220***
  germination, 253, 261
Temperature coefficient, 219
Temperature optima, 220, 253, 261
Temporal control, 186, 257, 258
Terpenoids, **99**
  biosynthesis (general), **102**
  biosynthesis (particular), **104**
  chemical constitution, **99**
  place in metabolism, 116*
Tetrasaccharides, 66
Tetraterpenes, 100*, **111**
  (*see also* Carotenoids)
Tetroses, 57
Thebaine, 154*
Theobromine, 156, 157*
Theophylline, 156, 157*
Thiamine, 79, 80*
Thiamine pyrophosphate, 79*, 80*, 81
Thioclastic cleavage, 95
2-Thiouracil, 20, 21*, 294, 295, 296*
Threonine, 142*, 143*, 192*, 193
Threonine deaminase, 192*, 193
Thylacoids, 54, 55*
Thymidilate synthetase, 20, 21*, 171*, 192
Thymidine, 168, 185*
Thymidine kinase, 21*, 185*
d-Thymidine -5-phosphate, 185
Thymine, 3*, 185
Thymol, 100*, 105*, 118
Tissue cultures
  cytokinins, 206, 207
  differentiation of shoot and root, 246
  equilibrium between division and growth, 235
  IAA, 198, 205
  technique, 245
Tobacco alkaloids
  see Nicotiana alkaloids
Totipotency, 172
TPP
  see Thiamine pyrophosphate
Transaminases, 141
Transamination, 141*
Transcription, 10, **12**
  in a cell-free system, 29
  phytohormones, 204*

regulation, **183**
Transfection, **9**
Transformation, **7***, 8*
Translation, 10, **14**
  on the polyribosome, **17**, 19*
  on the ribosome, **16**, 18*
  in a cell-free system, 29
  phytohormones, 204*
  regulation, **191**
Transpiration, 271, 275
  cuticular, 271
  stomatal, 271
Trichoblast, 229*
3,4,5-Trihydroxycinnamic acid, 122*
Triose phosphate, 51, 52*, 63, 75, 76*, 77*, 94
Triose phosphate isomerase
  see Isomerases
Trioses, 57
Trisaccharides, 66
Triterpenes, 100*, **106***
Tropane alkaloids, 145*, **149**
Tropic acid, 150*
Tropine, 150*
Tryptamine, 155*
Tryptophan, 118, 119*, 142*, 143*, 144, 145*, 148*, 154, 194*, 197
Turgescence, 233, 271
Tumors, **169** (*see also* Crown galls)
Tumor-inducing principle (TIP), 171
Two-electron transition, 41
Tyramine, 151*, 152*
Tyrosine, 23*, 118, 119*, 122, 123*, 142*, 143*, 144, 145*, 150, 151*, 152*, 153*, 154*, 194*
Tyrosine-ammonium lyase, 122, 123*, 127

Ubiquinone, 84*
UDP-Fructose, 67
UDPG, 59*, 60, 69, 72, 128*, 129
UDP-Galacturonic acid, 60, 61*, 73
UDP-Glucuronic acid, 60, 61*, 73
UDP-Xylose, 60, 61*
Umbelliferone, 125*
Unequal cell division, 225, **226**, 227*
  pollen, 229, 230*
  root hairs, 228, 229*
  stomata, 227, 228*
Unit membrane, 86, **108**
Uracil, 3*, 11, 20
Uridine, 3*, 177
Uridine diphosphate glucose
  see UDPG

Uroporphyrinogen, 160*

Valine, 142*, 143*
Vanillic acid, 127*, 129
Vascular system, **263**
  differentiation, **264**
  elements of, 264
  function, **302**
Verbascose, 278
Vernalization, **287**
  definition, 288
  duration, 290
  gibberellins, 291, 293*
  graft experiments, 289, 292*
  hypothesis, 296, 297*
  selective inhibition, 294, 296*
  site, 289
  stage of development, 288
Vitamin A, 112, 113

VitaminB₁ see Thiamine
Volatile oils, **101**

Wall pressure, 233, 234
Watson-Crick model, 5, 6*
Wood-Werkman reaction, 273
Wound callus,170, 171* (see also Callus)
Wound phelloderm, 188

Xanthophylls, 100*, **112***
Xanthosine-5'-phosphate, 156, 157*
Xylans, 71
Xylem, 125, 127*, **264**
  differentiation, **264**
  transport, 269
Xylose, 58*, 71
Xylulose-5-phosphate, 52*, 59, 60*, 62

Zeatin, 205, 206*
Zeaxanthin, 112*